Methods in Enzymology

Volume 223
PROTEOLYTIC ENZYMES IN
COAGULATION, FIBRINOLYSIS,
AND COMPLEMENT ACTIVATION
Part B
Complement Activation,
Fibrinolysis, and
Nonmammalian Blood Coagulation
Factors and Inhibitors

METHODS IN ENZYMOLOGY

EDITORS-IN-CHIEF

John N. Abelson   Melvin I. Simon

DIVISION OF BIOLOGY
CALIFORNIA INSTITUTE OF TECHNOLOGY
PASADENA, CALIFORNIA

FOUNDING EDITORS

Sidney P. Colowick and Nathan O. Kaplan

Methods in Enzymology

Volume 223

Proteolytic Enzymes in Coagulation, Fibrinolysis, and Complement Activation

Part B

Complement Activation, Fibrinolysis, and Nonmammalian Blood Coagulation Factors and Inhibitors

EDITED BY

Laszlo Lorand

DEPARTMENT OF BIOCHEMISTRY,
MOLECULAR AND CELL BIOLOGY
NORTHWESTERN UNIVERSITY
EVANSTON, ILLINOIS

Kenneth G. Mann

DEPARTMENT OF BIOCHEMISTRY
THE UNIVERSITY OF VERMONT
COLLEGE OF MEDICINE
BURLINGTON, VERMONT

ACADEMIC PRESS, INC.

A Division of Harcourt Brace & Company

San Diego   New York   Boston   London   Sydney   Tokyo   Toronto

Academic Press, Inc.
1250 Sixth Avenue, San Diego, California 92101-4311

United Kingdom Edition published by
Academic Press Limited
24–28 Oval Road, London NW1 7DX

International Standard Serial Number: 0076-6879

International Standard Book Number: 0-12-182124-2

PRINTED IN THE UNITED STATES OF AMERICA
93 94 95 96 97 98   MM   9 8 7 6 5 4 3 2 1

Table of Contents

Section I. Complement Activation

Section II. Fibrinolysis

Section III. Nonmammalian Blood Coagulation Factors and Inhibitors

Contributors to Volume 223

Article numbers are in parentheses following the names of contributors.
Affiliations listed are current.

NOBUO AOKI (11), *The First Department of Internal Medicine, Tokyo Medical and Dental University, Tokyo 113, Japan*

GÉRARD J. ARLAUD (4), *Institut de Biologie Structurale, Laboratoire d'Enzymologie Moléculaire, 38027 Grenoble Cedex 1, France*

KULWANT S. AULAK (6), *Division of Nephrology, Children's Hospital Research Foundation, Cincinnati, Ohio 45229, and Department of Pediatrics, Univeristy of Cincinnati College of Medicine, Cincinnati, Ohio 45229*

SCOTT R. BARNUM (5), *Department of Microbiology, University of Alabama at Birmingham, Birmingham, Alabama 35294*

DETLEV BEHNKE (9), *Zentralinstitut für Mikrobiologie und Experimentelle Therapie, D-6900 Jena, Germany*

NIELS BEHRENDT (13, 14), *The Finsen Laboratory, Rigshospitalet, DK-2100 Copenhagen Ø, Denmark*

JOHN J. BISSLER (6), *Division of Nephrology, Children's Hospital Research Foundation, Cincinnati, Ohio 45299*

FRANCIS J. CASTELLINO (8, 10), *Thrombolytics Venture Group, Abbott Laboratories, Abbott Park, Illinois 60044*

DÉSIRÉ COLLEN (Introduction, 12), *Center for Molecular and Vascular Biology, University of Leuven, Leuven B-3000, Belgium*

KELD DANØ (13, 14), *The Finsen Laboratory, Rigshospitalet, DK-2100 Copenhagen Ø, Denmark*

DONALD J. DAVIDSON (8, 10), *Department of Chemistry and Biochemistry, University of Notre Dame, Notre Dame, Indiana 46556*

ALVIN E. DAVIS III (6), *Division of Nephrology, Children's Hospital Research*

Foundation, Cincinnati, Ohio 45229, and Department of Pediatrics, University of Cincinnati College of Medicine, Cincinnati, Ohio 45229

A. J. DAY (1), *Department of Biochemistry, University of Oxford, Oxford OX1 3QU, United Kingdom*

ALISTER W. DODDS (3), *Medical Research Council, Immunochemistry Unit, Department of Biochemistry, University of Oxford, Oxford OX1 3QU, United Kingdom*

CHRISTOPHER T. DUNWIDDIE (18), *Department of Cardiovascular Biology, Rhone-Poulenc Rorer, Collegeville, Pennsylvania 19426*

JAY M. EDELBERG (17), *Department of Medicine, Massachusetts General, Boston, Massachusetts 02114*

VINCENT ELLIS (14), *The Finsen Laboratory, Rigshospitalet, DK-2100 Copenhagen Ø, Denmark*

JAN J. ENGHILD (7), *Pathology Department, Duke University Medical Center, Durham, North Carolina 27710*

M. FONTAINE (1), *INSERM-Unité 78, 76230 Bois-Guillaume, France*

PAUL A. FRIEDMAN (15, 18), *Merck Research Laboratories, West Point, Pennsylvania 19486*

STEPHEN J. GARDELL (15), *Department of Biological Chemistry, Merck Research Laboratories, West Point, Pennsylvania 19486*

DIETER GERLACH (9), *Zentralinstitut für Mikrobiologie und Experimentelle Therapie, D-6900 Jena, Germany*

RICHARD A. HARRISON (6), *Molecular Immunopathology Unit, Medical Research Council, Cambridge CB2 2QH, United Kingdom*

RYUJI HASHIMOTO (22), *Department of Biology, Faculty of Science, Kyushu University, Fukuoka-812, Japan*

SHINSAKU HIROSAWA (11), *The First Department of Internal Medicine, Tokyo Medical and Dental University, Tokyo 113, Japan*

GUNILLA HØYER-HANSEN (13), *The Finsen Laboratory, Rigshospitalet, DK-2100 Copenhagen Ø, Denmark*

SADAAKI IWANAGA (20, 21, 22, 23, 24, 25), *Department of Biology, Faculty of Science, Kyushu University, Fukuoka 812, Japan*

J. MICHAEL KILPATRICK (5), *Department of Medicine, University of Alabama at Birmingham, Birmingham, Alabama 35294*

YOUNG-JOON LEE (17), *Department of Pathology, Duke University Medical Center, Durham, North Carolina 27710*

H. ROGER LIJNEN (Introduction, 12), *Center for Molecular and Vascular Biology, University of Leuven, Leuven B-3000, Belgium*

EDWIN L. MADISON (16), *Department of Vascular Biology, Scripps Research Institute, La Jolla, California 92037*

JOHN M. MARAGANORE (19), *Department of Biological Research, Biogen, Inc., Cambridge, Massachusetts 02142*

JAMES McLINDEN (10), *Department of Molecular Biology, American Biogenetic Sciences, Notre Dame, Indiana 46556*

OSAMU MIURA (11), *The First Department of Internal Medicine, Tokyo Medical and Dental University, Tokyo 113, Japan*

TOSHIYUKI MIYATA (24), *Laboratory of Thrombosis Research, National Cardiovascular Center, Research Institute, Suita 565, Japan*

B. E. MOFFATT (1), *Medical Research Council, Immunochemistry Unit, Department of Biochemistry, University of Oxford, Oxford OX1 3QU, United Kingdom*

TAKASHI MORITA (20, 21, 22), *Department of Biochemistry, Meiji College of Pharmacy, Tokyo-188, Japan*

TATSUSHI MUTA (20, 21, 22), *Department of Biology, Faculty of Science, Kyushu University, Fukuoka-812, Japan*

TAKANORI NAKAMURA (20, 21, 22), *The Institute for Enzyme Research, The University of Tokushima, Tokushima 770, Japan*

NORIKAZU NISHINO (24), *Department of Applied Chemistry, Faculty of Engineering, Kyushu Institute of Technology, Kitakyushu 804, Japan*

KATHLEEN F. NOLAN (Introduction, 2), *Medical Research Council, Immunochemistry Unit, Department of Biochemistry, University of Oxford, Oxford OX1 3QU, United Kingdom*

TOSHIO ODA (21), *Department of Biology, Faculty of Science, Kyushu University, Fukuoka-812, Japan*

TAMOTSU OMORI-SATOH (24), *Department of Biochemistry and Cell Biology, National Institute of Health, Toyama 1-23-1, Shinjuku-ku, Tokyo 162, Japan*

SALVATORE V. PIZZO (17), *Department of Pathology, Duke University Medical Center, Durham, North Carolina 27710*

MICHAEL PLOUG (13), *The Finsen Laboratory, Rigshospitalet, DK-2100 Copenhagen Ø, Denmark*

JAMES T. RADEK (8), *Department of Biochemistry, Molecular Biology and Cell Biology, Northwestern University, Evanston, Illinois 60208*

KENNETH B. M. REID (Introduction, 2), *Medical Research Council, Immunochemistry Unit, Department of Biochemistry, University of Oxford, Oxford OX1 3QU, United Kingdom*

EBBE RØNNE (13), *The Finsen Laboratory, Rigshospitalet, DK-2100 Copenhagen Ø, Denmark*

ELLIOT ROSEN (10), *Department of Gene Expression, American Biogenetic Sciences, Notre Dame, Indiana 46556*

GUY SALVESEN (7), *Pathology Department, Duke University Medical Center, Durham, North Carolina 27710*

JOSEPH F. SAMBROOK (16), *Department of Biochemistry, University of Texas, Southwestern Medical Center at Dallas, Dallas, Texas 75235*

R. B. SIM (1), *Medical Research Council, Immunochemistry Unit, Department of Biochemistry, University of Oxford, Oxford OX1 3QU, United Kingdom*

STUART R. STONE (19), *MRC Centre, University of Cambridge, Cambridge CB2 2TS, United Kingdom*

YOSHIHIKO SUMI (11), *Biotechnology Research Laboratory, Teijin Limited, Tokyo 191, Japan*

HIROYUKI TAKEYA (24), *Department of Molecular Biology, Mie University School of Medicine, Tsu, Mie 514, Japan*

SHIGENORI TANAKA (23), *Research Laboratory, Seikagaku Corporation, Higashiyamato City, Tokyo-207, Japan*

JORDAN TANG (9), *Protein Studies Program, Oklahoma Medical Research Foundation, Oklahoma City, Oklahoma 73104*

NICOLE M. THIELENS (4), *Institut de Biologie Structurale, Laboratoire d'Enzymologie Moléculaire, 38027 Grenoble Cedex 1, France*

FUMINORI TOKUNAGA (20, 25), *Department of Life Science, Faculty of Science, Himeji Institute of Technology, Harima Science Park City, Kamigori, Hyogo 678-12, Japan*

THANG TRIEU[1] (9), *Protein Studies Program, Oklahoma Medical Research Foundation, Oklahoma City, Oklahoma 73104*

GEORGE P. VLASUK (18), *Corvas International, Inc., San Diego, California 92121*

JOHN E. VOLANAKIS (5), *Department of Medicine, University of Alabama at Birmingham, Birmingham, Alabama 35294*

LLOYD WAXMAN (18), *Merck Research Laboratories, West Point, Pennsylvania 19486*

TIMOTHY N. YOUNG (17), *Department of Pathology, Duke University Medical Center, Durham, North Carolina 27710*

KAMYAR ZAHEDI (6), *Division of Nephrology, Children's Hospital Research Foundation, Cincinnati, Ohio 45229, and Department of Pediatrics, University of Cincinnati College of Medicine, Cincinnati, Ohio 45229*

---

[1] Deceased.

Preface

Originally, a single volume was planned to cover the field of Proteolytic Enzymes in Coagulation, Fibrinolysis, and Complement Activation, but it soon became evident that—related though these topics may be—the material would have to be published in two parts. In this volume of *Methods in Enzymology* (223, Part B) the important fields of complement activation and fibrinolysis and the esoteric area of increasing pharmacological interest nonmammalian blood coagulation factors and inhibitors are covered. Its companion Volume 222, Part A deals with mammalian coagulation factors and inhibitors.

LASZLO LORAND
KENNETH G. MANN

METHODS IN ENZYMOLOGY

Introduction

Complement Activation*

Activation of the complement system can involve the utilization of the 12 serum complement components in a cascade fashion, via triggering of either the classical or the alternative pathways (Fig. 1). The system is one of the major immune effector mechanisms in the blood and its activation leads to the generation of inflammation, the killing and clearance of pathogenic microorganisms, and the elimination of immune complexes. There are five enzymes (C1r, C1s, C2, factor D, and factor B) directly involved with the activation and generation of the enzyme complexes of the two pathways of complement (Table I). All these enzymes, including the regulatory enzyme factor I, are serine proteases, and C1r, C1s, C2, and factor B circulate as proenzymes in the blood. A number of plasma proteins and membrane proteins act as negative regulators of complement activation (Table I) by accelerating the decay of the $\overline{C4b2a}$ or $\overline{C3bBb}$ complexes (which contain the activated forms of the C2 or factor B enzymes) or by acting as cofactors in the limited proteolysis of C4b, or C3b, by factor I. Properdin, which is the only known positive regulator of complement activation, stabilizes the $\overline{C3bBb}$ and $\overline{C3bBb3b}$ complexes, mainly via its interaction with C3b. A variety of other regulators/receptors and proteases are involved with the modulation of the biological activities mediated by the fragments of the activated complement components, and this raises the number of components, regulators/receptors, and enzymes within the complement system to approximately 30.

Both pathways of complement can be activated by antibody-dependent and antibody-independent mechanisms; however, the classical pathway is considered to be activated primarily via the Fc regions in immune complexes containing IgM antibody, or certain subclasses of IgG antibody. The interaction of immune complexes with the C1q portion of the Ca^{2+}-dependent C1q–C1r$_2$C1s$_2$ complex is considered to bring about a conformational change leading to the autocatalytic activation of proenzyme C1r, which, in turn, activates proenzyme C1s. The activated C1s, within the C1 complex, cleaves single bonds within component C4 and proenzyme C2, resulting in the assembly of the $\overline{C4b2a}$ complex. The C2a portion of C2 within the $\overline{C4b2a}$ complex is involved in the cleavage of the major serum complement protein, component C3, and the formation of the

---

* By Kenneth B. M. Reid and Kathleen F. Nolan.

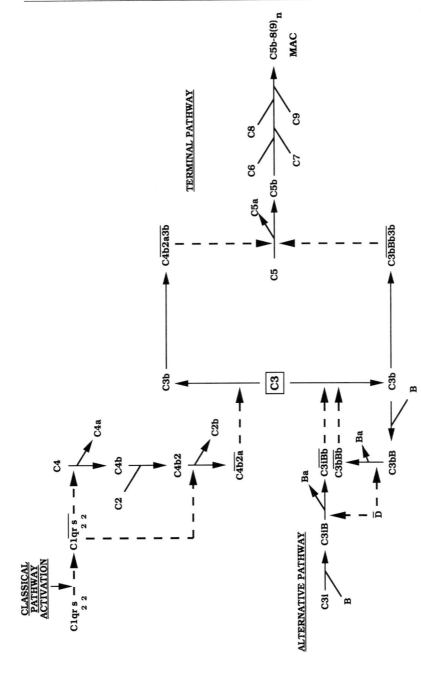

Fig. 1. The classical and alternative pathways of the serum complement system. The classical pathway is shown in the upper part of the diagram and the amplification loop of the alternative pathway is shown in the lower part of the diagram. Complexes containing enzymatic activity are denoted by overbars. Dashed lines represent enzymatic cleavages.

TABLE I

ENZYMES AND REGULATORY PROTEINS INVOLVED IN ACTIVATION AND CONTROL OF THE COMPLEMENT SYSTEM

Substance/involvement	Role within enzyme complex
(a) Enzymes involved in activation	
C1r	Forms complex with C1s, cleaves and activates proenzyme C1s
C1s	Forms complex with C1r, cleaves component C4 and proenzyme C2
C2	Forms complex with C4, cleaves components C3 and C5
Factor D	Activation of proenzyme factor B within C3iB or C3bB complexes
Factor B	Forms complex with C3i or C3b, cleaves C3
(b) Regulation of enzyme complexes	
Plasma proteins	
C1 inhibitor	Forms 1:1 complex with C1r or C1s and removes them from $\overline{C1}$ complex
C4-binding protein	Accelerates decay of $\overline{C4b2a}$ cofactor in cleavage of C4b by factor I
Factor H	Accelerates decay of $\overline{C3bBb}$ cofactor in cleavage of C3b by factor I
Factor I	Protease that, with the aid of cofactors, inactivates C4b and C3b
Properdin	Stabilizes $\overline{C3bBb}$ and $\overline{C3bBb3b}$ complexes
Membrane proteins	
Complement receptor 1 (CR1)	Accelerates decay of C3/C5 convertases, cofactor in cleavage of C3b or C4b by factor I
Membrane cofactor protein (MCP)	Cofactor in cleavage of C3b or C4b by factor I
Decay accelerating factor (DAF)	Accelerates decay of C3/C5 convertases

$\overline{C4b2a3b}$ complex, in which the C2a portion of C2 is now involved in the cleavage of C5, to yield C5a plus C5b. This results, without any further proteolytic cleavage, in the self-assembly of the membrane attack complex (MAC) from C5b, C6, C7, C8, and between 1 and 18 molecules of C9 (Fig. 1), thus causing membrane damage to the original target identified by antibody IgM, or IgG, as being foreign. The alternative pathway provides an efficient, antibody-independent means of triggering the complement cascade and thus provides an immediately available means of host defense that does not rely on immunological memory. The alternative pathway is activated by a wide range of surfaces and involves the C3b fragment of activated C3 or a "C3b-like" form of C3, known as $C3(H_2O)$ or C3i, associating with proenzyme factor B (Fig. 1). C3 plays a central role during complement activation because the two pathways merge at the C3 cleavage step. The C3bBb and C4b2a complexes split C3 at the same position in its α chain to yield the 77-amino acid residue-long C3a anaphylatoxin and the 180-kDa C3b fragment. The removal of C3a causes a conformational change in the C3b portion of C3, leading to the exposure of a thiol ester bond that is buried, and inaccessible to nucleophiles, within the native C3. Once the thiol ester is exposed it readily undergoes nucleophilic attack by water or the hydroxyl or amino groups of other molecules present in solution or on cell surfaces. This procedure allows C3b to become covalently attached to a variety of soluble proteins and membranes via ester or amide bonds. In the blood, C3 is continuously activated at a slow rate by serum proteases that convert C3 to C3b, or by water or other small nucleophiles, gaining access to the internal thiol ester bond. A molecule of C3 in which the thiol ester has been hydrolyzed, without the loss of C3a, is designated C3i or $C3(H_2O)$. The C3i, or $C3(H_2O)$, appears to have a molecular conformation similar to that of C3b, and therefore, in view of its C3b-like conformation, is able to form a Mg^{2+}-dependent C3 convertase, C3iBb, in the presence of factors B and D (Fig. 1). The C3iBb complex can generate C3b, which, in the presence of factors B and D, yields the C3bBb complex. The generation of C3b and C3i normally takes place at a very low level and consequently very little freshly activated C3b is deposited on host cell surfaces. If, however, C3b is deposited on an activator of the alternative pathway then it escapes normal control by the regulatory proteins (Table I) and serves as a seed for a positive-amplification loop (Fig. 1), resulting in massive conversion of C3 to C3b, coating of the target with C3b, and efficient generation of the membrane attack complex on the surface of the target. Properdin acts as a positive regulator of the C3bBb complex. All the other complement regulatory proteins carry out down-regulatory roles (Table I), because if C3bBb and C3iBb were left unchecked all the C3 in the blood would be rapidly consumed.

Down-regulation of the activated enzyme complexes of the classical and alternative pathways is mediated primarily by four plasma proteins and three membrane-bound proteins (Table I,b). The $C1qr_2s_2$ complex is controlled by the C1 inhibitor, which rapidly forms a covalent 1 : 1 complex with both activated C1r and C1s, probably via their catalytic sites, which results in the release of a C1 inhibitor–C1r–C1s–C1 inhibitor complex. The removal of C1r and C1s from C1q leaves the activator–C1q complex free to interact with the cell surface C1q receptor, resulting in the triggering of a variety of biological effects that may be used in the destruction/clearance of the target activator. The C1 inhibitor belongs to the serpin family of inhibitors and it can inhibit other activated plasma proteases (such as kallikrein, plasmin, Hageman factor, and factor XI), but it is the only plasma inhibitor directed against C1r and C1s. The $\overline{C4b2a}$, $\overline{C4b2a3b}$, $\overline{C3bBb}$, and $\overline{C3bBb3b}$ enzyme complexes are controlled by proteins that interact with them and simply accelerate the decay of the particular complex, or by proteins that, on interaction with an appropriate complex, also act as cofactors for limited proteolysis of C4b or C3b by the serine protease factor I (Table I,b).

The C4-binding protein has multiple binding sites for C4b and this property inhibits the formation of C4b2a and also accelerates its decay. When C4-binding protein is bound to C4b it acts as a cofactor in the cleavage of C4b at two positions by factor I. Factor H binds to C3b, which greatly accelerates the decay of $\overline{C3bBb}$ and $\overline{C3bBbP}$, and it also probably regulates the C5 convertase by competing with C5 with respect to its binding to C3b. Factor H also acts as cofactor for factor I in the limited cleavage of C3b (and C3i) at two positions in the α' chain to yield iC3b.

Three membrane-bound proteins play a major role in the regulation of the complement proteases on cell surfaces. Decay acceleration factor (DAF) and membrane cofactor protein appear to display activity only as regulators of the C3 convertases, whereas complement receptor 1 (CR1) plays a more versatile role, e.g., acting as a receptor for immune complexes, bearing C3b or C4b, and becoming involved in the triggering of receptor mediated-clearance of the complexes. Thus, it is not unexpected that DAF and MCP are found on a wide range of cells because their only purpose appears to be to serve as regulators of the C3 convertases and to protect host cells from damage by autologous complement.

References discussing the activation and control of specific areas of the complement system are given in this volume in the chapters that follow and in some reviews of the entire system.[1–5]

Fibrinolysis**

The blood fibrinolytic system comprises an inactive proenzyme, plasminogen, that can be converted to the active enzyme, plasmin, that de-

---

[1] T. Kinoshita, *Immunol. Today* **12**, 291 (1991).

[2] S. K. A. Law and K. B. M. Reid, in "In Focus" (D. Male, ed.), p. 1. IRL Press, Oxford, UK, 1988.

[3] R. P. Levine and A. W. Dodds, *Curr. Top. Microbiol. Immunol.* **153**, 73 (1989).

[4] B. P. Morgan, in "Complement: Clinical Aspects and Relevance to Disease." Academic Press, London, 1990.

[5] H. J. Müller-Eberhard, *Annu. Rev. Biochem.* **57**, 321 (1988).

** By H. Roger Lijnen and Désiré Collen.

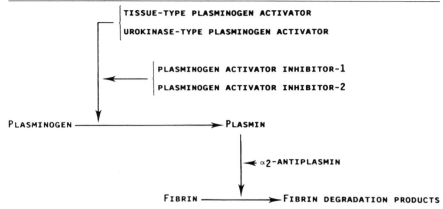

FIG. 2. Schematic representation of the fibrinolytic system. The proenzyme plasminogen is activated to the active enzyme plasmin by tissue-type or urokinase-type plasminogen activator. Plasmin degrades fibrin into soluble fibrin degradation products. Inhibition of the fibrinolytic system may occur at the level of the plasminogen activators, by plasminogen activator inhibitors, or at the level of plasmin, mainly by α_2-antiplasmin.

grades fibrin into soluble fibrin degradation products. In addition, the fibrinolytic system may play a role in several other biological phenomena such as tissue repair, malignant transformation, macrophage function, ovulation, and embryo implantation. Two immunologically distinct physiological plasminogen activators (PAs) have been identified: the tissue-type PA (tPA) and the urokinase-type PA (uPA), which may be obtained both as a single-chain form (scuPA) or a two-chain derivative (tcuPA, urokinase). The main role of uPA, however, may consist of proteolytic events other than fibrinolysis in the circulation. Inhibition of the fibrinolytic system may occur either at the level of the PA, by specific plasminogen activator inhibitors (PAIs), or at the level of plasmin, mainly by α_2-antiplasmin (Fig. 2). Regulation and control of the fibrinolytic system are mediated by specific molecular interactions among its main components,[6] and by a controlled release of plasminogen activators and plasminogen activator inhibitors from endothelial cells.[7,8]

The physiological plasminogen activators, tPA and scuPA, activate plasminogen preferentially at the fibrin surface. Plasmin, associated with the fibrin surface, is protected from rapid inhibition by α_2-antiplasmin and may thus efficiently degrade the fibrin of a thrombus. Similarities between

[6] D. Collen, *Thromb. Haemostasis* **43,** 77 (1980).

[7] V. W. M. Van Hinsbergh, T. Kooistra, J. J. Emeis, and P. Koolwijk, *Int. J. Radiat. Biol.* **60,** 261 (1991).

[8] D. J. Loskutoff, *Fibrinolysis* **5,** 197 (1991).

the roles of fibrin and roles of cell surfaces in plasminogen activation have been recognized.[9] Cells may express specific receptors for both tPA and scuPA and for plasminogen. At the cell surface, plasminogen is converted to plasmin and cell surface-bound plasmin is protected from rapid inhibition by α_2-antiplasmin. Plasmin may convert scuPA to tcuPA, resulting in markedly enhanced plasminogen activation. Cellular receptors for plasminogen, tPA, and uPA have been identified and partially characterized, and their function in cell surface proteolysis is being elucidated. The cell surface receptor for uPA (uPAR), a glycoprotein of M_r 55,000–60,000, has been identified on many cell types. uPAR binds both tcuPA and scuPA via the growth factor domain in the NH_2-terminal part of uPA. Assembly of plasminogen and PA at the endothelial cell surface thus provides a focal point for plasmin generation and may play an important role in maintaining blood fluidity and nonthrombogenicity.

Lipoprotein (a) [Lp(a)], a protein with multiple copies of a plasminogen kringle 4-like domain, a single copy of a kringle 5-like domain, and an inactive protease domain, binds to endothelial cells with an affinity and binding capacity similar to that of plasminogen. Competition of Lp(a) with plasminogen for binding to fibrin or to endothelial cells results in strongly down-regulated activation of surface-bound plasminogen by tPA, suggesting that Lp(a) may play a role in the regulation of fibrinolysis at the fibrin and the endothelial cell surface.[10]

The physiological importance of the fibrinolytic system is demonstrated by the association between abnormal fibrinolysis and a tendency toward bleeding or thrombosis.[11] Excessive fibrinolysis due to increased levels of tPA, or to α_2-antiplasmin or PAI-1 deficiency, may result in bleeding tendency. Impairment of fibrinolysis represents a commonly observed hemostatic abnormality associated with thrombosis. It may be due to a defective synthesis and/or release of tPA from the vessel wall, to a deficiency or functional abnormality in the molecular interactions regulating plasminogen activation, or to increased levels of inhibitors of tPA or of plasmin.

One approach to the treatment of thrombosis consists of a pharmacological dissolution of the blood clot via the intravenous infusion of plasminogen activators. Currently, five thrombolytic agents are either approved for clinical use or are under clinical investigation in patients with acute myocardial infarction. These are streptokinase, two-chain uPA, anisoylated plasminogen streptokinase activator complex (APSAC, eminase),

[9] E. F. Plow, J. Felez, and L. A. Miles, *Thromb. Haemostasis* **66**, 32 (1991).

[10] R. L. Nachman and K. A. Hajjar, *Ann. N. Y. Acad. Sci.* **614**, 240 (1991).

[11] B. Wiman and A. Hamsten. *Semin. Thromb. Hemostasis* **16**, 207 (1990).

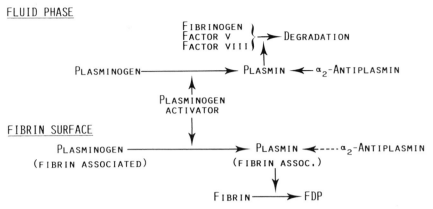

FIG. 3. Molecular interactions determining the fibrin specificity of plasminogen activators. Non-fibrin-specific plasminogen activators (streptokinase, urokinase, APSAC) activate both plasminogen in the fluid phase and fibrin-associated plasminogen. Fibrin-specific plasminogen activators (tPA and scuPA) preferentially activate fibrin-associated plasminogen; FDP, fibrin degradation product.

recombinant tPA (rtPA, alteplase), and recombinant single-chain uPA (rscuPA, prourokinase, saruplase). Staphylokinase, a protein obtained from *Staphylococcus aureus*, was recently shown to induce more fibrin-specific clot lysis in human plasma than streptokinase, and its thrombolytic properties are being evaluated in animal models of thrombosis and preliminarily in patients with acute myocardial infarction.

Whereas tPA and uPA activate plasminogen directly to plasmin by cleavage of the Arg^{561}-Val^{562} peptide bond, both streptokinase and staphylokinase form an equimolar complex with plasmin(ogen) that activates other plasminogen molecules. Streptokinase, APSAC, and urokinase (tcuPA) activate both circulating and fibrin-bound plasminogen; this causes systemic activation of the fibrinolytic system, which may result in depletion of α_2-antiplasmin, and generation of free plasmin that will degrade several plasma proteins, including fibrinogen, factor V, and factor VIII (Fig. 3). The more fibrin-specific mechanism of action of tPA and scuPA has triggered great interest in the use of these agents as thrombolytic agents.

Despite their widespread use, all currently available thrombolytic agents suffer from a number of limitations, including resistance to reperfusion, coronary reocclusion, and bleeding complications in a significant number of patients with acute myocardial infarction.[12] Furthermore, the

[12] D. Collen and H. R. Lijnen, *Blood* **78,** 3114 (1991).

fibrin specificity of rtPA and rscuPA in man is not as pronounced as was anticipated from animal models, and both agents have a very short plasma half-life, whereby their therapeutic use probably requires continuous intravenous infusion of relatively large amounts of material (50 to 100 mg). Therefore, the quest for improved thrombolytic agents continues. Several lines of research toward improvement of thrombolytic agents are being explored, including the construction, by site-specific mutagenesis, of mutants and variants of rtPA or rscuPA, chimeric (tPA/uPA) plasminogen activators, and conjugates of plasminogen activators with monoclonal antibodies.[13] Alternatively, plasminogen activators from animal origin, i.e., vampire bat salivary plasminogen activator, or from bacterial origin, i.e., staphylokinase, are being evaluated.

---

[13] H. R. Lijnen and D. Collen, *Thromb. Haemostasis* **66,** 88 (1991).

Section I

Complement Activation

[1] Complement Factor I and Cofactors in Control of Complement System Convertase Enzymes

By R. B. Sim, A. J. Day, B. E. Moffatt, and M. Fontaine

Complement factor I (EC 3.4.21.45) was first described and partially characterized in 1966–1968 as an enzyme involved in the physiological degradation of the major complement protein C3. In the 1970s it was also shown to act in the physiological breakdown of C4, a homolog of C3. It was previously named conglutinogen-activating factor (KAF), because its action on C3 exposed a binding site on a C3 fragment for the bovine protein conglutinin. Other former names include C3b inactivator, C3b INA, and C3b/C4b inactivator. The basic properties of factor I have been summarized previously in this series.[1,2] It is a serine protease of approximate molecular weight 88,000, and it circulates in plasma in activated, rather than proenzymatic, form at a concentration of about 35 μg/ml (0.4 μM). It is not inhibited by any of the protein protease inhibitors in plasma and has a very narrow substrate range.

Activation of Complement System and Formation of Substrates for Factor I

The physiological substrates of factor I are produced only when the complement system is activated. Activation of the classical pathway of the complement system is generally mediated by binding of the C1 complex to a target, such as immune complexes or microorganisms (for summary, see Refs. 3–5). The serine protease components of C1, namely C1r and C1s, become activated by this interaction, and activated C1s cleaves complement component C4, forming the proteolytic fragments C4a (9 kDa) and C4b (185 kDa). A proportion of the C4b that is formed binds covalently to the surface of the complement activator (immune complex or microorganism). This occurs by reaction of mildly nucleophilic surface groups (e.g., —NH$_2$, —OH) with a highly electrophilic carbonyl group within a thiol ester in C4b. Water also attacks the thiol ester in C4b, so that a high

[1] L. G. Crossley, this series, Vol. 80, p. 112.
[2] I. Gigli and F. A. Tausk, this series, Vol. 162, p. 626.
[3] S. K. A. Law and K. B. M. Reid, "Complement." IRL Press, Oxford, UK, 1988.
[4] R. B. Sim and K. B. M. Reid, *Immunol. Today* **12**, 307 (1991).
[5] R. B. Sim, *in* "Encyclopaedia of Immunology" (I. M. Roitt and P. J. Delves, eds.), p. 373. Academic Press, London, 1992.

METHODS IN ENZYMOLOGY, VOL. 223

proportion of the C4b formed does not bind to the activator surface, but remains in solution after reaction with water. Complement component C2 then binds to the C4b. If the C4bC2 complex is surface bound, and correctly oriented with respect to the surface-bound C1s, C1s cleaves C2 to form two fragments, C2a (60 kDa) and C2b (30 kDa). The C4bC2a2b complex (commonly denoted as C4b2a) is a serine protease, with its active site in the C2a fragment. C4b2a is termed the classical pathway C3 convertase, and cleaves complement component C3. In physiological conditions, the C4b2a enzyme is formed attached to the surface of a complement activator, and, because of the steric limitations for activation of C2 by surface-bound C1s, there is no significant formation of soluble, freely diffusing C4b2a. *In vitro,* however, soluble C4b2a can be formed in artificial conditions, by supplying soluble activated C1s to activate both C4 and C2.

Once C4b2a has formed on the surface of the complement activator, it cleaves C3, a homolog of C4, forming the fragments C3a (9 kDa) and C3b (178 kDa). C3b, like C4b, contains a thiol ester, and a small proportion of the C3b formed binds covalently to the surface of the complement activator; a larger proportion reacts with water and remains in solution. One molecule of C3b becomes covalently attached to C4b within the C4b2a complex, forming C4b2a3b, the classical pathway C5 convertase. The C4b and C3b within this complex form a binding site for complement component C5 and orient it for cleavage by the active site within C2a. C5, also a homolog of C4, is cleaved into the fragments C5a and C5b, similar to C4a and C4b. C5b does not contain a thiol ester and so does not bind covalently to the complement activator surface, but instead associates noncovalently with complement components C6, C7, C8, and C9, which insert into lipid bilayer membranes and cause membrane disruption.

The role of factor I is to control the activities of the convertase enzymes, C4b2a and C4b2a3b, by proteolytic cleavage of the C4b and C3b subunits of these enzymes. Before C3b or C4b can be cleaved by factor I, they must bind to one of several complement control proteins to form a noncovalent complex. C3b or C4b within such a complex is susceptible to cleavage by factor I (Fig. 1). It is assumed that these "cofactors" for factor I alter the conformation of the substrate so that it can be cleaved by factor I; alternatively, the cofactor proteins may provide a weak binding site for factor I to orient it toward the substrate. There are two major soluble (plasma) cofactors for factor I: these are C4b-binding protein, which forms complexes with C4b, and factor H, which binds to C3b. There are also two cell surface-expressed cofactor proteins, complement receptor type 1 (CR1, CD35), and membrane cofactor protein (MCP, CD45).

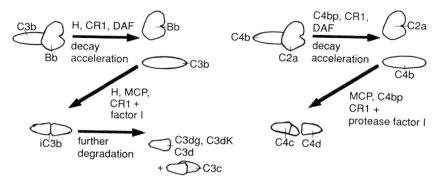

FIG. 1. Control of the convertase enzymes. The two subunits of the classical pathway convertase, C4b2a, dissociate with a half-life of <5 min. Dissociation is accelerated by interaction of C4bp, CR1, or DAF with the enzyme. This activity of C4bp, CR1, or DAF is referred to as "decay acceleration" activity. Once C2a has dissociated, C4b (185 kDa) forms a complex with one of the factor I–cofactor proteins, CR1, C4bp, or MCP, and is attacked by the protease factor I, which cleaves C4b at two sites, forming C4d (45 kDa), and C4c (140 kDa). If the C4b was covalently bound to a surface, C4d remains bound, while C4c diffuses away. In a similar way, the C3bBb enzyme of the alternative pathway is destroyed by control proteins. C3b (178 kDa), when it is bound to a factor I–cofactor protein, is cleaved in two places by factor I, releasing a 3-kDa fragment (C3f) and forming iC3b (175 kDa) (sometimes written as C3bi). iC3b is the ligand for the major phagocytosis receptor, CR3 (and CR4). iC3b is further degraded to form two separate fragments, like C4d and C4c. This further degradation is slow and may be mediated by factor I (using only CR1 as cofactor) or by a variety of proteases, including kallikrein or leukocyte elastase. The products are C3c (~135 kDa) and 28- to 38-kDa fragments called C3dg, C3d, or C3dK. If the C3b is covalently bound to a surface, iC3b and C3dg/C3d/C3dK remain bound, while C3c diffuses away. Reproduced with permission from Kluwer Publishers.

Complement is also activated by the alternative pathway.[3,6–8] In this pathway, convertase enzymes homologous to those in the classical pathway are formed. The alternative pathway is also activated by immune complexes and by a wide range of microorganisms. In this pathway, the initial event in activation is covalent deposition of a C3b molecule on the surface of a potential activator. Once deposited, C3b can bind factor B, a homolog of C2, and factor B in the C3bB complex is cleaved by the serine protease factor D, forming Ba (30 kDa) and Bb (60 kDa). Bb remains associated with C3b, forming C3bBb, the alternative pathway C3 convertase: this is similar in properties and structure to C4b2a. Because the enzyme that activates C3bB to form C3bBb is soluble, rather than surface

[6] M. K. Pangburn, this series, Vol. 162, p. 639.
[7] M. A. McAleer and R. B. Sim, in "Activators and Inhibitors of Complement" (R. B. Sim, ed.), p. 1. Kluwer Academic Publ., Dordrecht, The Netherlands, 1993.
[8] S. Meri and M. K. Pangburn, Proc. Natl. Acad. Sci. U.S.A. 87, 3982 (1989).

bound, significant soluble C3bBb can be generated in physiological conditions, in contrast to the classical pathway convertase.

C3b is, in the alternative pathway, a subunit of the enzyme that forms C3b, and the positive feedback in this series of reactions represents an amplification of C3 conversion, which results in covalent deposition of multiple C3b molecules in clusters on the complement activator. Multiplication of deposition of C3b is important in that C3b and its breakdown product, iC3b, are opsonins, and the clustered C3b/iC3b are important in interaction with C3 receptors on phagocytic cells. As in the classical pathway, activation of C3 by C3bBb results in covalent fixation of one further molecule of C3b to the C3b in the C3bBb complex. This alters the enzyme so that it will cleave C5; C5 is bound and oriented for cleavage by the two C3b molecules in the complex, and the activated C5 proceeds to form a lytic complex, as described above. Factor I and its cofactors also control the activity of the alternative pathway C3 and C5 convertases, and so control the extent of turnover and deposition of C3, the major opsonin of the complement system.

For initiation of the alternative pathway, as noted above, C3b must be deposited onto the surface of a potential activator. This first molecule of C3b can arise via classical pathway activation, or by random proteolysis of C3 by, e.g., kallikrein or plasmin. Mostly, however, it is likely to arise from a constant low-grade turnover of C3 mediated by another form of C3 convertase, the initiating convertase. This arises as follows: the thiol ester within C3 can be attacked and cleaved by low-molecular-weight nucleophiles, such as ammonia or water. This is thought to occur at a constant low rate in circulation and results in formation of a form of C3 known as C3u, $C3(H_2O)$, C3i, or C3b-like C3. This form of C3 has not been proteolytically activated and it cannot bind covalently to a surface because its thiol ester has already been cleaved. However, it is like C3b in that it can bind to factor B, and is also susceptible to the action of factor I and its cofactors. Once the $C3(H_2O)B$ complex is formed, it is activated by factor D to form $C3(H_2O)Bb$, which is a soluble (non-surface-bound) C3 convertase. This enzyme in turn will activate C3, generating normal C3b, which can bind randomly to any surface containing mild nucleophilic groups (e.g., $—NH_2$ or $—OH$). Once C3b has been deposited on a surface, it may be destroyed rapidly by the action of factor I and cofactors, or it may survive long enough to form a C3bBb complex, which will amplify deposition of C3b on the surface. On surfaces that are not alternative pathway activators, C3b is rapidly destroyed, but on activating surfaces, C3b appears to be protected from the action of factor I and cofactors, and amplification of C3 deposition occurs. It is not yet clear what structural differences there are between surfaces that activate or do not activate the alternative pathway, although surface charge clustering appears to be important. The cofactor protein, factor H, has a polyion-binding site[8] that is likely to modify the interaction between factor H and C3b, and this site is at least partly responsible for distinguishing between alternative pathway activators and nonactivators. Thus factors H and I play an important role in the initiation of the alternative pathway.

Control of Complement System Convertases

The activity of the convertase enzymes is controlled in three ways (Fig. 1): first, they are unstable, and the proteolytic subunits (C2a or Bb) dissociate irreversibly from C4b or C3b with a half-life of 5 min or less. C3bBb is, however, stabilized by the protein properdin, which increases the half-life of dissociation 4- to 10-fold. Second, the dissociation (or decay) of the enzymes is accelerated by interaction of the convertases with the control proteins factor H (which dissociates Bb from C3b), C4bp (which dissociates C2a from C4b), and also CR1 and decay-accelerating factor (DAF, CD55) (which act on both types of convertases). Because factor H, C4bp, CR1, and DAF influence dissociation of the convertases, they also in general inhibit formation of the convertases. Third, once the proteolytic subunits have been dissociated, factor H, CR1, or MCP binds to C3b, and the C3b is degraded to iC3b by factor I; similarly, C4b bound to C4bp, CR1, or MCP is degraded by factor I to form C4c plus C4d. The degraded forms of C4b and C3b can no longer interact with C2 or factor B, and so take no further part in the activation of the complement system. C3b is an opsonin and is recognized by CR1 on phagocytic cells; its degradation product iC3b is a more effective opsonin and is a ligand for the complement receptor type 3 (CR3) and possibly also type 4 (CR4) on phagocytes.[9] Factor I therefore has a role in opsonization.

The different cofactors for factor I are structurally related, as discussed below, and, when isolated, have superficially similar properties as cofactors. Their physiological roles are, however, different.[10] Factor H, the major soluble control protein for C3b, controls the activities of all the convertases that contain C3b and has the unique property of being involved in distinguishing between alternative pathway activators and nonactivators. Factor H is the major cofactor involved in factor I-dependent degradation of soluble (non-surface-bound) C3b and C3bBb. C4bp, also an abundant plasma protein, controls C4b-containing convertases and mediates factor I-dependent degradation of soluble C4b. It can also act as a cofactor for C3b degradation, but not in physiological conditions.[11] It has

[9] E. J. Brown, *Curr. Opin. Immunol.* **3,** 76 (1991).

[10] T. Kinoshita, *in* "Progress in Immunology VII" (F. Melchers, E. D. Albert, and L. Nicklin, eds.), p. 178. Springer-Verlag, New York, 1989.

[11] E. Sim, A. B. Wood, L.-M. Hsiung, and R. B. Sim, *FEBS Lett.* **132,** 55 (1981).

a further role in binding of the coagulation protein, protein S[12]: this appears unrelated to its function in the complement system. CR1 is found on erythrocytes, most leukocytes, tissue macrophages, and on a few other tissues. It acts as cofactor for degradation of both C3b and C4b. Unlike C4bp, it may distinguish between the major isotypes of C4, C4A, and C4B. It is much less abundant than factor H or C4bp, but is locally concentrated on cell surfaces. It may therefore have a special role in regulating C3 activation adjacent to the cell surface. It also acts as a receptor (the immune adherence receptor), mediating the transport of C3b- and C4b-bearing particles by erythrocytes, and participating in uptake of such particles by phagocytes and also by B lymphocytes. During transport of C3b-bearing particles attached to CR1 on erythrocytes, it is likely that CR1 has a major role in factor I-dependent conversion of C3b to iC3b, for subsequent transfer of the particles to tissue phagocytes bearing CR3. MCP is also involved in processing both C3b and C4b. It has a wide tissue distribution, and it is thought that it is likely to be involved in control of convertases that are attached to the surface of the same cell as the MCP. Thus MCP may have a particular role in protecting host tissues from complement attack, although the other cofactors are also important in this respect. Factor H, C4bp, and CR1, but not MCP, all have decay acceleration activity (i.e., the ability to accelerate dissociation, or inhibit assembly, of the subunits of the convertases), independent of factor I. The cell surface protein DAF also has this activity, but lacks factor I–cofactor activity. DAF has a wide tissue distribution and is thought to have a particular role in dissociating convertases attached to the same cell as the DAF. Descriptions[13,14] of inherited DAF deficiencies, however, show that erythrocytes lacking DAF are not abnormally susceptible to complement attack, and so the protective effect of DAF may be minor.

Protein and Gene Structures of Factor I and Cofactors

Factor I and Its Modules

The cDNA sequence of human factor I codes for a 583-residue single prepropeptide. Factor I is secreted in an active form, consisting of disul-

[12] A. Hillarp, *Scand. J. Clin. Lab. Invest.* **51,** Suppl. 204, 57 (1991).
[13] A. H. Merry, V. I. Rawlinson, M. Uchikawa, M. R. Daha, and R. B. Sim, *Br. J. Haematol.* **73,** 248 (1989).
[14] M. E. Reid, G. Mallinson, R. B. Sim, J. Poole, V. Pausch, A. H. Merry, Y. W. Liew, and M. J. A. Tanner. *Blood* **78,** 3291 (1991).

fide-linked heavy and light chains.[15,16] The overall polypeptide molecular weight of factor I is about 63,000. Each chain contains three potential sites for N-linked glycosylation, and if all these sites are occupied, the total molecular weight for the heavy and light chains would be 44,000–47,000 and 37,000–40,000, respectively. This is in reasonable agreement with estimates of molecular weight by SDS–PAGE, which indicate that the whole molecule is about 88,000 mol. wt., and the chains are approximately 50,000 and 38,000 mol. wt. The light chain (243 amino acids) is derived from the C-terminal end of the prepropeptide and is composed of a serine protease domain that is most similar to those of tissue plasminogen activator and urokinase. These serine proteases have an extra pair of cysteine residues (Cys-355, Cys-426 in factor I) that are presumably disulfide bonded.[15,16]

The heavy chain (318 amino acids), like the majority of complement proteins,[17] is composed of a number of protein modules[18,19] (i.e., factor I is a mosaic protein). There is an I/C6/C7 repeat (residues 23–89) at the N-terminal end followed by a CD5 module and two low-density apolipo-protein receptor (LDLr) modules as shown in Fig. 2a.

The I/C6/C7 repeat is approximately 70 amino acids in length and is based on a consensus sequence of 8 or 10 cysteines, a tryptophan, and other highly conserved residues[18] as shown in Fig. 2b. The terminal com-plement components C6[20] and C7[21] both contain two of these repeats, which may be involved in C5 binding[22], at their C-terminal ends.

The CD5 module, which is approximately 100 residues long (Fig. 2b), is found in single copy in factor I and in the macrophage scavenger receptor (murine and bovine).[23] Three modules of this type are also found in human CD5 (Ly-1 in mouse). The repeat was first identified in the sea urchin

[15] C. F. Catterall, A. Lyons, R. B. Sim, A. J. Day, and T. J. R. Harris, *Biochem. J.* **242**, 849 (1987).

[16] G. Goldberger, G. A. P. Burns, M. Rits, M. D. Edge, and D. J. Kwiatkowski, *J. Biol. Chem.* **262**, 10065 (1987).

[17] A. W. Dodds and A. J. Day, *in* "Complement in Health and Disease" (K. Whaley, ed.). Kluwer Academic Publ., Dordrecht, The Netherlands, 1993, in press.

[18] A. J. Day and M. Baron, *J. Biomed. Sci.* **1**, 153 (1991).

[19] M. Baron, D. G. Norman, and I. D. Campbell, *Trends Biochem. Sci.* **16**, 13 (1991).

[20] R. G. DiScipio and T. E. Hugli, *J. Biol. Chem.* **264**, 16197 (1989).

[21] R. G. DiScipio, D. N. Chakravarti, H. J. Müller-Eberhard, and G. H. Fey, *J. Biol. Chem.* **263**, 549 (1988).

[22] B. A. Fernie, M. J. Hobart, R. DiScipio, and P. J. Lachmann, *Complement Inflammation* **8**, 148 (1991).

[23] M. Freeman, J. Ashkenas, D. J. G. Rees, D. M. Kingsley, N. G. Copeland, N. A. Jenkins, and M. Krieger, *Proc. Natl. Acad. Sci. U.S.A.* **87**, 8810 (1990).

sperm peptide "speract" receptor.[24] In the speract receptor there are four CD5 modules that comprise almost all of the extracellular domain of the protein. At present the function of this module is not known, but a role in protein–protein interaction seems likely.

Factor I contains two LDLr modules (Fig. 2a). The LDLr module was first identified in the low-density lipoprotein receptor,[25] which has seven contiguous LDLr modules at its N terminus, and is based on a consensus sequence of six invariant cysteines and other highly conserved residues (Fig. 2b). This module is also found in the α_2-macroglobulin (α_2M) receptor as well as in the terminal complement components C6, C7, C8α, C8β, and C9.[17]

There are no tertiary structural data available for the LDLr, CD5, or I/C6/C7 modules and the disulfide bond organizations are not known. However, in the first LDLr module of factor I, consensus Cys-1 and Cys-3 are missing and this suggests that these cysteines are disulfide bonded in other LDLr domains.

In Catterall et al.[15] amino acids 12–59 of factor I were incorrectly identified as an epidermal growth factor (EGF) domain. This region is now known to be part of the I/C6/C7 module. Similarly, Goldberger et al.[16] inappropriately assigned amino acids 71–136 of factor I as a complement control protein (CCP) module (also called a short consensus repeat or SCR). This region is now known to be part of the CD5 module. At the time these papers were published neither the CD5 module nor the I/C6/C7 had been described.

The structural gene for factor I has been mapped to chromosome 4 in humans using somatic cell hybrids.[16] It has been further localized to band 4q25 and has been shown to be fairly close to genes for alcohol dehydrogenase (ADH1, ADH2) and has been physically linked to epidermal growth factor.[26] The gene is approximately 25 kilobases (kb) in size[26]. There are

[24] L. J. Dangott, J. E. Jordan, R. A. Bellet, and D. L. Garbers, Proc. Natl. Acad. Sci. U.S.A. 86, 2128 (1989).
[25] T. Yamamoto, C. G. Davies, M. S. Brown, W. J. Schneider, M. L. Casey, J. L. Goldstein, and D. W. Russel, Cell (Cambridge, Mass.) 39, 27 (1984).
[26] K. Kölble, V. Buckle, and R. B. Sim, Cytogenet. Cell Genet. 51, 1024 (1989).

FIG. 2. (a) The modular composition of factor I and its cofactors is shown. The major polymorphic variant of CR1 (CR1-A) is made up of four long homologous repeats A–D (LHRA–D), each made up of seven CCPs, followed by two additional CCPs. The C3b- and C4b-binding sites are indicated. Probable ligand-binding sites in MCP, DAF, factor H, and C4bp are also indicated. (b) Consensus sequences for the modules that are shown in (a). Highly conserved residues are shown, and x indicates nonconserved positions. (c) A ribbon diagram of the three-dimensional structure of the sixteenth CCP module of factor H. The arrows denote β-strand structure. The N terminus is at the bottom.

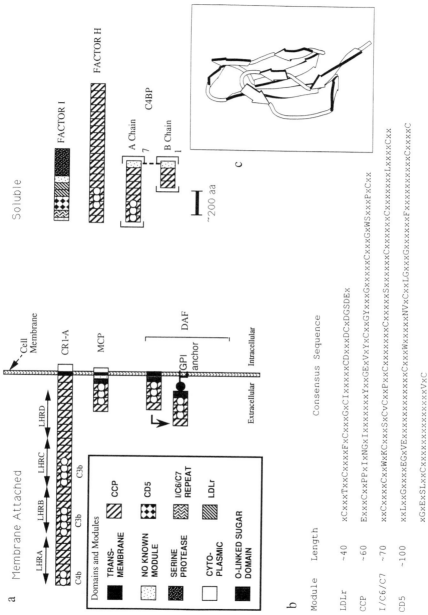

a Membrane Attached

Soluble

Cell Membrane

FACTOR I

FACTOR H

CR1-A

LHRA LHRB LHRC LHRD

C4b C3b C3b C3b

MCP

A Chain
7

B Chain

C4BP

~200 aa

DAF

GPI anchor

Extracellular | Intracellular

c

Domains and Modules

■ TRANS-MEMBRANE	▨ CCP	
░ NO KNOWN MODULE	◈ CD5	
▓ SERINE PROTEASE	▨ I/C6/C7 REPEAT	
□ CYTO-PLASMIC	▨ LDLr	
■ O-LINKED SUGAR DOMAIN		

b

Module	Length	Consensus Sequence
LDLr	~40	xCxxxTxxCxxxxFxCxxxGxCIxxxxxCDxxxDCxDGSDEx
CCP	~60	ExxxCxxPPxIxNGxIxxxxxxxYxxxGExVxYxCxxGYxxxGxxxxxCxxxGxWSxxxPxCxx
I/C6/C7	~70	xxCxxxxCxxWxxKCxxxSxCvCxxPxxCxxxxxxSxxxxxCxxxxxxCxxxxxxLxxxxCxx
CD5	~100	xxLxxGxxxxEGxVExxxxxxxxxxxxxCxxxWxxxxxxNVxCxxLGxxxGxxxxxFxxxxxxxxxxCxxxxC
		xGxExSLxxCxxxxxxxxxxxxxxxxVxC

at least two allelic variants of factor I that have different charges,[27] but this polymorphism has not been defined at the sequence level. At present there is no evidence for alternative splicing in the factor I gene and no genomic sequence is published.

Factor I Cofactors and Decay Accelerators

The other proteins discussed above, namely factor H, CR1, MCP, C4bp, and DAF, are diverse in molecular weight, but are all made up of the same type of structural unit, or module. They are therefore homologous to each other and are evolutionarily related.

Factor H and Structure of Complement Control Protein Module. Factor H is an abundant, single polypeptide chain plasma glycoprotein with a molecular weight of approximately 155,000. It is present in plasma at 200–500 μg/ml (1.3–3.2 μM). The amino acid sequences of mouse and human factor H have been derived by cDNA sequencing studies.[28,29] Human factor H is 1213 amino acids in length and the entire sequence is made up of 20 contiguous complement control protein (CCP) modules (also called short consensus repeats) (Fig. 2a). The CCP module is the most common structural unit of the cofactors of factor I, and the complement system as a whole, and has been extensively reviewed elsewhere.[17,30] The CCP module is approximately 60 amino acids in length and has a consensus sequence based on four generally invariant cysteines, disulfide bonded in the pattern Cys^1-Cys^3 and Cys^2-Cys^4 (see Ref. 31), and other highly conserved residues (Fig. 2b). A general feature of protein modules is that they are encoded at the gene level by discrete and symmetrical exons or cassettes of exons.[18,32] This is illustrated by the murine factor H gene, where CCP repeat 1 and CCP repeats 3–20 are encoded by single exons and CCP repeat 2 is split into two exons.[33] The human factor H gene has been mapped to chromosome 1q[34] and has been physically linked to the gene for factor XIIIb subunit using yeast artificial chromosomes.[35]

[27] H. Nishimukai and Y. Tamaki, *Hum. Hered.* **36,** 195 (1986).

[28] T. Kristensen and B. F. Tack, *Proc. Natl. Acad. Sci. U.S.A.* **83,** 3963 (1986).

[29] J. Ripoche, A. J. Day, T. J. R. Harris, and R. B. Sim, *Biochem. J.* **249,** 593 (1988).

[30] K. B. M. Reid and A. J. Day, *Immunol. Today* **10,** 177 (1989).

[31] J. Janatova, K. B. M. Reid, and A. C. Willis, *Biochemistry* **28,** 4754 (1989).

[32] L Patthy, *Curr. Biol.* **1,** 351 (1991).

[33] D. P. Vik, J. B. Keeney, P. Munoz-Canoves, D. D. Chaplin, and B. F. Tack, *J. Biol. Chem.* **263,** 16720 (1988).

[34] S. Hing, A. J. Day, S. J. Lintin, J. Ripoche, R. B. Sim, K. B. M. Reid, and E. Solomon, *Ann. Hum. Genet.* **52,** 117 (1988).

[35] D. Hourcade, J. M. Moulds, A. D. Garcia, P. Taillon-Miller, L. M. Wagner, T. W. Post, V. M. Holers, N. Bora, and J. P. Atkinson, *Complement Inflammation* **8,** 163 (1991).

The protein sequence of factor XIIIb subunit,[36] of the blood coagulation system, is composed entirely of 10 CCP modules that are highly related to those in factor H.

There are five allelic variants of factor H that differ by charge.[37] The difference between the two major alleles (FH1 and FH2 with gene frequencies of 0.685 and 0.301, respectively, in caucasoids) has been established and corresponds to a single point mutation at base 1277.[38] This leads to tyrosine/histidine polymorphism at amino acid 384.

Alternative splicing of the factor H gene gives rise to a truncated form of factor H (49 kDa), which corresponds to the N-terminal seven CCP modules and is active as a cofactor for factor I.[39–42] There are two further "truncated" forms of factor H; one is composed of five CCP modules (39–43 kDa) and is the product of a separate, but highly related, gene.[42,43] Its function is unknown. A further smaller protein of about 29 kDa appears to be the product of another related gene. The 29- and 39- to 43-kDa proteins are closely related in sequence to the C-terminal region of factor H. The truncated forms of factor H are present in plasma, but at low levels (approximately 2–20 μg/ml).

There is considerable interest in defining the exact binding sites for C3b and C4b in the complement control proteins. In factor H preliminary evidence suggests the C3b-binding site is located within CCP modules 3–5 as shown in Fig. 2a.[44]

The three-dimensional structures of several CCP modules from human factor H have been determined in solution by protein expression and proton nuclear magnetic resonance (^{1}H NMR) spectroscopy.[45,46] The CCP

[36] A. Ichinose, B. A. McMullen, K. Fujikawa, and E. W. Davie, *Biochemistry* **25**, 4633 (1986).

[37] S. Rodriquez de Cordoba and P. Rubinstein, *Immunogenetics* **25**, 267 (1987).

[38] A. J. Day, A. C. Willis, J. Ripoche, and R. B. Sim, *Immunogenetics* **27**, 211 (1988).

[39] M. A. McAleer, J. Ripoche, A. J. Day, B. E. Moffatt, M. Fontaine, and R. B. Sim, *Complement Inflammation* **6**, 366 (1989).

[40] R. Misasi, H. P. Huemer, W. Schwaeble, E. Solder, C. Larcher, and M. P. Dierich, *Eur. J. Immunol.* **19**, 1765 (1989).

[41] C. Estaller, V. Koistinen, W. Schwaeble, M. P. Dierich, and E. H. Weiss, *J. Immunol.* **146**, 3190 (1991).

[42] M. Fontaine, M. J. Demares, V. Koistinen, A. J. Day, C. Davrinche, R. B. Sim, and J. Ripoche, *Biochem. J.* **258**, 927 (1989).

[43] W. Schwaeble, E. Feifel, C. Estaller, A. Barbier, M. Molgg, V. Koistinen, E. H. Weiss, and M. P. Dierich, *Immunobiology* **182**, 307 (1991).

[44] A. Steinkasserer, P. Barlow, D. G. Norman, Z. Kertesz, I. D. Campbell, A. J. Day, and R. B. Sim, *Complement Inflammation* **8**, 225 (1991).

[45] P. N. Barlow, M. Baron, D. G. Norman, A. J. Day, A. C. Willis, R. B. Sim, and I. D. Campbell, *Biochemistry* **30**, 997 (1990).

[46] D. G. Norman, P. N. Barlow, M. Baron, A. J. Day, R. B. Sim, and I. D. Campbell, *J. Mol. Biol.* **219**, 717 (1991).

module is based on a β-sandwich arrangement where both faces of the sandwich contribute hydrophobic side chains that form a compact core. This is shown in Fig. 2c. An analysis of the sequence alignments of approximately 150 CCP repeats reveals a high degree of conservation of residues of structural importance (i.e., hydrophobic core residues), whereas insertions and deletions in various repeats are found in loops between elements of the secondary structure. This indicates that many members of the CCP module superfamily can be accurately modeled on the basis of these structures. The dimensions of an individual CCP module are approximately 3.8 × 2.0 × 2.0 nm.[46] This is consistent with physical studies on proteins containing many CCP modules (i.e., factor H, C4bp), which are very elongated proteins with structures resembling a string of beads.[46a,47]

C4b-Binding Protein. Human C4b-binding protein (C4bp) is a multimeric protein composed of seven identical A chains (each composed of eight CCP modules)[48] and a single B chain (containing three CCP modules)[49] that are covalently linked by their C-terminal regions as shown in Fig. 2a. The overall molecular weight is approximately 500,000, and C4bp is present in plasma at about 250 μg/ml (0.5 μM). C4bp has a spiderlike structure with flexible tentacles extending from a central core when visualized in the electron microscope.[47,50] The A chains are responsible for C4b binding and cofactor activity. It has been shown that in murine C4bp (which has only six CCP modules per A chain) the three N-terminal CCP modules are required for C4b binding, as shown in Fig. 2a.[51] The B chain has been implicated in binding of the vitamin K-dependent protein S, which is a regulator of coagulation.[12,49]

The human C4bpα gene has been estimated to be about 30 kb in length[52] and has been mapped to chromosome 1q.[34] There are three allelic variants of the human C4bp gene with gene frequencies of 0.986, 0.010, and 0.004 in caucasoids.[38] The A and B chain genes have been physically linked by pulsed-field gel electrophoresis and are less than 5 kb apart.[53]

[46a] P. N. Barlow, A. Steinkasserer, D. G. Norman, B. Kieffer, A. P. Wiles, R. B. Sim, and I. D. Campbell, *J. Mol. Biol.* in press (1993).

[47] R. B. Sim and S. J. Perkins, *Curr. Top. Microbiol.* **153,** 209 (1990).

[48] L. P. Chung, D. R. Bentley, and K. B. M. Reid, *Biochem. J.* **230,** 133 (1985).

[49] A. Hillarp and B. Dahlback, *Proc. Natl. Acad. Sci. U.S.A.* **87,** 1183 (1990).

[50] B. Dahlback, C. A. Smith, and H. J. Müller-Eberhard, *Proc. Natl. Acad. Sci. U.S.A.* **80,** 3461 (1983).

[51] R. T. Ogata, P. Mathias, B. Bradt, and N. R. Cooper, *Complement Inflammation* **8,** 202 (1991).

[52] S. J. Lintin and K. B. M. Reid, *FEBS Lett.* **204,** 77 (1986).

[53] F. Pardo-Manuel, J. Rey-Campos, A. Hillarp, B. Dahlback, and S. Rodriguez de Cordoba, *Proc. Natl. Acad. Sci. U.S.A.* **87,** 4529 (1990).

Complement Receptor Type 1. Complement receptor type 1 (CR1) is the largest of the cofactors for factor I and exists as four distinct size variants, CR1-A, -B, -C, and -D, which have apparent molecular weights on SDS–PAGE (under reducing conditions) of about 220,000, 250,000, 190,000, and 280,000, respectively. The major allelic variant, CR1-A, is composed of 30 CCP modules that are organized in higher order structures called long homologous repeats (LHRs), each consisting of seven CCP modules.[54,55] In CR1-A there are four LHRs (A, B, C, and D) extending from the N terminus, which are followed by two CCP modules (which are not part of a LHR), a membrane-spanning sequence, and a 70-amino acid cytoplasmic domain (see Fig. 2a). The basis for the size polymorphism between the four allelic variants of CR1 is thought to correspond to the number of LHRs present in each gene. For example in CR1-D there is an extra N-terminal LHR.[56] The intron–exon organization within each LHR is conserved with repeats 1, 5, and 7 being encoded in single exons, repeats 2 and 6 each being encoded by two exons, and CCP repeats 3 and 4 being contained in a common exon.[56] The structural gene for CR1 has been mapped by *in situ* hybridization to band q32 of the human chromosome 1.[57] There is evidence for alternative splicing in the CR1 gene giving rise to a truncated secreted form consisting of $8\frac{1}{2}$ CCP modules,[58] which is reminiscent of the truncated form of factor H (see above).

The N-terminal two CCP modules of LHRA have been shown to contain a site determining C4b specificity and the N-terminal two CCP modules of LHRB and LHRC have a site that determines C3b specificity.[55] More recently it has been shown that for full C3b binding and cofactor activity, four CCP modules are required as shown in Fig. 2a (i.e., modules 1–4 of LHRB and LHRC).[59] A short linear sequence in CCP module 9 (LHRB-2) has been shown to be important in C3b binding,[60] and this is consistent with its position in the CCP module tertiary structure.[46]

Membrane Cofactor Protein. Membrane cofactor protein (MCP) con-

[54] L. B. Klickstein, W. W. Wong, J. A. Smith, J. H. Weis, J. A. Wilson, and D. T. Fearon, *J. Exp. Med.* **165,** 1095 (1987).

[55] L. B. Klickstein, T. J. Bartow, V. Miletič, L. D. Rabson, J. A. Smith, and D. T. Fearon, *J. Exp. Med.* **168,** 1699 (1988).

[56] D. P. Vik, B. F. Tack, and W. W. Wong, *Complement Inflammation* **8,** 238 (1991).

[57] J. H. Weis, C. C. Morton, G. A. Burns, J. J. Weis, L. B. Klickstein, W. W. Wong, and D. T. Fearon, *J. Immunol.* **138,** 312 (1987).

[58] D. Hourcade, D. R. Miesner, J. P. Atkinson, and V. M. Holers, *J. Exp. Med.* **168,** 1255 (1988).

[59] K. R. Kalli, T. J. Bartow, P. Hsu, L. B. Klickstein, J. M. Ahearn, A. K. Matsumoto, and D. T. Fearon, *Complement Inflammation* **8,** 171 (1991).

[60] M. Krych, D. Hourcade, and J. P. Atkinson, *Proc. Natl. Acad. Sci. U.S.A.* **88,** 4353 (1991).

sists of four N-terminal CCP modules, followed by a region that is heavily O-glycosylated, a transmembrane segment, and a short cytoplasmic domain (see Fig. 2a).[61] The protein was formerly called gp45–70, to reflect its heterogeneity in molecular mass (45–70 kDa), which arises from glycosylation differences and from multiple alternative splicing. The gene for MCP has been mapped to band q32 of human chromosome 1.[61] Deletion mutants of MCP have been constructed differing in the number of CCP modules present.[62] Mutants lacking the third and fourth CCP module did not bind C4b or C3b and lacked cofactor activity. The mutant deleted of repeat 2 bound to, but lacked cofactor activity for, C3b and did not bind C4b. Deletion of the first CCP module did not affect C3b binding or cofactor activity but diminished the efficiency of C4b binding. Therefore, for full functional activity, all four CCP modules are probably required, with the last three being particularly important (see Fig. 2a).

Decay-Accelerating Factor. Decay-accelerating factor (DAF) is somewhat like MCP in primary structure and size, having four N-terminal CCP modules followed by a region that is rich in O-linked sugars and a transmembrane segment.[63,64] The protein sequence ends with a membrane-spanning sequence and has no basic residues following this, which are important for anchoring the protein to the membrane. Therefore, the transmembrane sequence is removed posttranslationally and a C-terminal glycophospholipid anchor is attached as shown in Fig. 2a (see Ref. 65). Studies to identify the functionally important CCP modules of DAF indicate that repeats 2–4 are essential for DAF activity (see Fig. 2a).[66]

Regulation of Complement Activation Gene Cluster. The complement control proteins CR1, C4bp, MCP, DAF, and factor H are composed almost entirely of CCP modules as shown in Fig. 2a and as discussed above. Complement receptor type 2 (CR2) also has this structure, but is not thought to share the regulatory activities discussed above. These proteins are all encoded in the "regulation of complement activation" (RCA) gene cluster on human chromosome 1q.[37] The genes for MCP,

[61] D. M. Lublin, M. K. Liszewski, T. W. Post, M. A. Arce, M. M. Le Beau, M. B. Rebentisch, R. S. Lemons, S. Tsukasa, and J. P. Atkinson, *J. Exp. Med.* **168,** 181 (1988).

[62] E. M. Adams, M. C. Brown, M. Nunge, M. Krych, and J. P. Atkinson, *Complement Inflammation* **8,** 123 (1991).

[63] I. W. Caras, M. A. Davitz, L. Rhee, G. Weddell, D. W. Martin, and V. Nussenzweig, *Nature (London)* **325,** 545 (1987).

[64] M. E. Medof, D. M. Lublin, V. M. Holers, D. J. Ayers, R. R. Getty, J. F. Leykam, J. P. Atkinson, and M. L. Tykocinski, *Proc. Natl. Acad. Sci. U.S.A.* **84,** 2007 (1987).

[65] D. M. Lublin and J. P. Atkinson, *Annu. Rev. Immunol.* **7,** 35 (1989).

[66] D. M. Lublin, T. Kinoshita, T. Fujita, D. J. Anstee, and W. F. Rosse, *Complement Inflammation* **8,** 184 (1991).

CR1, CR2, DAF, C4bpα, and C4bpβ are all located within a megabase of DNA and map to 1q32, with the factor H locus being 5–10 centimorgans (cM) away from this cluster.[37,53,57,67] Another CCP module-containing protein, coagulation factor XIIIb subunit, which consists of 10 CCP repeats, has been closely linked to factor H and therefore can be considered as a member of the RCA gene family. Therefore, there are at least 112 CCP modules from seven proteins encoded in the extended RCA gene cluster. This close association of many CCP modules indicates that these proteins have coevolved by a combination of gene duplication and exon shuffling.[17]

Isolation of Factor I

By Conventional Chromatography

Isolation of factor I by conventional chromatography methods involves multiple steps, and the yield is relatively low. A satisfactory method for isolation of factor I has been described in this series.[1]

By Binding to Anti-factor I Antibodies

In general, immunoaffinity methods for isolation of factor I are more rapid and give higher yield than multiple-step chromatography. Methods using polyclonal antibodies have been described briefly in this series.[2,6] The more detailed description below is based on the monoclonal antibody affinity method described by Hsiung et al.[68] All procedures are carried out at 4–6° unless stated otherwise.

Outdated or fresh-frozen human plasma or serum is used. Factor I is stable on storage in serum/plasma, so fresh material is not essential. The plasma/serum is first passed through a column of lysine-Sepharose, to remove plasmin/plasminogen: plasmin/plasminogen binds weakly to immunoglobulin columns, and will contaminate the final product if it is not removed in advance. The plasma/serum is then passed through a column of nonimmune IgG immobilized on Sepharose: this removes proteins that bind to immunoglobulin or to modified agarose (e.g, C1q, fibronectin, rheumatoid factors). The plasma/serum is finally passed through a column of monoclonal anti-factor I immobilized on Sepharose, and bound factor I is eluted in a chaotropic buffer.

[67] J. Rey-Campos, P. Rubinstein, and S. Rodriguez de Cordoba, J. Exp. Med. 167, 664 (1988).
[68] L.-M. Hsiung, A. N. Barclay, M. R. Brandon, E. Sim, and R. R. Porter, Biochem. J. 203, 293 (1982).

Preparation of Monoclonal Antibody Columns. The hybridoma MRC OX21[68,69] (IgG$_1$ subclass) is passaged in mice and 50 ml of ascites fluid obtained. The monoclonal antibody MRC OX21 is purified by triple 18% (w/v) sodium sulfate precipitation, followed by DEAE-Sephacel chromatography. Briefly, these procedures are as follows: ascites fluid is made 18% (w/v) sodium sulfate by addition of anhydrous sodium sulfate, and incubated at 37° for 1 hr, then centrifuged for 30 min at 3000 g at 20°. The precipitate is redissolved in 30 ml of water, again made 18% (w/v) sodium sulfate, and the procedure is repeated. The second precipitate is dissolved in 30 ml of water, again made 18% (w/v) sodium sulfate, and the precipitation is done a third time. The final pellet is dissolved in water to a volume of 50 ml and is dialyzed twice against 2 liters of 10 mM potassium phosphate, pH 7.0. The monoclonal antibody (about 85% pure at this stage) is loaded on to a column (22 × 2.2 cm diameter) of DEAE-Sephacel (Pharmacia, Uppsala, Sweden) equilibrated in the dialysis buffer. The column is washed with 1 liter of the same buffer, then the antibody is eluted, as the only major peak, with a linear gradient, total volume 600 ml, of 0–300 mM NaCl in the starting buffer. The final yield of MRC OX21 is ~150 mg.

The purified MRC OX21 (150 mg in 80–100 ml) is dialyzed against 50 mM potassium phosphate, 150 mM KCl, pH 7.2–7.5, and mixed on a slow rotary stirrer (2 hr, 20°) with 15–20 ml packed volume (5–6 g dry weight) of CNBr-activated Sepharose 4B (Pharmacia). It is particularly important that the pH for coupling of the antibody to the resin is not greater than pH 7.5, as higher pH leads to loss of the antigen-binding capacity of the immobilized antibody. The resin is then washed with 150 ml of 2 M NaCl and mixed as above with 100 ml of 100 mM ethanolamine hydrochloride, 150 mM NaCl, pH 8.5, to block reactive sites. The coupling efficiency is generally 95–98%, and ~8–10 mg of antibody should be bound per milliliter (packed volume) of Sepharose.

Preparation of Guard (Preadsorption) Column. Nonimmune IgG is prepared from rabbit or human serum by triple precipitation with sodium sulfate, as above, but using 14% (w/v) sodium sulfate at each step. Rabbit or human IgG preparations are coupled to CNBr-activated Sepharose as described above, to a concentration of 10 mg IgG/ml of packed Sepharose.

---

[69] MRC OX21, MRC OX23, and MRC OX24 hybridomas are available from the European Collection of Animal Cell Cultures (ECACC), Division of Biologics, PHLS Centre for Applied Microbiology & Research, Porton Down, Salisbury, SP4 0JG, U.K. Preparations of the antibodies are available commercially from Serotec Ltd, 22 Bankside, Station Approach, Kidlington, Oxford OX5 1JE, U.K.

Removal of Plasminogen/Plasmin. Up to 2 liters of pooled outdated human plasma (with citrate anticoagulant) or serum, made 5 mM with EDTA, is made 0.5 mM with the protease inhibitor Pefabloc-SC[70] and passed through a lysine-Sepharose column (8.5 × 5.2 cm diameter) equilibrated in 100 mM sodium phosphate, 150 mM NaCl, 15 mM EDTA, pH 7.4, to remove plasminogen/plasmin.[71] The eluted plasma, now diluted to a volume of 2.5 liters, is again made 0.5 mM with Pefabloc-SC and stored (1–2 days) at 4° for further use. The lysine-Sepharose column is washed with 200 mM ε-aminocaproic acid in 100 mm sodium phosphate, 150 mM NaCl, 15 mM EDTA, pH 7.4, to elute bound plasminogen/plasmin, and reequilibrated in the same buffer without ε-aminocaproic acid.

Purification Procedure. Plasma/serum depleted of plasminogen/plasmin (400 ml) is dialyzed against 25 mM Tris-HCl, 140 mM NaCl, 0.5 mM EDTA, pH 7.4 (running buffer), and is made 0.5 mM with Pefabloc-SC. The guard column (8.5 × 2 cm diameter) and affinity column (MRC OX21-Sepharose, 10 × 1.5 cm diameter) are connected in series, equilibrated in the running buffer, and the plasma/serum run through at a rate of 25 ml/hr. The guard column is then disconnected and the affinity column washed with 500 ml of the running buffer. Factor I is then eluted with 200 ml of the chaotrope, 3 M MgCl$_2$, adjusted to pH 6.8–7.0 by addition of Tris base. Factor I emerges as a single sharp peak with a trailing edge. Analysis of reduced or nonreduced fractions by SDS–PAGE shows the characteristic 88-kDa band (nonreduced factor I), and the 50- and 38-kDa bands of reduced factor I. The MRC OX21 column is reequilibrated in the running buffer and the guard column is regenerated by washing with 200 ml of 3 M MgCl$_2$, followed by reequilibration in the running buffer. C1q can be prepared readily from the guard column eluate.

Factor I eluted from the antibody column is generally >98% pure. The material is then dialyzed against 25 mM Tris-HCl, 140 mM NaCl, 0.5 mM EDTA, pH 7.4, and the concentration adjusted, as desired, by ultrafiltration (e.g., Amicon, Beverly, MA, PM10 membrane) or other convenient method such as precipitation with ammonium sulfate at 60% saturation.[68] Factor I is quite soluble and can readily be concentrated up to about 3 mg/ml. For long-term storage, factor I should be dialyzed against 10 mM potassium phosphate, 140 mM NaCl, 0.5 mM EDTA, pH 7.0–7.5, and stored at −20°. Factor I eluted from the column in 3 M MgCl$_2$ cannot be

---

[70] The broad-spectrum serine protease inhibitor Pefabloc-SC [4-(2-aminoethyl)benzenesulfonylfluoride hydrochloride] is a less toxic substitute for DFP (diisopropyl fluorophosphate). It is generally more effective than PMSF (phenylmethylsulfonyl fluoride), and is available from Pentapharm, Engelgasse 109, CH-4002, Basel, Switzerland.

[71] B. F. Tack, J. Janatova, M. L. Thomas, R. A. Harrison, and C. H. Hammer, this series, Vol. 80, p. 64.

dialyzed directly against the phosphate buffer because of precipitation of magnesium phosphate and hydroxide. Factor I can be stored at $-20°$ for >1 year without degradation or loss of activity.

Minor contamination of the purified factor I with human IgG is sometimes visible. If it is necessary to remove traces of IgG, the factor I preparation should be dialyzed against 25 mM Tris-HCl, 140 mM NaCl, 0.5 mM EDTA, pH 7.4, concentrated to about 10 ml, and run on a gel-filtration column (Sephacryl S-200, Pharmacia, Uppsala, Sweden), 100 × 2.5 cm diameter, in the same buffer.[68]) Factor I is readily separated from IgG, and is eluted as the major peak.

The yield of factor I by this method is about 70% (i.e., about 10 mg from 400 ml of plasma). The guard column and monoclonal antibody affinity column can be reused up to about 30 times, depending on how thoroughly they are washed with 3 M MgCl$_2$. The binding capacity of these columns diminishes gradually. Elution of bound protein with 3 M MgCl$_2$ is more efficient, and preserves the binding capacity of the columns for much longer than do the acid or alkaline elution techniques often used with antibody columns.

Isolation of Factor H

By Binding to Anti-factor H Antibodies

In general, immunoaffinity methods for isolation of factor H are also more rapid and give higher yield than multiple-step chromatography. Factor H can conveniently be prepared using the same method as described above for factor I. The major differences are (1) the monoclonal anti-factor H antibodies, MRC OX23 or MRC OX24,[69,72] are used instead of MRC OX21 and (2) fresh-frozen plasma or fresh plasma should be used, rather than serum or outdated plasma. Factor H undergoes gradual cleavage of the single 155-kDa polypeptide chain into a form with two disulfide-linked chains (approximately 38 and 120 kDa) on storage in plasma, probably by the action of plasmin. This form has slightly modified activity.[73] As for the factor I preparation above, the plasma is first passed through a column of lysine-Sepharose, then through a column of nonimmune IgG immobilized on Sepharose, and finally through a column of monoclonal anti-factor H immobilized on Sepharose, and bound factor H is eluted in 3 M MgCl$_2$.

Preparation of Monoclonal Antibody Columns. The antibodies MRC

---

[72] E. Sim, M. S. Palmer, M. Puklavec, and R. B. Sim, *Biosci. Rep.* **3,** 1119 (1983).
[73] R. B. Sim and R. G. DiScipio, *Biochem. J.* **205,** 285 (1982).

OX24 or MRC OX23 are prepared from ascites fluid and bound to CNBr-activated Sepharose exactly as described above for MRC OX21. Again, it is particularly important that the pH for coupling of the antibody to the resin is not greater than pH 7.5, as higher pH leads to loss of the antigen-binding capacity of the immobilized antibody. Both anti-factor H antibodies perform equally well in this procedure, although they are directed against different epitopes.[72] Therefore either may be used, and the comments below apply to purification procedures with either of the two antibodies. The binding capacity of the MRC OX23 or OX24 columns for factor H is approximately equivalent to the weight of antibody bound to the column (i.e., a column containing 150 mg of immobilized MRC OX23 will bind ~150 mg of factor H).

Preparation of Guard (Preadsorption) Column. The same guard column as described above for factor I purification is used.

Purification Procedure. Plasma depleted of plasminogen/plasmin (400 ml), as described above, is dialyzed against 25 mM Tris-HCl, 140 mM NaCl, 0.5 mM EDTA, pH 7.4 (running buffer), and made 0.5 mM with Pefabloc-SC. The guard column (8.5 × 2 cm diameter) and affinity column (MRC OX23- or OX24-Sepharose, 10 × 1.5 cm diameter) are connected in series, equilibrated in the running buffer, and the plasma run through at a rate of 25 ml/hr. The guard column is then disconnected and the affinity column washed with 500 ml of the running buffer. Factor H is then eluted with 200 ml of the chaotrope, 3 M MgCl$_2$, adjusted to pH 6.8–7.0 by addition of Tris base. Factor H emerges as a single sharp peak with a pronounced trailing edge. Analysis of reduced or nonreduced fractions by SDS–PAGE shows the characteristic single band of factor H. Factor H migrates anomalously in SDS–PAGE, generally migrating with an apparent molecular weight, versus globular protein standards, of 120,000–145,000 (unreduced) or 150,000–170,000 (reduced). The antibody column and the guard column are regenerated and and reequilibrated as described above.

Factor H eluted from the antibody column is generally >98% pure, and overall yield is about 80%. The material is then dialyzed against water, then against 10 mM potassium phosphate, 140 mM NaCl, 0.5 mM EDTA, pH 7.0–7.5, and stored at $-20°$, after addition of Pefabloc-SC to 0.5 mM. Factor H can be stored at $-20°$ for >1 year without degradation or loss of activity. Like factor I, factor H is relatively soluble and can be concentrated to about 3 mg/ml by ultrafiltration, or to 10–20 mg/ml by precipitation with 14% (w/v) polyethylene glycol 3500 (Sigma, St. Louis, MO) and redissolving in the same phosphate buffer. Factor H can also be concentrated easily by Zn^{2+} ions: the protein, in a nonchelating buffer at physiological salt strength and pH 7.0–8.0, is made 1 mM with ZnCl$_2$.

Precipitation occurs over a period of 24 hr at 4°. The precipitate is harvested by centrifugation at 10,000 g for 10 min, and redissolves readily in a physiological buffer containing 5 mM EDTA, at pH 7.0–8.0.

The monoclonal antibodies MRC OX24 and MRC OX23 also recognize one of the truncated forms of factor H mentioned above. A form containing the first seven (N-terminal) modules of factor H, followed by the short unique amino acid sequence SFTL, arises by alternative splicing, and is present in plasma, at a level of about 1–5 μg/ml.[40,42] This protein, H-49, which also has factor I cofactor activity, binds to the monoclonal antibodies and is eluted with factor H. H-49 is evidently not stable in plasma, as it is detectable only in preparations made from fresh plasma. In factor H preparations eluted from the antibody columns, H-49 appears as a faint band ($<$1% of the concentration of factor H) migrating with an apparent molecular weight on SDS–PAGE of 42,000–49,000 (reduced) or 32,000–33,000 (unreduced). H-49 adsorbs to surfaces and is generally lost on further handling of the factor H preparation.

The guard column and monoclonal antibody affinity columns, as with factor I, can be reused up to about 30 times, depending on how thoroughly they are washed with 3 M MgCl$_2$. The binding capacity of these columns diminishes gradually. The acid or alkaline elution techniques often used with antibody columns do not work well with MRC OX23 and OX24, as antigen-binding capacity is rapidly destroyed by acid elution, and alkaline elution is relatively ineffective, and, at pH $>$ 11, causes precipitation of factor H.

By Conventional Chromatography

Isolation of factor H by conventional chromatography methods involves multiple steps, and the yield is relatively low. Satisfactory methods for isolation of factor H have been published.[1,73,74] A brief updating of the method described by Ripoche et al.[74] is given below. All procedures are at 4–6° unless stated otherwise.

Step 1: Preparation of Euglobulins. Fresh serum, or serum made from fresh-frozen plasma (500 ml), is made 20 mM with sodium EDTA (pH 7.2) and 50 mM with ε-aminocaproic acid. Serum is then dialyzed twice (16 hr, 4°), against 20 volumes of 5 mM EDTA, 2 mM ε-aminocaproic acid, pH 5.3. The final pH at equilibrium is 5.4. After centrifugation (11,000 g, 30 min), the precipitate is harvested and washed twice with the cold dialysis buffer, then redissolved in ~50 ml of 150 mM NaCl, 50 mM ε-amino caproic acid, 20 mM EDTA, pH 7.2.

---

[74] J. Ripoche, A. Al-Salihi, J. Rousseaux, and M. Fontaine, *Biochem. J.* **221**, 89 (1984).

Step 2: DEAE-Sephacel Chromatography. The redissolved euglobulins are diluted with distilled water to a conductivity of 2 mS/cm (20°) and run on a column of DEAE-Sephacel (Pharmacia) (90 × 2.6 cm diameter) equilibrated in 5 mM Tris, 5 mM EDTA, 10 mM ε-aminocaproic acid, pH 7.2. The column is washed with starting buffer until the OD_{280} is close to zero, then two gradient elutions are performed: the first is from 0 to 100 mM NaCl (total volume 250 ml) in the starting buffer. The second is from 100 to 250 mM NaCl (total volume 750 ml) in the starting buffer. Two major peaks emerge in the gradient. Factor H is the first of these, and factor H is eluted at approximately 90 mM NaCl. Factor H is 90% pure at this stage, with minor contamination by immunoglobulins and C4bp. The material is concentrated to 10–15 ml by ultrafiltration.

Step 3: Gel Filtration. The factor H preparation is finally purified on a column (90 × 2.6 cm diameter) of Sephacryl S-300 (Pharmacia) in 150 mM NaCl, 50 mM ε-aminocaproic acid, 20 mM EDTA, pH 7.2. Factor H emerges as the single major peak, with contaminants running at the leading and trailing edges. The yield of factor H by this method is 30–37%.

As noted in Ref. 74, factor H, purified by either of the methods described above, can be further fractionated, on phenyl-Sepharose, into two forms, termed φ1 and φ2. These are indistinguishable on SDS–PAGE, and do not appear to differ in cofactor activity, but only the φ2 form binds specifically to B lymphoblastoid cells (i.e., binds to a factor H receptor). These two forms are likely to differ by a posttranslational modification, possibly involving tyrosine sulfation. Other properties known to differ between the two forms include the monoclonal antibody BGH1, which binds preferentially to φ2, and the aggregation of φ2, but not φ1, at zero salt strength.[74,75]

Isolation of C4bp

C4bp is a large protein of limited solubility, and isolation procedures in general do not provide a high yield. It is particularly difficult to separate from IgM. Methods for preparation of C4bp, with yields of 20 mg/liter, using multiple chromatographic steps have been described in this series.[2,76] C4bp can also be prepared[77] as a by-product of C1 purification,[78] using affinity chromatography on C4c-Sepharose.

[75] J. Ripoche, A. Erdei, D. Gilbert, A. Al-Salihi, R. B. Sim, and M. Fontaine, *Biochem. J.* **253**, 475 (1988).
[76] V. Nussenzweig and R. Melton, this series, Vol. 80, p. 124.
[77] E. Sim and R. B. Sim, *Biochem. J.* **210**, 567 (1983).
[78] R. B. Sim, this series, Vol. 80, p. 6.

Isolation of Membrane-Bound Cofactors and Regulatory Proteins

The isolation of CR1 from erythrocytes, using monoclonal antibody and dye–ligand affinity, has been described in detail in this series.[79] Other large-scale purification methods for CR1 include monoclonal antibody[80] and C3b affinity[81] methods. MCP can also be purified from cultured cells (e.g., U937 cells) by affinity for C3b or C3(H_2O).[82] An immunoaffinity method for DAF purification from erythrocytes, with a yield of 20–25%, is available.[83]

Enzymatic Properties of Factor I

The enzymatic properties of factor I have been described in detail previously.[1] Factor I is relatively stable to denaturation and is not inhibited by protein protease inhibitors or by the common protease inhibitors such as DFP or PMSF. It has recently been suggested that factor I may react with DFP when factor I is bound to its substrate, C3b,[84] but this is not in agreement with previous work[1,85] and requires further investigation. Factor I, in combination with appropriate cofactors, cleaves two Arg-Ser bonds, at positions 1281 and 1299, in C3b, converting it to iC3b. It is generally believed that factor I, with CR1 as cofactor, cleaves a third arginyl bond (Arg-Glu) in iC3b, cleaving it to C3c (140 kDa) plus C3dg (35 kDa). There are some reservations about this specificity, however, as discussed in Refs. 84 and 85. In C4b, factor I also cleaves two arginyl bonds to form C4c and C4d. The specificity of the enzyme is therefore very restricted, and factor I can be assayed conveniently only by using its natural substrates and in the presence of cofactors. Convenient assay methods for assessing factor I and the cofactor-dependent breakdown of C3b and C4b have been described previously in this series. Such methods include the three following approaches: (1) Functional assay of erythro-cyte-bound C3b breakdown in the presence of factors H and I.[1] This can be adapted to assess either factor I activity, in the presence of a constant quantity of cofactor, or cofactor activity, in the presence of constant factor I. (2) Functional assay of the decay acceleration of C3bBb bound

[79] W. W. Wong and D. T. Fearon, this series, Vol. 150, p. 579.

[80] V. M. Holers, T. Seya, E. Brown, J. J. O'Shea, and J. P. Atkinson, *Complement* **3**, 63 (1986).

[81] R. B. Sim, *Biochem. J.* **232**, 883 (1985).

[82] T. Seya, J. R. Turner, and J. P. Atkinson, *J. Exp. Med.* **163**, 837 (1986).

[83] M. A. Davitz, D. Schlesinger, and V. Nussenzweig, *J. Immunol. Methods* **97**, 71 (1987).

[84] J. D. Becherer, J. Alsenz, and J. D. Lambris, *Curr. Top. Microbiol. Immunol.* **153**, 45 (1989).

[85] V. Malhotra and R. B. Sim, *Biochem. Soc. Trans.* **12**, 781 (1984).

to erythrocytes.[6] This assay is suitable for assessing factor I-independent decay acceleration activity, if factor I is absent, or for measuring factor I and factor I cofactor activity under appropriate circumstances. (3) Assay of the breakdown of fluid-phase purified radioiodinated C3b, in the presence of factor I and cofactors, has also been described in detail.[77] This type of assay relies on fluid-phase incubation of C3b, factor I, and cofactor, followed by analysis, by SDS–PAGE and autoradiography, of the extent of breakdown, by factor I, of the α' chain of C3b (110 kDa) to 68- and 43-kDa fragments. This assay is again suitable for measuring either factor I activity or the activity of cofactors.

[2] Properdin

By Kathleen F. Nolan and Kenneth B. M. Reid

Properdin was first described by Pillemer et al. in 1954[1] when it was proposed that plasma contained a nonspecific defense mechanism that was mediated by properdin, acting in conjunction with complement, in a Mg^{2+}-dependent manner. Although these studies initially introduced the concept of an antibody-independent, alternative pathway of complement activation, these views were only fully accepted in the late 1960s after Pensky et al.[2] had rigorously demonstrated properdin to be a unique serum protein, distinct from immunoglobulins and components of the classical pathway (for review, see Lepow[3]).

Structure of Properdin

Properdin is present in plasma in the form of oligomers, mainly dimers (P_2), trimers (P_3), and tetramers (P_4), of an approximately 53-kDa monomer. The apparent molecular mass of the monomer, when examined under dissociating conditions, has been reported to lie between 46 and 57 kDa.[4-6] This wide range is probably due to the asymmetric nature of the monomer

[1] L. Pillemer, L. Blum, I. H. Lepow, O. A. Ross, E. W. Todd, and A. C. Wardlaw, Science 120, 279 (1954).
[2] J. Pensky, C. F. Hinz, E. W. Todd, R. J. Wedgwood, J. T. Boyer, and I. H. Lepow, J. Immunol. 100, 142 (1968).
[3] I. H. Lepow, J. Immunol. 125, 471 (1980).
[4] K. B. M. Reid and J. Gagnon, Mol. Immunol. 18, 949 (1981).
[5] J. O. Minta and I. H. Lepow, Immunochemistry 11, 361 (1974).
[6] R. G. DiScipio, Mol. Immunol. 19, 631 (1982).

and its relatively high (9.8%, w/w) carbohydrate content.[5] The molecular mass of the unglycosylated monomer of human properdin (a 442-residue-long polypeptide chain), as calculated from the cDNA-derived amino acid composition,[7] is 48,949 Da. Assuming 9.8% (w/w) carbohydrate (in the form of 3.8% hexose, 3.8% sialic acid, 1.5% hexosamine, and 0.8% fucose[5]), the estimated molecular mass of the glycosylated monomer of human properdin is 53,246 Da.[8]

In plasma, the oligomeric forms of properdin have been shown to exhibit a natural distribution of $26:54:20$, $(P_2:P_3:P_4)$.[9] When these oligomeric forms are viewed by replica electron microscopy[10] they appear as cyclic structures, formed by the presumably "head-to-tail" interaction of the N- and C-terminal regions of the monomers. The monomers appear extended, each ~26 nm long and 2.5–3.0 nm wide, with a reproducible bend toward the center and pronounced thickening, to ~4 nm, toward each end. The variation in intersubunit contact angle, from a few degrees in dimers to 120° in hexamers, could confer flexibility on the cyclic structures necessary for them to cross-link and stabilize clusters of C3 convertases on activating surfaces. The specific activities of these oligomers, in a functional assay, differ quite markedly, with P_4, on a molar basis, being 10-fold more active than P_2.[9] It has been suggested that this difference may reflect an increased affinity resulting from the presence of multiple binding sites in the P_4 oligomers. The binding site for properdin, on C3, has been localized to residues 1402–1435 in the α chain of C3[11] as judged by peptide inhibition studies, and the analysis of overlapping peptides indicated that the site could be further refined to residues 1424–1432.[12–14] It has been shown that properdin also interacts with factor B[15] and this interaction appears to take place through sites located on both the Ba and Bb portions of the molecule.[16]

[7] K. F. Nolan, W. Schwaeble, S. Kaluz, M. P. Dierich, and K. B. M. Reid, *Eur. J. Immunol.* **21**, 771 (1991).

[8] K. F. Nolan and K. B. M. Reid, *Biochem. Soc. Trans.* **18**, 1161 (1990).

[9] M. K. Pangburn, *J. Immunol.* **142**, 202 (1989).

[10] C. A. Smith, M. K. Pangburn, C. W. Vogel, and H. J. Müller-Eberhard, *J. Biol. Chem.* **259**, 4582 (1984).

[11] M. E. Daoudaki, J. D. Becherer, and J. D. Lambris, *J. Immunol.* **140**, 1577 (1988).

[12] J. D. Becherer, J. Alsenz, and J. D. Lambris, *Curr. Top. Microbiol. Immunol.* **153**, 45 (1989).

[13] J. Alsenz, J. D. Becherer, I. Esparza, M. E. Daoudaki, D. Avila, S. Oppermann, and J. D. Lambris, *Complement Inflammation* **6**, 307 (abstr.) (1989).

[14] D. Grossberger, A. Marcuz, L. Du Pasquier, and J. D. Lambris, *Proc. Natl. Acad. Sci. U.S.A.* **86**, 1323 (1989).

[15] R. G. DiScipio, *Biochem. J.* **199**, 485 (1981).

[16] T. C. Farries, P. J. Lachmann, and R. A. Harrison, *Biochem. J.* **252**, 47 (1988).

Monomeric and open circular forms of properdin are not considered to be present in nondissociating conditions, but the observation that the monomers are generated under low pH conditions suggests that the interactions between them are probably mainly ionic.[9,10] If, after acid denaturation, the monomers are allowed to renature, then a distribution of oligomers closely resembling that found in serum is seen.[9] This suggests that the assembly of oligomers occurs spontaneously following biosynthesis of monomers and that this distribution may be a thermodynamically derived ratio, similar to that discussed for the distribution of cyclic polymers formed by the interactions of IgG with a small bivalent hapten.[17,18]

The majority of the properdin amino acid sequence[7,19,20] is composed of five tandemly repeating motifs, each approximately 60 amino acids long and showing similarity to the type 1 repeats of the cell adhesion protein thrombospondin, i.e., thrombospondin repeat (TSR) modules.[19,21,22] Motifs of this type are also seen in the terminal components of complement (C6, C7, C8α, C8β, and C9)[23–29] and in the malaria parasite proteins TRAP and SSP2[30,31] (Fig. 1). In properdin these motifs, or TSR modules, are flanked by a 49-amino acid N-terminal region and a 38-amino acid segment that resembles the N-terminal region of a TSR module and which is followed by a distinct, 54-residue-long C-terminal region.[20] Secondary structure predictions for human properdin, based on Fourier transform infrared

[17] V. N. Schumaker, G. W. Seegan, C. A. Smith, S. K. Ma, J. D. Rodwell, and M. F. Schumaker, *Mol. Immunol.* **17,** 413 (1980).

[18] B. G. Archer and H. Krakauer, *Biochemistry* **16,** 618 (1977).

[19] D. Goundis and K. B. M. Reid, *Nature (London)* **335,** 82 (1988).

[20] K. F. Nolan, S. Kaluz, J. M. G. Higgins, D. Goundis, and K. B. M. Reid, *Biochem. J.* **287,** 291 (1992).

[21] D. Goundis, Ph.D. Thesis, Oxford University (1988).

[22] J. Lawler and R. O. Hynes, *J. Cell Biol.* **103,** 1635 (1986).

[23] R. G. DiScipio and T. E. Hugli, *Complement Inflammation* **6,** 330 (1989).

[24] R. G. DiScipio, D. N. Chakravarti, H. J. Müller-Eberhard, and G. H. Fey, *J. Biol. Chem.* **263,** 549 (1988).

[25] A. G. Rao, O. M. Z. Howard, S. C. Ng, A. S. Whitehead, H. R. Colten, and J. M. Sodetz, *Biochemistry* **26,** 3556 (1987).

[26] J. A. Haefliger, J. Tschopp, D. Naedelli, W. Wahli, H. P. Kocher, M. Tosi, and K. K. Stanley, *Biochemistry* **26,** 3551 (1987).

[27] O. M. Z. Howard, A. G. Rao, and J. M. Sodetz, *Biochemistry* **26,** 3565 (1987).

[28] R. G. DiScipio, M. R. Gehring, E. R. Podack, C. C. Kan, T. E. Hugli, and G. F. Fey, *Proc. Natl. Acad. Sci. U.S.A.* **81,** 7298 (1984).

[29] K. K. Stanley, H. P. Kocher, J. P. Luzio, P. Jackson, and J. Tschopp, *EMBO J.* **4,** 375 (1985).

[30] K. J. H. Robson, J. R. S. Hall, M. W. Jennings, T. J. R. Harris, K. Marsh, C. I. Newbold, V. E. Tate, and D. J. Weatherall, *Nature (London)* **335,** 79 (1988).

[31] R. C. Hedström, J. R. Campbell, M. L. Leef, Y. Charoenvit, M. Carter, M. Sedegah, R. L. Beaudoin, and S. L. Hoffman, *Bull. W.H.O.,* **68,** Suppl, 152 (1990).

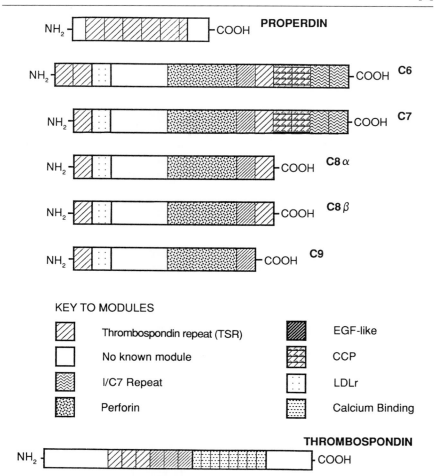

FIG. 1. Schematic representation of the primary structures of the complement proteins properdin (53 kDa), C6 (120 kDa), C7 (110 kDa), C8α (64 kDa), and C8β (64 kDa), and C9 (71 kDa), which contain TSR modules, and the cell adhesion protein thrombospondin (420 kDa) in which TSR modules were first described (see text for references).

spectroscopy analysis, indicate substantial quantities of β-sheet and β-turn structure.[32] Using neutron and X-ray scattering analysis,[33] good agreement was obtained with the electron micrograph structures of properdin.

[32] S. J. Perkins, A. S. Nealis, P. I. Haris, D. Chapman, D. Goundis, and K. B. M. Reid, *Biochemistry* **28,** 7176 (1989).

[33] K. F. Smith, K. F. Nolan, K. B. M. Reid, and S. J. Perkins, *Biochemistry* **30,** 8000 (1991).

The mouse properdin sequence contains two potential N-linked glyco-sylation sites at asparagine residues located at positions 370 and 400, which lie within the partial TSR module and the C-terminal region, respectively.[19] Although both the asparagine sites are conserved in human properdin, the first is unlikely to be glycosylated due to the presence of a prolyl residue at position X of the Asn-X-Ser/Thr glycosylation sequence.[7,34] All the carbohydrate present on human properdin is probably N linked and appears not to be required for properdin secretion, polymerization, or binding to C3iBb.[35]

There are 44 cysteine residues in each 442-residue-long chain of human properdin, and because no free sulfhydryl groups have been detected it is probable that there are 22 disulfide bonds per molecule; however, no information is yet available regarding the arrangement of the disulfide bridges.

Properdin Properties and Function

Properdin is a basic glycoprotein with an isoelectric point of >9.5[36] and is now known to be present in normal human plasma at a concentration that is significantly lower than the previous, widely cited value of 20–25 mg/liter. This concentration value was determined by a radioimmunoassay in which a molar absorption coefficient (A_{280}, 1%, 1 cm) of 14.8 was assumed to calculate the concentration of the standard properdin. Pang-burn[9] previously revised the concentration value to 5.7 ± 1 mg/liter, using a molar absorption coefficient of 18.00; however, the correct molar absorption coefficient for human properdin is now known to be 23.4 (based on a molecular mass of 53,246 and a precise knowledge of the Trp, Tyr, and Cys contents), and this knowledge has yielded a further revised value for the concentration of properdin in normal human plasma of approxi-mately 4.3–5.7 mg/liter (although high normal values of approximately 10 mg/liter have been reported).[9]

Properdin functions as a positive regulator of the alternative pathway by binding to and stabilizing the inherently labile C3/C5 convertase com-plexes, C3bBb and C3b2Bb, on activating surfaces. Binding of properdin brings about an approximately 10-fold increase in the half-lives of the convertases and also protects them from the negative regulatory action of factor I.[15,16,36,37] Properdin does not interfere with the binding of factor

[34] E. Bause, *Biochem. J.* **209,** 331 (1983).

[35] T. C. Farries and J. P. Atkinson, *J. Immunol.* **142,** 842 (1989).

[36] D. T. Fearon and K. F. Austen, *J. Exp. Med.* **142,** 856 (1975).

[37] R. G. Medicus, O. Gotze, and H. J. Müller-Eberhard, *J. Exp. Med.* **144,** 1076 (1976).

H to the convertase complexes,[15] is not necessary for the initial recognition of an activating surface, and does not appear to have any influence on the fluid-phase "tick-over" of the alternative pathway.[38]

It has been proposed that properdin exists in two interconvertible forms, native and activated,[37,39,40] but studies have indicated that activated properdin is composed of amorphous aggregates and arises as an artifact of the isolation procedures.[41] At physiological ionic strength a weak interaction has been demonstrated of native properdin with cell-bound C3b, which was significantly enhanced in the presence of factor B and further enhanced (approximately fivefold) with cell-bound C3bBb.[16,38] At low ionic strength, native properdin shows weak binding to fluid-phase C3b, iC3b, and C3c, but no interaction with fluid-phase C3 or C3i, which is consistent with the inability of properdin to accelerate C3 consumption in serum in the absence of an activating surface.[38,42]

Gene Structure, Chromosome Localization, and Genetic Deficiency

The human properdin gene spans approximately 6.0 kilobases (kb) of DNA and is organized into 10 exons.[20] Identification and characterization of a polymorphic dinucleotide repeat region, located within 16 kb of the 3′ end of the properdin gene, has enabled the gene to be positioned on the genetic map of the short arm of the X chromosome <1 centimorgan (cM) distal to TIMP and >10 cM proximal to MAO-A and DXS228, in the same genetic interval as DXS426.[43] These mapping studies, in combination with restriction analysis of yeast artificial chromosome (YAC) clones, have enabled the physical localization of the properdin gene to be refined to the region Xp11.23–Xp11.3, <100 kb distal to TIMP.[43]

Deficiency of properdin function is inherited as an X-linked recessive disorder and, although the condition is rare (with only 70 cases reported in the literature up to July, 1991), affected individuals are predisposed to life-threatening bacterial—particularly meningococcal—infections.[44] Al-

[38] T. C. Farries, P. J. Lachmann, and R. A. Harrison, *Biochem. J.* **253,** 677 (1988).

[39] O. Gotze and H. J. Müller-Eberhard, *J. Exp. Med.* **139,** 44 (1974).

[40] R. G. Medicus, A. F. Esser, H. N. Fernandez, and H. J. Müller-Eberhard, *J. Immunol.* **124,** 602 (1980).

[41] T. C. Farries, J. T. Finch, P. J. Lachmann, and R. A. Harrison, *Biochem. J.* **243,** 507 (1987).

[42] O. Gotze, R. G. Medicus, and H. J. Müller-Eberhard, *J. Immunol.* **118,** 525 (1977).

[43] M. P. Coleman, J. C. Murray, H. F. Willard, K. F. Nolan, K. B. M. Reid, D. J. Blake, S. Lindsay, S. S. Bhattacharaya, A. Wright, and K. E. Davies, *Genomics* **11,** 991 (1991).

[44] J. E. Figueroa and P. Densen, *Clin. Microbiol. Rev.* **4,** 359 (1991).

though the precise genetic defects are unknown, three properdin-deficiency phenotypes have been described: type I, total deficiency, in which serum properdin levels are very low ($<2\%$ of normal)[45–47]; type II, partial deficiency, characterized by properdin that is only $\sim 10\%$ as abundant as in normal serum, but is apparently functionally active[48,49]; and type III, in which a dysfunctional form of the protein is produced at normal antigenically determined levels.[50] The X-linked nature reported for the three forms of properdin deficiency[44,51] and the colocalization of deficiency and structural gene loci in a type II-deficient family[43,52,53] are both suggestive that the underlying genetic lesions responsible for the properdin deficiency phenotype probably exist in, or are very close to, the properdin structural gene sequence.

Immunochemical and functional investigations of serum properdin levels allow for the accurate diagnosis of properdin-deficient males who can then be immunized, and these procedures could potentially be used to screen selected groups of susceptible individuals.[54] Female carriers of the properdin-deficient allele do not appear predisposed to bacterial infections, and it is difficult to unambiguously diagnose these individuals within deficient families based solely on their properdin serum concentration levels. The polymorphic CA repeat, located less than 16 kb from the 3' end of the properdin structural gene, has provided a rapid, technically simple, nonradioactive method for detecting carrier status within deficient kindreds, thus facilitating the early detection and immunization of affected male offspring. This marker has now been used in genetic linkage analysis to unambiguously assign female carrier status in three unrelated properdin-deficient families.[55]

[45] P. Densen, J. M. Weiler, J. M. Griffiss, and L. G. Hoffmann, *N. Engl. J. Med.* **316,** 922 (1987).
[46] A. G. Sjoholm, J.-H. Braconier, and C. Soderström, *Clin. Exp. Immunol.* **50,** 291 (1982).
[47] E. R. Holme, J. Veitch, A. Johnston, G. Hauptmann, B. Uring-Lambert, M. Seywright, V. Docherty, W. N. Morley, and K. Whaley, *Clin. Exp. Immunol.* **76,** 76 (1989).
[48] A. G. Sjoholm, C. Soderström, and L.-Å. Nilsson, *Complement* **5,** 130 (1988).
[49] H. E. Nielsen and C. Koch, *Clin. Immunol. Immunopathol.* **44,** 134 (1987).
[50] A. G. Sjoholm, E. J. Kuijper, A. Jansz, P. Bol, L. Spanjaard, and H. C. Zanen, *N. Engl. J. Med.* **319,** 33 (1988).
[51] A. G. Sjoholm, *APMIS* **98,** 861 (1990).
[52] P. Goonewardena, A. G. Sjoholm, L.-Å. Nilsson, and U. Petterson, *Genomics* **2,** 115 (1988).
[53] D. Goundis, S. M. Holt, Y. Boyd, and K. B. M. Reid, *Genomics* **5,** 56 (1989).
[54] C. A. P. Fijen, E. J. Kuijper, A. J. Hannema, A. G. Sjoholm, and J. P. M. van Putten, *Lancet* 585 (1989).
[55] K. Kolble, K. F. Nolan, and K. B. M. Reid, *Complement Inflammation* **8,** 176 (1991).

Purification of Human Properdin

Properdin has been successfully purified from serum by a variety of methods. The principal features of the three main procedures that have been used are (1) preabsorption of starting serum on Sepharose followed by affinity chromatography on an antiproperdin column, as the main step, with further purification by ion-exchange chromatography (QAE-Sephadex) and/or gel filtration (Sephacryl S-300); (2) with euglobulin precipitation as the first step followed by ion-exchange chromatography (CM-cellulose and DEAE-Sephadex) and removal of traces of IgG by antihuman IgG-Sepharose; (3) with precipitation by polyethylene glycol as the first step followed by ion-exchange chromatography (QAE-Sephadex), then removal of traces of IgG by antihuman IgG-Sepharose. The first method requires the availability of a monoclonal or polyclonal antibody against human properdin, whereas the second and third methods require no special reagents [other than antihuman IgG-Sepharose, which may be replaced by commercially available protein A-Sepharose, or protein G-Sepharose (Sigma Chemical Company Ltd., Poole, Dorset, UK)]. All procedures are carried out at 4°.

Method 1.[9,32] A Sepharose antiproperdin monoclonal or polyclonal antibody column is prepared by coupling a purified mouse IgG monoclonal antibody (Dr. C. Koch, Statens Seruminstitut, Denmark) or the IgG fraction, from a rabbit polyclonal antihuman properdin antiserum, to CNBr-activated Sepharose 4B. The column (1 × 5 cm, containing 5 mg of monoclonal IgG antibody/ml of Sepharose 4B; or 2 × 12 cm, containing 8 mg of polyclonal IgG antibody fraction/ml of Sepharose 4B) is equilibrated with phosphate-buffered saline (PBS; 10 mM sodium phosphate, 150 mM NaCl, pH 7.4). Serum (2 liters) that has been extensively dialyzed against PBS is preabsorbed by application to a column (2 × 12 cm) of Sepharose 4B prior to being run through the antibody column (adjustment of the ionic strength and pH of the sample to the same values as those of the column buffers reduces the possibility of precipitation of proteins taking place on the column). After extensive (1 liter) washing of the column with PBS the column is further washed with PBS made 3 M with respect to NaCl (500 ml) and then the bound proteins are eluted from the column with 0.2 M glycine–0.5 M NaCl, pH 2.5,[32] or 0.1 M glycine, pH 2.3.[9] The low-pH eluate is immediately neutralized by the addition of 1 M NaOH[32] or 0.5 M Na$_2$HPO$_4$.[9] This fraction is reapplied to the affinity column and eluted using the same conditions as before. After dialysis of the sample against 20 mM Tris-HCl, 30 mM NaCl, pH 8.5 (3.9 mS at 4°), and removal of any precipitate by centrifugation, the properdin can be purified further by application to a column (4 × 15 cm) of QAE-Sephadex A-50 equilibrated with 20 mM Tris, 30 mM NaCl, pH 8.5.[9,42] The column is run at ~60 ml/hr and the properdin eluted over the region between 60–120 ml. Traces of IgG can be removed on an antihuman IgG-Sepharose column. The final yield of properdin is ~7 mg from 2 liters of serum.

Method 2.[2,56,57] Full details of the precipitation of the euglobulins at low ionic strength for up to 2 liters of human serum, and descriptions of the ion-exchange chromatography on DEAE-Sephadex A-50 and CM-32, are given in a previous volume.[57,58] The properdin is finally purified on an antihuman IgG-Sepharose column, with the final yield being ~6.7 mg from 2 liters of serum.

Method 3.[15,59] Serum, obtained by clotting 5 liters of plasma, is adjusted to pH 7.5 and made 15 mM with respect to EDTA. Polyethylene glycol 4000 is added to give a 5% (w/v) final concentration and the mixture is centrifuged (4500 g for 30 min). The precipitate is dissolved by stirring overnight in 20 mM Tris HCl, 300 mM NaCl, pH 8.5 (250 ml). After centrifugation, to remove insoluble material, the supernatant is dialyzed against 20 mM Tris-HCl, 30 mM NaCl, pH 8.5 (3.9 mS at 4°). After centrifugation the sample is applied to a column (6 × 72 cm) of QAE-Sephadex equilibrated with 20 mM Tris-HCl, 30 mM NaCl, pH 8.5. The column is run at 100 ml/hr and properdin eluted between 500 and 1100 ml.[59] Further purification is achieved by immunoadsorption on rabbit antihuman IgG-Sepharose, column (2.5 × 20 cm) equilibrated in 25 mM Tris-HCl, pH 8.0, 70 mM NaCl, 10 mM EDTA, 0.01% (w/v) NaN$_3$. This procedure gives a yield of ~2 mg properdin/liter of plasma.[59]

Properdin is best stored at 4° because freezing and thawing causes aggregation of the molecule, yielding nonphysiological material that activates complement when added to normal human serum.[9,41] Concentration of properdin can be carried out by ultrafiltration under compressed air using Amicon Diaflo ultrafiltration cells and PM10 membranes (Amicon BV, Oosterhout, Holland). If concentration is to be carried out in a buffer such as 12 mM sodium phosphate, 150 mM NaCl, 0.2 mM EDTA, pH 7.4, it has been found that to achieve high concentrations of properdin (above 0.8 mg/ml), without losses due to precipitation, it is necessary for 0.2 M glycine to be added to the sample buffer.[32] The amount of properdin present in the sample can be estimated using a molar absorption coefficient (A_{280}, 1%, 1 cm) of 23.4.[8]

[56] J. O. Minta and E. S. Kunar, *J. Immunol.* **116**, 1099 (1976).
[57] K. B. M. Reid, this series, Vol. 80, p. 143.
[58] K. B. M. Reid, this series, Vol. 80, p. 16.
[59] M. K. Pangburn, this series, Vol. 162, p. 639.

Separation of Oligomeric Forms of Properdin

The P_2, P_3, P_4, and higher oligomers, which are present in quite constant proportions in preparations of purified properdin, can be separated from each other by conventional gel filtration on Sephacryl S-300[10,58] or by HPLC on either a TSK-400 gel-filtration column (Tosohaas, Montgomeryville, PA) or Mono S cation-exchange column (Pharmacia LKB Biotechnology, Milton Keynes Burks, UK).[9] Although finer resolution of the oligomers is achieved by use of HPLC, the Sephacryl S-300 separation has the advantage of not requiring the availability of high-performance liquid chromatography equipment.

The isolated oligomers preparations have been shown to be very stable and eluted at the expected position when rerun on Sephacryl S-300 or HPLC, but could be redistributed into the different oligomeric forms by denaturation–renaturation cycles.

Separation on Sephacryl S-300

The purified properdin sample (~8 mg) is concentrated to ~8 ml and applied to a column (2.5 × 85 cm) of Sephacryl S-300 (Pharmacia Fine Chemicals, Piscataway, NJ), which is equilibrated with 0.15 M NaCl–12 mM sodium phosphate–0.2 mM EDTA buffer, pH 7.4. The properdin is eluted in three distinct main peaks,[57] between 160 and 225 ml, corresponding to the P_4, P_3, and P_2 forms (as identified by electron microscopy).[10] Each peak from the Sephacryl S-300, when concentrated and reapplied to the same column, is eluted in its original position.

Separation on HPLC

Good separation of the oligomers has been achieved by gel filtration on TSK-400.[59] Purified properdin (260 μg in 200 μl) is applied to a column (7.5 × 600 mm) of TSK-400 (Bio-Rad, Richmond, CA, equilibrated with 50 mM sodium phosphate–100 mM sodium acetate, pH 6.8, and eluted at a flow rate of 1 ml/min). The P_4, P_3, and P_2 oligomers are separated from each other between 16 and 24 ml after sample application.[9] Better resolution is achieved by cation-exchange chromatography on Mono S. Purified properdin (1 mg in 770 μl) is diluted with 230 μl of 50 mM sodium phosphate, pH 6, and loaded onto a column (5 × 50 mm) of Mono S (Pharmacia, Fine Chemicals, Piscataway, NJ). The column is washed with 50 mM sodium phosphate–100 mM NaCl, pH 6.0, and then eluted with a 20-ml gradient from 50 mM sodium phosphate–100 mM NaCl, pH 6.0. The P_2, P_3, and P_4 oligomers are separated between 9 and 17 ml of the gradient.[9] This procedure yields mixtures of 26 : 54 : 20 for $P_2 : P_3 : P_4$,

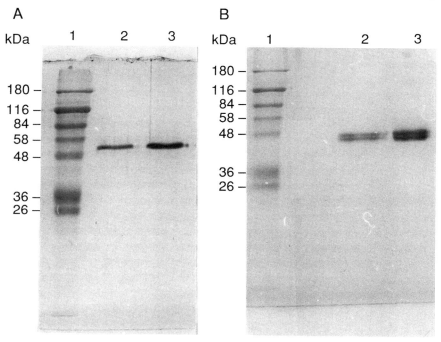

FIG. 2. SDS–PAGE on a 10% (w/v) gel of two preparations of human properdin run under (a) reducing and (b) nonreducing conditions. Lane 1 contains standard proteins as molecular weight markers. Lanes 2 and 3 contain the human properdin preparations.

which is the same as that found in normal human serum or plasma[9] or in purified preparations of properdin prior to separation of the oligomers.

The purity of a properdin preparation is readily assessed by SDS–PAGE because a sample of highly purified properdin (whether as an unfractionated mixture of oligomers or as individual oligomers) yields two or three very closely spaced bands of approximately 504,000 apparent molecular weight on 10% (w/v) SDS–PAGE, in nonreducing conditions and a single band of approximately 54,000 apparent molecular weight in reducing conditions (Fig. 2).

Functional Assay for Properdin

Assay of properdin function has been discussed in a previous volume in this series.[58] The most widely used functional assay involves the ability of properdin samples to promote the lysis of rabbit red blood cells,[58] or neuraminidase-treated sheep red blood cells,[9] in properdin-depleted human serum. Details on how to carry out the long-term storage of rabbit red blood cells at $-80°$, thus providing a readily available source of cells for use in the assay, are given by Pangburn.[59]

[3] Small-Scale Preparation of Complement Components C3 and C4

By ALISTER W. DODDS

Introduction

C3 is the central and most abundant protein of the human complement system.[1-3] Its intrachain thioester bond allows it to bind covalently to activating surfaces,[4,5] a property that it shares with C4.[6] The two molecules share a common evolutionary origin and show approximately 25% conservation of primary structure after alignment. They have parallel roles in the formation of the classical and alternate pathway C3 convertases.[7,8] A volume of "Current Topics in Microbiology and Immunology" has been devoted entirely to the chemistry and biology of C3.[9]

For the large-scale preparation of C3 and C4, the procedures described in previous volumes of this series[10-12] are still the methods of choice. However, these methodologies are expensive, both in the volumes of starting plasma used and in the amounts of chromatographic media required. Although large quantities of pure protein can be obtained by these methods, storage of the thioester-containing proteins over long periods is

[1] J. E. Volanakis, *Curr. Top. Microbiol. Immunol.* **153,** 1 (1990).
[2] J. D. Becherer, J. Alsenz, and J. D. Lambris, *Curr. Top. Microbiol. Immunol.* **153,** 45 (1990).
[3] Z. Fishelson, *Mol. Immunol.* **28,** 545 (1991).
[4] S. K. A. Law, *Ann. N.Y. Acad. Sci.* **421,** 246 (1983).
[5] R. P. Levine and A. W. Dodds, *Curr. Top. Microbiol. Immunol.* **153,** 73 (1990).
[6] E. Sim and A. W. Dodds, *in* "Complement in Health and Disease" (K. Whaley, ed.), p. 99. MTP Press, Lancaster, UK, 1986.
[7] S. K. A. Law and K. B. M. Reid, *in* "Complement" (D. Male, ed.). IRL Press, Oxford, UK, 1988.
[8] H. J. Müller Eberhard, *Annu. Rev. Biochem.* **57,** 321 (1988).
[9] J. D. Lambris, ed., *Curr. Top. Microbiol. Immunol.* **153,** (1990).
[10] B. F. Tack, J. Janatova, M. L. Thomas, R. A. Harrison, and C. H. Hammer, this series, Vol. 80, p. 64.
[11] J. Janatova, this series, Vol. 162, p. 579.
[12] I. Gigli and F. A. Tausk, this series, Vol. 162, p. 626.

difficult. Major conversion to the thioester-hydrolyzed forms of these proteins can occur on freezing and thawing even under optimal conditions. In the case of a freezer malfunction or power outage most of the active protein can be destroyed during a slow thawing process.[11]

Here we describe rapid and reproducible methods for the production of small amounts of C3 and C4. There are a number of situations when this can be useful: (1) when proteins are required only occasionally and in small amounts, for example, as hemolytic or immunochemical assay reagents; (2) when proteins are required for functional studies from small samples of plasma from individual patients or donors; (3) when proteins are required from other species, especially of small animals, where the starting volumes required for the large-scale purifications described are impractical; and (4) when recombinant proteins are recovered from tissue culture supernatants.

Two types of purification procedure will be described: affinity methods and ion-exchange chromatography. These methods have been developed using a Pharmacia (Piscataway, NJ) fast protein liquid chromatography (FPLC) apparatus and a Waters (Milford, MA) advanced protein purification system (APPS). Any other comparable setup can be substituted. Apart from the chromatographic steps utilizing Mono Q, the procedures can be carried out using peristaltic pumps and manual gradient formers, or even under gravity flow.

Detection of Proteins

Hemolytic Assay

Standard hemolytic assays using sensitized sheep erythrocytes and serum from guinea pigs genetically deficient in C4[12] or human serum depleted of C3 by immunoadsorption[11] as a source of other complement components are the simplest and most reproducible of the many methods available. When C3 or C4 from other animals is prepared, the standard assays, which are optimized for the detection of human complement components, may not be applicable due to interspecies incompatibilities between the components.[13]

Immunochemical Methods

Polyclonal and monoclonal antibodies to C3 and C4 from humans and a number of other species, for use in immunochemical detection systems,

---

[13] W. D. Linscott, *Prog. Vet. Microbiol. Immunol.* **2**, 54 (1986).

are available from various companies.[14] When C3 or C4 from novel species is studied it may be possible to detect cross-reactions with an antibody to the protein from a phylogenetically related species.

SDS–PAGE Analysis

Human C3 is a major plasma protein and has a distinctive two-chain structure following reduction. It should be possible to identify it from the earliest stages of a purification scheme (see Fig. 1B). All C4 proteins studied thus far are eluted very late in the salt gradient from anion-exchange columns, being among the most negatively charged plasma proteins. The characteristic three-chain structure of C4 following reduction should be detectable on SDS–PAGE from the early stages of a purification scheme (see Fig. 1B). Under most circumstances, standard SDS–PAGE systems can be utilized, but for the differentiation of the C4A and C4B isotypes a system based on that of Laemmli,[15] but with a reduced level of cross-linking bisacrylamide (10% acrylamide; 0.06% bisacrylamide), is required.[16]

Thioester Detection

Column fractions can be tested for the presence of C3 and C4 using either direct radiolabeled methylamine incorporation[17] or by incorporation of radiolabeled iodoacetamide into the free SH released after nucleophile treatment,[11] followed by SDS–PAGE and autoradiography. The methods are not applicable to whole plasma or serum, where the methylamine-treated proteins are rapidly broken down by factor I and its cofactors together with other proteases. With human components, problems have not been encountered with proteolytic breakdown in column fractions. However, this can be a problem with some other species, presumably due to the coelution of proteases in the thioester protein-containing fractions.

Small-Scale Purification of C3 and C4 by Ion-Exchange Chromatography

The procedure described is for the purification of C3 and C4 from 5 ml of human EDTA plasma, which should be fresh or stored at $-70°$. However, it is suitable for volumes of plasma from 0.5 to 50 ml, with

---

[14] W. D. Linscott, ed., "Linscott's Directory of Immunological and Biological Reagents," 6th ed. Linscott's Directory, Mill Valley, CA, 1990.

[15] U. K. Laemmli, *Nature (London)* **227**, 680 (1970).

[16] M. H. Roos, E. Mollenhauer, P. Demant, and C. Rittner, *Nature (London)* **298**, 854 (1982).

[17] H. S. Auerbach, R. Burger, A. W. Dodds, and H. R. Colten, *J. Clin. Invest.* **86**, 96 (1990).

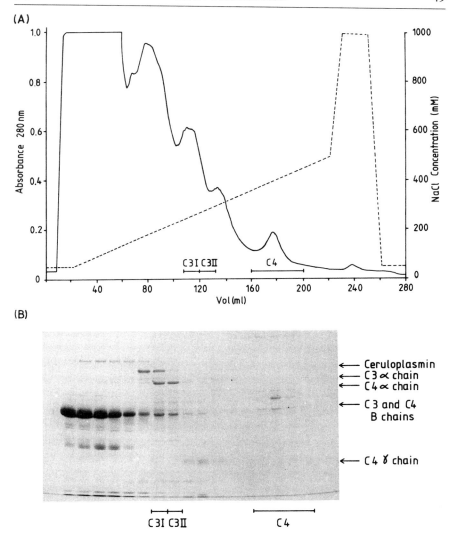

FIG. 1. Human EDTA plasma on Q-Sepharose fast flow. (A) The elution profile obtained when 5 ml of human EDTA plasma is loaded onto a 10-ml column of Q-Sepharose and eluted with a 200-ml linear gradient from 50 to 500 m*M* NaCl. The absorbance at 280 nm is indicated by the solid line, and the molarity of NaCl is shown by the dashed line. Fractions (4 ml) were collected. (B) SDS–PAGE analysis, under reducing conditions, of alternate fractions of the material eluted between 50 and 210 ml in the profile.

slight adjustment of the size of columns and the volumes of gradients. The method starts with a polyethylene glycol (PEG) precipitation step, which may be omitted if only C4 is required. If C3 is to be purified the PEG precipitation is important to remove an unidentified contaminant, which causes severe blocking of the Mono Q column used in the final stage of purification. The amount of C3 produced from 5 ml of plasma is too large for a single run on the Mono Q 5/5 column used in the final stage of purification. Larger columns of this resin are available (but are extremely expensive) and could be substituted for the multiple small runs described here. The following buffers are used and are mixed to give buffers of the required ionic strength.

Buffer A: 20 mM Tris, 50 mM ε-aminocaproic acid (EACA), 5 mM EDTA, 0.02% (w/v) NaN$_3$, 0.2 mM phenylmethylsulfonyl fluoride (PMSF), pH 7.5, with HCl

Buffer B: Buffer A with 1 M NaCl

Method

Human plasma (5 ml) containing 10 mM EDTA and 2.5 mM diisopropyl fluorophosphate (DFP) is made 5% (w/v) with polyethylene glycol (PEG 3350 molecular weight, Sigma, St. Louis, MO) by adding 2.5 ml of a 15% (w/v) solution of PEG in buffer A. The mixture is stirred gently at 0° for 30 min and spun at 10,000 rpm for 20 min to remove the resulting precipitate. The supernatant is loaded, at a flow rate of 2 ml/min, onto a column (50 × 16 mm diameter) of Q-Sepharose Fast Flow (Pharmacia, Piscataway, NJ), equilibrated with 95% buffer A, 5% buffer B (i.e., 50 mM NaCl), and eluted with a 200-ml linear salt gradient to a final buffer containing 500 mM NaCl. The column can be run at 4° or at room temperature but in the latter case fractions should be collected into tubes on ice, if possible, or immediately removed from the fraction collector and placed on ice. The resulting elution profile is shown in Fig. 1A, and SDS–PAGE analysis of the fractions is shown in Fig. 1B. C4 is eluted in the final peak. C3 is eluted approximately halfway through the gradient immediately after the protein ceruloplasmin, the blue coloration of which should be visible in the fractions immediately preceding it. If C3 is required free of contamination with ceruloplasmin, a cut should be made avoiding this protein, as indicated (C3II in Figs. 1A and 1B). Ceruloplasmin can be seen on SDS–PAGE analysis eluting slightly before C3 and above the C3 α chain on the gel (Fig. 1B).

The pool containing C3 is diluted with half a volume of water to lower the ionic strength sufficiently to allow binding to the Mono Q column. If a single pool has been made containing C3 and ceruloplasmin, the diluted sample should be run in four or five separate batches to avoid overloading the column and reducing resolution. If a pool has been taken that avoids ceruloplasmin and hence approximately halves the yield of C3, the column is large enough for the sample to be run in two batches. As mentioned earlier, if a large column of Mono Q (1 × 10 cm or 1.6 × 10 cm) is available, the C3 could be chromatographed in a single run.

A column (0.5 × 5 cm) of Mono Q is equilibrated with 90% buffer A, 10% buffer B (i.e., 100 mM NaCl). The C3-containing sample is loaded and immediately eluted with a 20-ml linear gradient to 300 mM NaCl at a flow rate of 1 ml/min. Figure 2A shows the elution profile obtained from 50% of the pool lacking ceruloplasmin (C3II in Fig. 1) and Fig. 2B shows

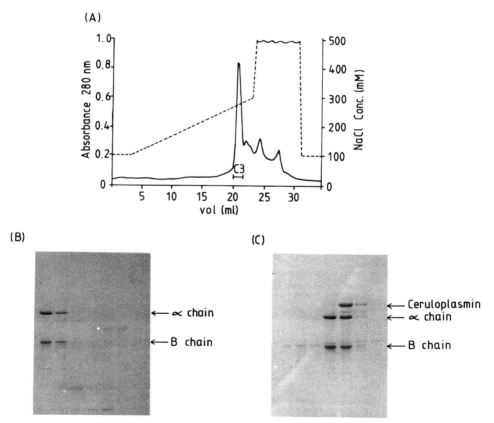

FIG. 2. Human C3 from Q-Sepharose on Mono Q. Half of the material in pool C3II from the Q-Sepharose column was loaded onto a 1-ml column of Mono Q and eluted with a 20-ml linear gradient from 100 to 300 mM NaCl. (A) The resultant elution profile. (B) SDS–PAGE analysis of the 1-ml fractions from 20 to 27 ml in the profile. (C) SDS–PAGE analysis of the fractions obtained when half of pool C3I was run on Mono Q under the same conditions; 1-ml fractions, from 17 to 23 ml, are shown in the profile.

SDS–PAGE analysis of the fractions under reducing conditions. If the C3 pool that includes the contaminating ceruloplasmin is run under the same conditions, then the ceruloplasmin is eluted slightly after the C3, but is very poorly resolved from it, as shown in Fig. 2C.

The C4 pool is diluted with half a volume of water and loaded at a flow rate of 1 ml/min onto a column (0.5 × 5 cm) of Mono Q, equilibrated with buffer containing 200 mM NaCl and eluted with a 20-ml linear gradient to 500 mM NaCl. A typical elution profile is shown in Fig. 3A and SDS–PAGE analysis of the fractions is shown in Fig. 3B. C4 is contained in the second of the two major peaks to be eluted: the other major peak is prothrombin. The C4 contains only one major contaminant, inter-α-trypsin inhibitor, which can be seen running close to the top of the gel.

Affinity Purification of C3 and C4 Using Polyclonal Antibody

Rabbit antihuman C3 or C4 F(ab')$_2$ is prepared[18] and bound to CNBr-activated Sepharose 4B (Pharmacia) at approximately 6 mg per ml of gel and unreacted sites on the gel are blocked with glycine. A column containing 10 ml of gel is equilibrated with 50 mM Tris, 150 mM NaCl, 50 mM ε-aminocaproic acid, 5 mM EDTA, 0.2 mM PMSF, 0.02% sodium azide, pH 7.4. EDTA plasma (10 ml) is loaded onto the column at 0.1 ml/min and the column is washed with starting buffer until the absorbance at 280 nm approaches zero. The column is then washed with 30 ml of starting buffer containing 1 M NaCl. Bound C3 or C4 is eluted either with 50 mM glycine/HCl, pH 2.2, or 50 mM diethanolamine, pH 11.5. Fractions (2 ml) are collected into tubes containing 100 μl of 1 M Tris, pH 7.4, to neutralize the eluted protein immediately.

The acid-eluted C4 and C3 are totally devoid of activity whereas the material eluted under basic conditions retains approximately 25% of its activity, the remaining material having been converted to the thioester-hydrolyzed form. Minor contaminants can be removed by chromatography on Mono Q as described in the previous section. Although this method is not a good one for the preparation of active proteins, it is useful for the preparation of material for immunochemical work and also for the preparation of a reagent deficient in C3 or C4 for use in hemolytic assays. The use of Fab or F(ab')$_2$ fragments is important in this context because C1q will be considerably depleted if whole immunoglobulin is used on the affinity column.

[18] J. M. Wilkinson, *Biochem. J.* **112**, 173 (1969).

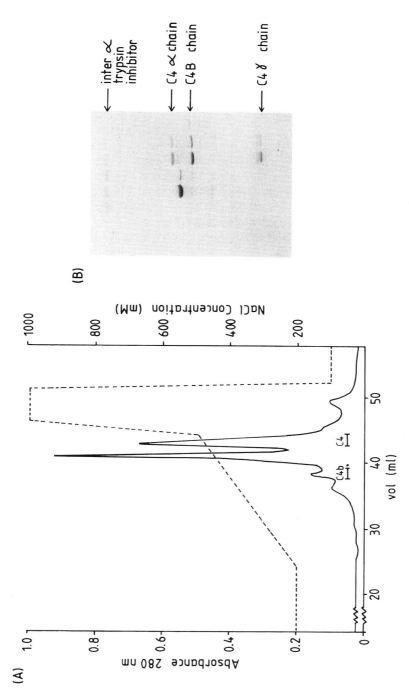

FIG. 3. Human C4 from Q-Sepharose on Mono Q. (A) The C4-containing pool from Q-Sepharose was loaded onto a 1-ml column of Mono Q and eluted with a 20-ml linear gradient from 200 m*M* NaCl to 500 m*M* NaCl. (B) SDS–PAGE analysis of the reduced 1-ml fractions from 39 to 46 ml.

Affinity Purification of C4 and C3 Using Monoclonal Antibodies

The methods described have been developed using monoclonal antibodies LOO1 and LOO3 (anti-C4),[19–21] which are available from Commonwealth Serum Laboratories (Parkville, Victoria, Australia), and WM1 (anti-C3),[22] which is available for noncommercial use from the European Collection of Animal Cell Cultures (accession number 92021211). The methods should be applicable to other monoclonal antibodies, provided that they are specific and have a high enough affinity.

Affinity Purification Buffer

For the monoclonal antibody affinity purifications described below a buffer system containing a number of different buffering components is used. A mixture containing equimolar amounts of phosphate (pK_a 7.20 and 12.38), borate (pK_a 9.24), and Tris (pK_a 8.08) gives moderately good buffering at all of the pH values required, from neutral to pH 11.5, and produces an approximately linear pH profile when buffers of the same composition, but adjusted to different pH values, are mixed. The composition of the basic stock buffer from which the others are prepared is 100 mM NaH$_2$PO$_4$, 100 mM Tris, 25 mM Na$_2$B$_4$O$_7$, 5 mM EDTA, 0.02% (w/v) NaN$_3$, 0.2 mM phenylmethylsulfonyl fluoride. This stock solution is referred to throughout the following sections as affinity purification buffer (APB). The pH of the stock solution is adjusted with 5 M HCl or 4 M NaOH and dilutions made as required. The buffers are described as the percentage of APB followed by the pH value, e.g., 50% APB, pH 7.5, means the basic stock buffer adjusted to pH 7.5 with HCl and diluted to half the original concentration with water. The buffer strength and pH described have been optimized for the monoclonal antibodies used; other antibodies may require slightly different conditions for optimal recovery and purification.

Affinity Purification of C4 on Monoclonal Antibody

Mouse antihuman C4 IgG (LOO1), which recognizes a determinant in the β chain of C4,[21] is prepared from ascites fluid and bound to CNBr-activated Sepharose 4B at 2.5 mg of antibody/ml of gel, and excess sites on the gel are blocked with glycine. A column containing 4 ml of affinity resin is equilibrated with 50% APB, pH 7.0. Human plasma (1–10 ml)

[19] A. W. Dodds, S. K. A. Law, and R. R. Porter, *EMBO J.* **4,** 2239 (1985).
[20] C. M. Giles and D. S. Ford, *Transfusion (Philadelphia)* **26,** 370 (1986).
[21] L. M. Hsiung, D. W. Mason, and A. W. Dodds, *Mol. Immunol.* **24,** 91 (1987).
[22] A. S. Whitehead, R. B. Sim, and W. F. Bodmer, *Eur. J. Immunol.* **11,** 140 (1981).

containing 10 mM EDTA and 2.5 mM DFP (0.2 mM PMSF could be substituted) is loaded onto the column at a flow rate of 0.25 ml/min. The column is washed at 0.25 ml/min with 50% APB, pH 7.0, until the bulk of the protein has passed through (approximately 5 ml). The flow rate is then increased to 1 ml/min and washing is continued until the absorbance at 280 nm approaches zero. The C4 is eluted with a 20-ml linear gradient from 50% APB, pH 7.0, to 50% APB, pH 11.5. The column can be run at 4° or at room temperature, but in the latter case fractions should be placed on ice as they are eluted. The eluted peak monitored by absorption at 280 nm is pooled and loaded without adjustment of pH or buffer concentration onto a column (0.5 × 5 cm) of Mono Q for concentration and removal of trace contaminants, as described in the section on C4 purification by ion-exchange chromatography. The C4 obtained by this procedure is more than 95% active, because very little of the thioester is disrupted by the gradient elution. In experiments where C4 is eluted by a stepped change in pH, it has been found that the recoveries of active C4 are considerably reduced. Purified C4 is dialyzed into 10 mM sodium phosphate, 125 mM NaCl, 0.5 mM EDTA, pH 7.4, and stored at 4°. The loss of activity under these conditions is slow, and approximately 90% of the activity remains after 1 month of storage.

Purification of C4 Isotypes and Allotypes

Human C4 is coded by two separate genes located in the class III region of the MHC.[23] The products of these two genes, termed C4A and C4B, are subtly different in their activities.[24,25] Both loci are highly polymorphic, and by a combination of electrophoretic mobility, serology, and restriction site polymorphism, 65 alleles and variants have been described at the two loci.[26] Individuals deficient at one of the two loci comprise approximately 4%[27] of the population and it is possible to purify C4A and C4B from these individuals by the ion-exchange or affinity methods described above. However, most of the less common C4 allotypes are never found in combination with a deficient allele at the other locus. A number of monoclonal antibodies have been described that can recognize

[23] R. D. Campbell, S. K. A. Law, K. B. M. Reid, and R. B. Sim, *Annu. Rev. Immunol.* **6,** 161 (1988).

[24] S. K. A. Law, A. W. Dodds, and R. R. Porter, *EMBO J.* **3,** 1819 (1984).

[25] D. E. Isenman and J. R. Young, *J. Immunol.* **132,** 3019 (1984).

[26] G. Mauff, M. Brenden, M. Braun-Stilwell, G. Doxiadis, C. M. Giles, G. Hauptmann, C. Rittner, P. M. Schneider, B. Stradmann-Bellinghausen, and B. Uring-Lambert, *Complement Inflammation* **7,** 193 (1990).

[27] G. Hauptmann, J. Goetz, B. Uring-Lambert, and E. Grosshans, *Prog. Allergy* **39,** 232 (1986).

antigenic determinants normally expressed on the different isotypes.[19,28–30] One of these antibodies (LOO3)[19] recognizes a determinant carried on the α chains of both C4 types but has different affinities, particularly at high pH for the two types. This antibody has been used to separate a number of C4 allotypes of each isotype.[19,31] The method has not been tried with any of the other monoclonal antibodies that have been described, but because most of these are more specific than LOO3, they should work equally well or better. If antibodies "specific" for C4A or C4B are used, samples will need to be passed independently through the two columns, or could be run in series for sample loading. However, pH gradient elution is still recommended, because there may be some weak cross-reaction of the antibody with the other C4 isotype: any contaminating C4 of the wrong isotype should be removed early in the gradient, along with other nonspecifically bound contaminants. Gradient elution also ensures that the protein is released under the mildest possible conditions of pH.

Affinity columns bearing mouse antihuman C4 IgG (LOO3) are prepared as described in the previous section. A column containing 4 ml of affinity resin is equilibrated with 25% APB, pH 7.0. Human plasma (5 ml) containing 10 mM EDTA and 2.5 mM DFP (0.2 mM PMSF could be substituted) is loaded onto the column at a flow rate of 0.25 ml/min. The column is washed with 25% APB, pH 7.0, until the absorbance at 280 nM approaches zero and the flow rate is increased to 1 ml/min after approximately 5 ml of washing. The C4A and C4B are eluted with a 30-ml linear gradient from 100% APB, pH 8.5, to 100% APB, pH 11.5. As shown in Fig. 4A, two peaks of absorbance are eluted. The first peak eluting at pH 8–9 is C4A and the second peak eluting at pH 9.5–10.5 is C4B. SDS–PAGE analysis of the eluted peaks was performed in gels containing 10% acrylamide and 0.06% bisacrylamide.[16] As shown in Fig. 5, the α chain of C4A has an apparently higher molecular mass than that of C4B under these conditions and this difference can be seen in the eluted fractions (Fig. 4B). Pools are made as indicated, ensuring that intermediate fractions containing a mixture of the two C4 types are discarded. The protein is concentrated and trace contaminants are removed on a column of Mono Q as described in the section on C4 purification by ion-exchange chromatography. The C4 obtained by this procedure is more than 95% active.

[28] G. J. O'Neill, *Vox Sang.* **47,** 362 (1985).
[29] J. Chrispeels, S. Bank, C. Rittner, and D. Bitter-Suermann, *J. Immunol. Methods* **125,** 5 (1989).
[30] B. D. Reilly, R. P. Levine, and V. M. Skanes, *J. Immunol.* **147,** 3018 (1991).
[31] A. W. Dodds, S. K. A. Law, and R. R. Porter, *Immunogenetics* **24,** 279 (1986).

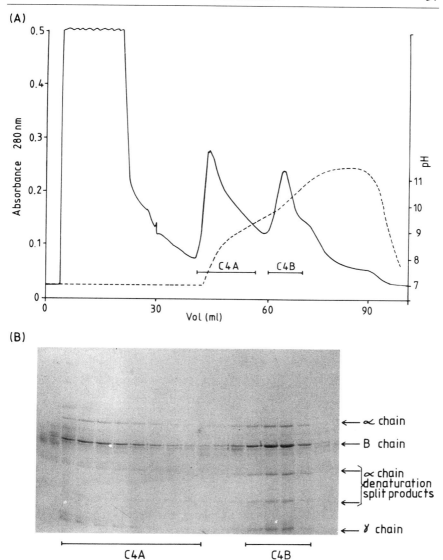

FIG. 4. Human C4A and C4B from an affinity column bearing monoclonal antibody L003. Human EDTA plasma (5 ml) was loaded onto a column bearing monoclonal antibody L003, and eluted with a 30-ml gradient from pH 8.5 to pH 11.5. (A) Profile obtained; the absorbance at 280 nm is shown as a solid line and the pH of the fractions is shown as a broken line. (B) SDS–PAGE analysis of 2-ml fractions eluting between 40 and 74 ml.

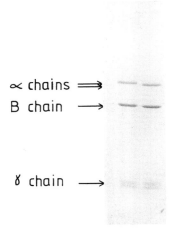

∝ chains ⟹

B chain ⟶

ɤ chain ⟶

Fɪɢ. 5. SDS–PAGE of C4A and C4B. The C4A and C4B pools from the LOO3 monoclonal antibody affinity column were loaded onto a 1-ml Mono Q column and eluted with a 20-ml linear gradient from 200 mM NaCl to 500 mM NaCl. The concentrated and purified proteins were reduced and electrophoresed on gels with a very low level of cross-linking bis-acrylamide.

Affinity Purification of C3 on Monoclonal Antibodies

C3 can be purified by affinity chromatography on monoclonal antibody WM1 using the same general method as has been described for C4. This antibody gives somewhat stronger binding of C3 than the LOO1 and LOO3 antibodies do for C4. The thioester bond of C3 is hydrolyzed rapidly above pH 10.5 and the monoclonal antibody releases the C3 rather slowly at this pH. The column is equilibrated with 25% APB, pH 7.0, and the plasma

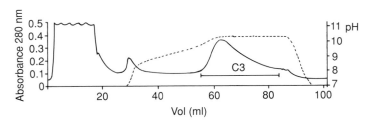

Fɪɢ. 6. Human C3 from an affinity column bearing monoclonal antibody WM1. Human EDTA plasma (2.5 ml) was loaded onto a WM1 affinity column and eluted with a 30-ml gradient from pH 8.5 to pH 10.5. Washing was continued with the buffer at pH 10.5 until all of the C3 had been eluted.

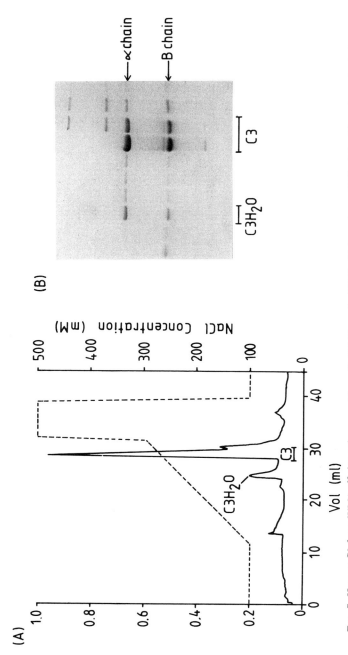

FIG. 7. Human C3 from WM1 affinity column on Mono Q. (A) From the WM1 affinity column, 25% of the C3 pool was loaded onto a 1-ml column of Mono Q and eluted with a 20-ml gradient from 100 to 300 m*M* NaCl. (B) SDS–PAGE analysis of the fractions eluting between 24 and 33 ml.

is loaded and washed as described for C4 on LOO3. The C3 is eluted with a 30-ml gradient from 100% APB, pH 8.5, to 100% APB, pH 10.5. Washing with 100% APB, pH 10.5, is continued until the absorbance at 280 nm approaches zero (Fig. 6). Under these conditions approximately 90% of the thioester bonds in the C3 remain intact. Using a final buffer with a higher pH gives a much sharper peak, but most of the C3 is inactivated. The C3 peak is pooled and concentrated on the Mono Q column as described in the section on ion-exchange purification, as shown in Fig. 7A. SDS–PAGE analysis (Fig. 7B) reveals that the C3 in which the thioester bond has been hydrolyzed is eluted earlier in the gradient than is C3. The high-molecular-mass components eluting slightly after the bulk of the C3 protein have not been fully characterized, but appear to be denatured C3 products.

Preparation of C3b and C4b

The activated forms of C3 and C4 are widely used in the functional evaluation of their control proteins, factors I and H, C4bBP, CR1, etc. C3b can be prepared from C3 using the classical or alternative pathway C3 convertases, but this requires that a number of other components are available. A simpler and more straightforward method is to use trypsin as the activating enzyme. C3 is treated for 15 min at 37° with 0.1% (w/w) trypsin. The reaction is stopped by the addition of an excess of soybean trypsin inhibitor (SBTI). The resultant mixture can be rerun on Mono Q to remove the SBTI and trypsin–SBTI complexes.

C4b can also be generated by trypsin digestion, but because its natural activator, C1s, is not difficult to prepare,[32] most investigators prefer to use it. C4 is treated with 1% (w/w) C1s for 15 min at 37°. Any unactivated material can be removed by chromatography on Mono Q under the conditions described in previous sections.

Conclusions

Two types of method have been described for the preparation of C3 and C4 from small volumes of plasma. The affinity methods using monoclonal antibodies give virtually 100% recovery of pure protein and are the methods of choice when small amounts of material are required. If monoclonal antibodies of suitable affinity are unavailable, the small-scale ion-exchange methods can be used. Although rapid and very reproducible, these methods do not give 100% purity, but the preparations are suitable for most purposes. These methods are also useful for the preparation of the proteins

---

[32] R. B. Sim, this series, Vol. 80, p. 26.

from novel species.[33,34] Both methods are very rapid and can be performed in a single day.

[33] A. W. Dodds and S. K. A. Law, *Biochem. J.* **265**, 495 (1990).
[34] M. W. Nicholls and J. P. Atkinson, *J. Immunol.* **144**, 4262 (1990).

[4] Human Complement Serine Proteases C$\bar{\text{1}}$r and C$\bar{\text{1}}$s and Their Proenzymes

By GÉRARD J. ARLAUD and NICOLE M. THIELENS

Introduction

The first component of the classical pathway of complement, C1, is a complex protease comprising two weakly interacting subunits, a noncatalytic protein (C1q) and C1s–C1r–C1r–C1s, a Ca^{2+}-dependent tetrameric complex of two homologous serine protease zymogens, C1r and C1s. C1q mediates C1 binding to various activators, either directly, in the case of certain bacteria or viruses, or through antibodies linked to antigenic determinants on a cell surface. C1q binding causes conformational changes within C1, triggering autocatalytic conversion of proenzyme C1r to its activated form C$\bar{\text{1}}$r,[1] which in turn activates C1s. This double-activation process occurs through limited proteolysis and results in the conversion of single-chain proenzymes (C1r, C1s) into active proteases (C$\bar{\text{1}}$r, C$\bar{\text{1}}$s) comprising two disulfide-linked chains. Activated C$\bar{\text{1}}$s then cleaves and activates the fourth (C4) and second (C2) components of complement, and thereby triggers activation of the classical pathway. The properties of the C1 complex and of its C1q subcomponent are described in recent reviews.[2-5] This article will describe the isolation and properties of the proenzyme and the activated forms of C1r and C1s.

[1] Activated complement components are designated by a superscript bar (e.g., C$\bar{\text{1}}$r is activated C1r).
[2] N. R. Cooper, *Adv. Immunol.* **37**, 151 (1985).
[3] V. N. Schumaker, P. Zavodszky, and P. H. Poon, *Annu. Rev. Immunol.* **5**, 21 (1987).
[4] G. J. Arlaud, M. G. Colomb, and J. Gagnon, *in* "New Comprehensive Biochemistry" (R. A. Harrison, ed.). Elsevier, Amsterdam, 1993 (in press).
[5] K. B. M. Reid, *Biochem. Soc. Trans.* **11**, 1 (1983).

Assays of C1r, C$\overline{1}$r, C1s, and C$\overline{1}$s

Esterolytic Assays

Proenzymes C1r and C1s do not cleave synthetic substrates at a significant rate, but the active forms C$\overline{1}$r and C$\overline{1}$s can both be assayed by hydrolysis of amino acid esters. Some of these assays are sensitive and provide a convenient means to obtain an estimate of the proteolytic activity of C$\overline{1}$r and C$\overline{1}$s. They are not specific, however, and should therefore be used only with purified proteins or intermediate purification fractions essentially free from other plasma proteases.

Assay of C$\overline{1}$r. C$\overline{1}$r hydrolyzes a restricted number of amino acid or peptide esters, including N-acetyl-L-arginine methyl ester (Ac-Arg-OMe),[6] N-acetylglycine-L-lysine methyl ester (Ac-Gly-Lys-OMe),[6] and N-carbobenzyloxyglycine-L-arginine thiobenzyl ester (Z-Gly-Arg-S-Bzl).[7] It should be noted, however, that C$\overline{1}$r cleaves these substrates at a slow rate, compared to other plasma proteases, including plasmin, thrombin, and C$\overline{1}$s. The following assay, using Ac-Gly-Lys-OMe as a substrate, and based on the spectrophotometric measurement of methanol released on hydrolysis,[8] is sensitive and can be used to determine the esterolytic activity of purified C$\overline{1}$r preparations. The reaction mixture, prepared in 100 mM K$_2$HPO$_4$, pH 7.5, contains 1 mM Ac-Gly-Lys-OMe (Sigma Chemical Co., St. Louis, MO), 1.6 U/ml[9] alcohol oxidase from *Pichia pastoris* (Sigma), 1 U/ml[9] horseradish peroxidase type VI (Sigma), and 0.5 mg/ml 2,2'-azinobis(3-ethylbenzthiazoline-6-sulfonic acid) (ABTS) (Sigma). The mixture (3 ml) is incubated at 30° in two spectrophotometer cuvettes. The reaction is initiated by addition to the test sample of 10–100 μl of the C$\overline{1}$r solution, and the increase in absorbance at 414 nm, due to the oxidation of ABTS, is monitored against the control sample for 5–10 min. Cleavage of Ac-Gly-Lys-OMe by C$\overline{1}$r yields methanol, which is oxidized by alcohol oxidase. This reaction produces H$_2$O$_2$, which in turn is used by the peroxidase for the conversion of the chromogenic substrate ABTS to its oxidized form. The molar extinction coefficient at 414 nm of the oxidized form of ABTS is 3.6×10^4 M^{-1} cm^{-1}. Under the conditions of the assay, the specific activity is 2.0–4.0 U/mg pure C$\overline{1}$r.[9,10] The increase in absorbance is linear up to a ΔA_{414}/min of 0.10, and the sensitivity limit of the assay is about 0.5 μg or 3 pmol of C$\overline{1}$r. An alternative sensitive

---

[6] J. M. Andrews and R. D. Baillie, *J. Immunol.* **123**, 1403 (1979).

[7] B. J. McRae, T.-Y. Lin, and J. C. Powers, *J. Biol. Chem.* **356**, 12362 (1981).

[8] G. R. Herzberg and M. Rogerson, *Anal. Biochem.* **149**, 354 (1985).

[9] One unit of enzyme is defined as the amount that transforms 1 μmol of substrate per minute. This definition is used throughout this section.

[10] As illustrated in Fig. 1, the specific esterolytic activity of C$\overline{1}$r decreases when its concentration is increased. This question is discussed later in this section.

procedure for C̄r assay by hydrolysis of Z-Gly-Arg-S-Bzl is described byMcRae *et al.*[7]

Assay of C̄s. C̄s exhibits a wider esterolytic activity than does C̄r and can be assayed using a number of amino acid and peptide esters, including those already mentioned for C̄r, as well as N^α-carbobenzyloxy-L-lysine *p*-nitrophenyl ester (Z-Lys-ONp),[11] *N*-carbobenzyloxy-L-tyrosine *p*-nitrophenyl ester (Z-Tyr-ONp),[12] N^α-benzoyl-L-arginine ethyl ester (Bz-Arg-OEt),[13] *p*-tosyl-L-arginine methyl ester (Tos-Arg-OMe),[13] and *N*-acetyl-L-tyrosine ethyl ester (Ac-Tyr-OEt).[14] Again, these substrates are not specific for C̄s.

Ac-Gly-Lys-OMe. The procedure using Ac-Gly-Lys-OMe, already described for C̄r, can also be used to assay C̄s esterolytic activity. The specific activity is 32.0 U/mg pure C̄s,[9] and the sensitivity limit of the assay is about 50 ng or 0.6 pmol of C̄s.

Bz-Arg-OEt. Bz-Arg-OEt can be used as a substrate in a spectrophotometric alcohol dehydrogenase-linked assay.[15] This assay is less sensitive than the previous one but, because Bz-Arg-OEt is not cleaved by C̄r, it can be used on fractions containing both proteases.

Reagents

> Buffer: 75 mM Tetrasodium diphosphate, 0.3 M semicarbazide dihydrochloride, 25 mM Tris, 50 mM glycine, pH adjusted to 8.5 with NaOH
> Bz-Arg-OEt solution: 6 mM N^α-benzoyl-L-arginine ethyl ester hydrochloride (Sigma, St. Louis, MO)
> Glutathione solution: 1.3 mM reduced glutathione
> NAD solution: 30 mM
> ADH solution: alcohol dehydrogenase from yeast (Sigma), 320 U/mg,[9] 30 mg/ml in 0.1% (w/v) bovine serum albumin (BSA)

Procedure

In two spectrophometer cuvettes, thermostatted at 30°, mix 2.30 ml of buffer, 0.50 ml of Bz-Arg-OEt solution, 0.10 ml of NAD solution, 0.01 ml of glutathione solution, and 0.02 ml of ADH solution. The reaction is initiated by addition to the test sample of 10–100 μl of the C̄s-containing fraction; monitor against the control sample for 5–10 min by measuring

[11] R. B. Sim, R. R. Porter, K. B. M. Reid, and I. Gigli, *Biochem. J.* **163**, 219 (1977).
[12] D. H. Bing, *Biochemistry* **8**, 4503 (1969).
[13] G. J. Arlaud, Ph.D. Thesis, University of Grenoble (1980).
[14] G. Valet and N. R. Cooper, *J. Immunol.* **112**, 339 (1974).
[15] I. Trautschold, E. Werle, and G. Schweitzer, *in* "Methods of Enzymatic Analysis" (H. U. Bergmeyer, ed.), p. 1031. Springer-Verlag, Berlin and New York, 1974.

the increase in absorbance at 340 nm. The molar extinction coefficient at 340 nm of NADH is $6.22 \times 10^3 \, M^{-1} \, cm^{-1}$. Under the conditions of the assay, the specific activity is 2.1 U/mg pure C$\overline{1}$s,[9] and the sensitivity limit is about 0.5 μg or 6 pmol of C$\overline{1}$s. The increase in absorbance is linear up to a ΔA_{340}/min of 0.06.

Other Substrates. Other assays using Z-Lys-ONp,[16] Z-Tyr-ONp,[12] Tos-Arg-OMe,[17] Ac-Tyr-OEt,[14] or various peptide thioesters[7] have been described.

Proteolytic Assays

The complex interaction and activation properties of C1r and C1s involved in the assembly and activation of the C1 complex can be measured by means of a C1 reconstitution assay. Other simple proteolytic assays can be used to test the activation properties of C1r and C1s, as well as the proteolytic activity of C$\overline{1}$r and C$\overline{1}$s. Purified C1r and C1s, or fractions containing only one of these proteins, should be used in these assays.

C1 Reconstitution Assay. The proenzyme C1 complex is reassembled in solution by incubating purified C1q,[17,18] C1r, and C1s in a C1q : C1r : C1s molar ratio of 1 : 2 : 2 for 5 min at 4° in 145 mM NaCl, 2.5 mM CaCl$_2$, 50 mM triethanolamine hydrochloride, pH 7.4. Efficient reconstitution of the complex occurs at C1q, C1r, and C1s concentrations higher than 230, 86, and 79 μg/ml, respectively. Reconstituted C1 activates spontaneously on incubation at 37°, and both C1r and C1s become fully activated within 30 min. Activation can be measured by the ability of C$\overline{1}$r and C$\overline{1}$s to hydrolyze synthetic esters (see above), or of C$\overline{1}$s to cleave its natural substrates, complement proteins C4 and C2 (see below). Conversion of proenzyme C1r and C1s to their activated counterparts can also be monitored by SDS polyacrylamide gel electrophoresis. Under reducing conditions, proenzyme C1r migrates as a single-chain protein of apparent M_r 85,000, whereas C$\overline{1}$r yields two chains, A and B, of apparent M_r 57,000 and 35,000, respectively. Under the same conditions, the apparent M_r of proenzyme C1s and of the A and B chains of C$\overline{1}$s are 84,000, 55,000, and 30,000, respectively. The sensitivity limit of the analysis by SDS polyacrylamide gel electrophoresis (about 1 μg protein under standard staining conditions) can be improved by using ^{125}I-labeled C$\overline{1}$s and subsequent quantitation of C1s activation by scanning of radioactivity. A detailed procedure for

[16] R. B. Sim, this series, Vol. 80, p. 26.
[17] G. J. Arlaud, R. B. Sim, A.-M. Duplaa, and M. G. Colomb, *Mol. Immunol.* **16**, 445 (1979).
[18] K. B. M. Reid, this series, Vol. 80, p. 16.

monitoring activation of C1 reconstituted from C1q, C1r, and ^{125}I-labeled C1s has been described.[19]

Other Assays. Purified proenzyme C1r autoactivates on incubation at 37° in the absence of Ca^{2+} ions. At this temperature, in 145 mM NaCl, 50 mM triethanolamine hydrochloride, pH 7.4, activation is complete in 1 hr. C1r activation can be monitored by SDS polyacrylamide gel electrophoresis as described above. Activated C̄1r can also be assayed by its ability to cleave proenzyme C1s (see below) or to hydrolyze synthetic esters (see above).

Activation of proenzyme C1s is achieved by incubation of the protein (1 mg/ml) with 5% (w/w) C̄1r for 1 hr at 37° in 145 mM NaCl, 50 mM triethanolamine hydrochloride, pH 7.4. Again, the activation process can be monitored by SDS polyacrylamide gel electrophoresis under reducing conditions. Activated C̄1s can also be assayed by its esterolytic activity or by limited proteolysis of its protein substrates C4 and C2. Purified C4,[20,21] at a concentration of 1 mg/ml, is incubated with 5% (w/w) C̄1s for 30 min at 37° in 145 mM NaCl, 1 mM EDTA, 50 mM triethanolamine hydrochloride, pH 7.4. Conversion of C4 (apparent M_r 200,000) into fragments C4a (M_r 9,000) and C4b (M_r 190,000) can be monitored by SDS polyacrylamide gel electrophoresis under nonreducing conditions. Identical incubation and analysis conditions are used for cleavage of purified C2[20,22] by C̄1s. Native C2 (apparent M_r 100,000) is converted into fragments C2a (M_r 70,000) and C2b (M_r 30,000).

Hemolytic Assay

The functional properties of purified C1r can also be measured by means of a hemolytic assay. Like the above-described C1 reconstitution assay, this assay provides a measurement of the ability of C1r or C1s to reconstitute a functionally active C1 complex in the presence of the other two C1 subcomponents. Once reconstituted, the C1 complex is assayed by its ability to trigger lysis of antibody-coated red blood cells in the presence of an excess of all other components of the classical pathway of complement. The specific reagents required for this type of assay (antibody-coated cell intermediates and complement components) are rela-

[19] N. R. Cooper and R. J. Ziccardi, *J. Immunol.* **119,** 1664 (1977).

[20] I. Gigli and F. A. Tausk, this series, Vol. 162, p. 626.

[21] A. Reboul, N. M. Thielens, M.-B. Villiers, and M. G. Colomb, *FEBS Lett.* **103,** 156 (1979).

[22] N. M. Thielens, M.-B. Villiers, A. Reboul, C. L. Villiers, and M. G. Colomb, *FEBS Lett.* **141,** 19 (1982).

tively complex and their preparation is time-consuming, but most are commercially available. This assay represents a reliable method for verifying the overall functional state of highly purified proenzymatic C1r and C1s, which can be detected at the 1- to 10-ng level. A detailed procedure for hemolytic assay of C1r and C1s has been described.[16,23]

Purification of C1r, C̄1r, C1s, and C̄1s

Several methods for the purification of human C1r and C1s have been reported. Most involve a first step consisting in isolation of the C1 complex from serum or plasma by either conventional (euglobulin) precipitation[16,24,25] or by means of affinity procedures involving selective C1 binding to insoluble antibody–antigen aggregates[17,26] or immunoglobulin G (IgG)-Sepharose.[27–30] The C1 complex is then disrupted by EDTA, and subsequent purification of C1r and C1s is achieved by ion-exchange and gel-filtration procedures. Unless protease inhibitors are used throughout the preparation, all methods result in partial activation of C1r and C1s. Purification methods for the activated or proenzyme forms of C1r and C1s,[17,26] based on the same affinity procedure, are described below. All steps are done at 4°, unless stated otherwise.

Isolation of C̄1r and C̄1s

Citrated human plasma pooled from normal donors is made 20 mM with CaCl$_2$ and left to clot for 4 hr at room temperature, then for 16 hr at 4°. The clot is removed by centrifugation and subsequent filtration through muslin, and the serum is stored at $-80°$ until required. Outdated plasma may be used as a starting material for this preparation, but plasma containing heparin or EDTA as anticoagulant is not suitable.

C1 Binding to IgG–Ovalbumin Aggregates. Human serum (1 liter) is thawed and centrifuged for 15 min at 12,500 g. Soybean trypsin inhibitor (STI) (Sigma) is added to a concentration of 50 μg/ml and the pH is adjusted

[23] R. B. Sim, this series, Vol. 80, p. 6.
[24] R. J. Ziccardi and N. R. Cooper, *J. Immunol.* **116**, 496 (1976).
[25.] I. Gigli, R. R. Porter, and R. B. Sim, *Biochem. J.* **157**, 541 (1976).
[26] G. J. Arlaud, C. L. Villiers, S. Chesne, and M. G. Colomb, *Biochim. Biophys. Acta* **616**, 116 (1980).
[27] P. A. Taylor, S. Fink, D. H. Bing, and R. H. Painter, *J. Immunol.* **118**, 1722 (1977).
[28] D. H. Bing, J. M. Andrews, K. M. Morris, E. Cole, and V. Irish, *Prep. Biochem.* **10**, 269 (1980).
[29] M. C. Peitsch, T. J. Kosacsovics, and H. Isliker, *J. Immunol. Methods* **108**, 265 (1988).
[30] P. D. Lane, V. N. Schumaker, Y. Tseng, and P. H. Poon, *J. Immunol. Methods* **141**, 219 (1991).

to 7.0. Insoluble IgG–ovalbumin aggregates formed at equivalence[17,31] (1.0 g) are suspended in the serum and the suspension is incubated for 45 min at 4° with occasional shaking. The resulting (IgG–ovalbumin)–C1 aggregates are then centrifuged (8 min at 10,000 g), suspended in 200 ml of 120 mM NaCl, 5 mM CaCl$_2$, 40 μg/ml STI, 20 mM Tris-HCl, pH 7.0, and recentrifuged. The washing step is repeated twice, and the aggregates are suspended in 150 ml of the washing buffer without STI. At this stage, activation of the C1 complex is only partial. To achieve complete activation, the suspension is incubated for 35 min at 30°, then cooled on ice and centrifuged (8 min at 10,000 g). The resulting pellet is suspended twice in 50 ml of 20 mM EDTA, 60 mM NaCl, 50 mM Tris-HCl, pH 7.0, and centrifuged (8 min at 10,000 g). Treatment with EDTA disrupts the CĪ complex, removes CĪr and CĪs, and leaves C1q bound to the IgG–ovalbumin aggregates. CĪr and CĪs represent about 95% of the material contained in the pooled EDTA supernatants (20–25 mg protein).

DEAE-Cellulose Chromatography. The EDTA extract is centrifuged (20 min at 25,000 g), dialyzed against 5 mM EDTA, 20 mM sodium phosphate, pH 7.4, and then applied, at a flow rate of about 50 ml/hr, to a DEAE-cellulose (Whatman, Clifton, NJ DE-52) column (2.5 × 15 cm) equilibrated with the same buffer. The column is washed with 150 ml of the starting buffer, and the bound proteins are eluted by a linear NaCl gradient formed from 350 ml of 5 mM EDTA, 20 mM sodium phosphate, pH 7.4, and 350 ml of the same buffer containing 0.25 M NaCl. Figure 1 shows a typical DEAE-cellulose elution profile. CĪr and CĪs are eluted in that order and pooled as indicated.

The CĪr pool is dialyzed against 145 mM NaCl, 50 mM triethanolamine hydrochloride, pH 7.4, and concentrated to 0.4–0.5 mg/ml by ultrafiltration on a PM10 membrane (Amicon, Danvers, MA). The protein may be stored for a few months at 4° without loss of activity, although slight aggregation occurs on storage. Freezing causes extensive precipitation of the protein and must therefore be avoided.

CĪs eluted from DEAE-cellulose contains significant amounts (up to 5%) of CĪr, and further purification by affinity chromatography on anti-CĪr IgG-Sepharose is required. The CĪs pool is dialyzed against 145 mM NaCl, 2 mM CaCl$_2$, 50 mM triethanolamine hydrochloride, pH 7.4, and passed at a flow rate of 20 ml/hr through a 1.5 × 10 cm column of anti-CĪr IgG-Sepharose 4B[32] equilibrated with the dialysis buffer. The eluate

---

[31] J. W. Prahl and R. R. Porter, *Biochem. J.* **107**, 753 (1968).
[32] G. J. Arlaud, A. Reboul, C. M. Meyer, and M. G. Colomb, *Biochim. Biophys. Acta* **485**, 215 (1977).

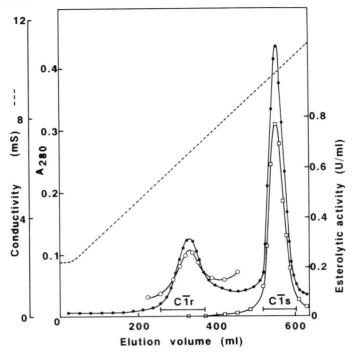

FIG. 1. Separation of C̄1r and C̄1s by ion-exchange chromatography on DEAE-cellulose. The EDTA extract obtained from 2 liters of serum was applied to the column and chromatographed as described in the text. ●, A_{280}. Esterolytic activity was determined using Ac-Gly-Lys-OMe (○) or Bz-Arg-OEt (□) as substrate. One unit is defined as the amount of protease that transforms 1 μmol of substrate per minute.

is then dialyzed against 145 mM NaCl, 50 mM triethanolamine hydrochloride, pH 7.4, and concentrated to 1.0 mg/ml by ultrafiltration. This material is very stable, and may be stored for several months at 4° without loss of activity. Unlike C̄1r, C̄1s is unaffected by freezing, and may therefore be stored at −20°.

Recycling of the IgG–Ovalbumin Aggregates. Removal of C̄1r and C̄1s from C̄1 by EDTA leaves C1q bound to the IgG–ovalbumin aggregates. In order to remove this subcomponent, the aggregates are washed twice with 50 ml of 0.7 M NaCl, 50 mM Tris, pH 10.0, and centrifuged (8 min at 10,000 g). The pooled supernatants contain almost exclusively C1q, which can be further purified by ion-exchange chromatography on CM-cellulose.[17,33] The aggregates contained in the pellet are suspended in 100 ml

[33] K. B. M. Reid, *Biochem. J.* **141**, 189 (1974).

of 150 mM NaCl, 20 mM Tris-HCl, pH 7.0, and incubated for 30 min at 30° in the presence of 5 mM diisopropyl fluorophosphate (DFP)[34] (Sigma), then cooled on ice and centrifuged (8 min at 10,000 g). Finally, the aggregates are washed twice with 150 mM NaCl, 20 mM Tris-HCl, pH 7.0, and suspended at about 10 mg/ml in this buffer. After each purification cycle, the average recovery of the aggregates is 85–90%. After addition of fresh aggregates to reconstitute the initial amount (1 g), the suspension can be used for a new purification cycle without significant loss of efficiency. The aggregates are very stable and can be stored for several months at 4° in the presence of 0.1% (w/v) sodium azide.

Isolation of C1r and C1s

Isolation of the proenzyme forms, C1r and C1s, is achieved by essentially the same method as for the active proteases, with several modifications designed to prevent activation. All manipulations, including centrifugations and DEAE-cellulose chromatography, must be performed at a temperature as close to 0° as possible, and serine protease inhibitors are present in all buffers. As described above, human serum (1 liter) is thawed, centrifuged, and adjusted to pH 7.0. Serum is then preincubated for 15 min at 0° in the presence of 1 mM DFP and 1 mM p-nitrophenyl-p'-guanidinobenzoate (NPGB)[35] (Merck, Darmstadt, Germany). After addition of the IgG–ovalbumin aggregates (1 g), the suspension is incubated for only 25 min at 0°. The (IgG–ovalbumin)–C1 aggregates are then centrifuged (8 min at 10,000 g) and washed twice with 200 ml of 120 mM NaCl, 5 mM CaCl$_2$, 20 mM Tris-HCl, pH 7.0, containing 1 mM DFP and 1 mM NPGB. For solubilization of C1r and C1s, the aggregates are suspended four times at 0° in 50 ml of 56 mM EDTA, 40 mM Tris-HCl, pH 7.0, containing 5 mM DFP and 1 mM NPGB, and centrifuged (8 min at 10,000 g). The pooled EDTA supernatants are centrifuged (15 min at 20,000 g) and dialyzed overnight at 0° against 5 mM EDTA, 20 mM sodium phosphate, pH 7.4, containing 1 mM DFP and 0.1 mM NPGB. After addition of 1 mM DFP, the dialyzate is applied to the DEAE-cellulose column and separation of C1r and C1s is achieved as described above for the active forms, except that 1 mM DFP is added to all buffers before use. The elution profile is essentially the same as in Fig. 1, except that an additional peak of serum amyloid protein (SAP) is eluted between C1r and C1s.

The C1r pool is dialyzed against 145 mM NaCl, 50 mM triethanolamine hydrochloride, pH 7.4, then concentrated by ultrafiltration and stored as

---

[34] DFP is prepared as a stock 0.5 M solution in anhydrous 2-propanol. It is highly toxic and should be used with extreme caution, as indicated by the suppliers.

[35] NPGB is prepared as a stock 0.1 M solution in dimethylformamide.

described for $C\overline{1}r$. C1s eluted from DEAE-cellulose contains, in addition to C1r, significant amounts of SAP. Both contaminants are removed by affinity chromatography on anti-$C\overline{1}r$ IgG-Sepharose 6B.[36] The procedure used is the same as described above, except that 1 mM DFP is added to the C1s pool before application to the column, as well as to the elution buffer. The eluate is finally dialyzed against 145 mM NaCl, 50 mM triethanolamine hydrochloride, pH 7.4, concentrated by ultrafiltration, and stored as described for $C\overline{1}s$.

The procedure used for recycling the aggregates is different from that described above for the purification of $C\overline{1}r$ and $C\overline{1}s$. The (IgG–ovalbumin)–C1q aggregates are first suspended in 200 ml of 150 mM NaCl, 20 mM Tris-HCl, pH 7.0, then incubated for 30 min at 30° and centrifuged (8 min at 10,000 g). This washing procedure is performed three times, and is essential for efficient subsequent solubilization of C1q. C1q is then removed from the aggregates at pH 10.0 by the same procedure as described above for the purification of $C\overline{1}r$ and $C\overline{1}s$. In this case, however, the supernatants also contain large amounts of contaminants and are therefore not suitable for C1q purification. Finally, the aggregates are treated with 5 mM DFP, washed twice with 150 mM NaCl, 20 mM Tris-HCl, pH 7.0, suspended at about 10 mg/ml in this buffer, and stored exactly as described above. The average recovery is the same as observed for the purification of $C\overline{1}r$ and $C\overline{1}s$, and the aggregates can also be used for new purification cycles, although a slight decrease in C1 binding efficiency is observed after each cycle.

Purity and Yield. A summary of the purification stages of $C\overline{1}r$ and $C\overline{1}s$ is shown in Table I. The yields of the active enzymes are $C\overline{1}r$, 6.1–8.5 mg/liter of serum (average 6.9 mg); $C\overline{1}s$, 6.5–8.6 mg/liter of serum (average 7.8 mg). Slightly higher values are usually obtained for the proenzyme forms. The concentrations of C1r and C1s in serum have been estimated at 34 and 31 mg/liter,[37] respectively. Final preparations of C1r, $C\overline{1}r$, C1s, and $C\overline{1}s$ are homogeneous as judged by SDS polyacrylamide gel electrophoresis. It should be noted that proenzyme C1r usually contains trace amounts (3–5%) of $C\overline{1}r$. In this respect, it should be kept in mind that IgG–ovalbumin aggregates are potent activators of the C1 complex. To keep C1r activation to a minimal level, it is essential to conform strictly to the purification protocol. In particular, all manipulations should be

---

[36] In contrast with C1r, which binds to the IgG antibody, SAP is removed through nonspecific, Ca^{2+}-dependent binding to the Sepharose matrix. Complete removal of C1r is crucial in the case of proenzyme C1s, because even trace amounts of C1r lead to C1s activation.

[37] R. J. Ziccardi and N. R. Cooper, *J. Immunol.* **118,** 2047 (1977).

TABLE I
PURIFICATION OF C1r AND C1s FROM HUMAN SERUM

Material	Total volume (ml)	Total protein (mg)	Esterolytic activity (units)[a]		Specific activity (units/mg)[a]	
			Bz-Arg-OEt	Ac-Gly-Lys-OMe	Bz-Arg-OEt	Ac-Gly-Lys-OMe
Serum	1000	70,000	—	—	—	—
EDTA supernatant	101	22.8[b]	27.2	643	1.2	26.3
C1r pool from DEAE-cellulose	120	6.8[b]	ND[c]	27	—	3.9
C1s pool from DEAE-cellulose	96	8.0[b]	16.8	232	2.1	29.0
C1s pool from anti-C1r IgG-Sepharose 4B	104	6.5[b]	13.6	206	2.1	31.7

[a] One unit is defined as the amount of protease that transforms 1 μmol of substrate per minute.

[b] Protein was determined from measurement of A_{280}, using extinction coefficients ε (1%, 1 cm) = 12.4 (C1r), 14.5 (C1s), and 13.4 (EDTA supernatants).

[c] Not detectable.

done as quickly as possible, and at a temperature as close to 0° as possible. It should be emphasized that $\overline{C1r}$ present in the C1r preparation is expected to be quantitatively inactivated by NPGB or DFP, because these inhibitors, known to react irreversibly with the $\overline{C1r}$ active site, are used at all stages of the purification.

C1r and C1s in Other Species. C1 subcomponents and/or C1 activity have been detected in a wide range of species.[4] Rabbit,[38] guinea pig,[39] and bovine[40] C1r and C1s have been isolated by methods similar to those used for human C1r and C1s. Insofar as analyzed, they have structural and functional properties similar to those of their human counterparts.

Isolation of Proteolytic Fragments

The fragments responsible for either the interaction or the catalytic properties of C1r ($\overline{C1r}$) and C1s ($\overline{C1s}$) can be isolated by limited proteolysis of the native proteins.

Catalytic (γ-B) Fragments. The dimeric $(\gamma\text{-B})_2$ catalytic fragments of $\overline{C1r}$, comprising the C-terminal (γ) region of the A chain disulfide-linked to the B chain, are generated by autolytic cleavage of $\overline{C1r}$ on incubation

[38] Y. Mori, M. Koketsu, N. Abe, and J. Koyama, *J. Biochem. (Tokyo)* **85,** 1023 (1979).

[39] R. M. Bartholomew and A. F. Esser, *Biochemistry* **19,** 2847 (1980).

[40] R. D. Campbell, N. A. Booth, and J. E. Fothergill, *Biochem. J.* **183,** 579 (1979).

for 7 hr at 37° in 145 mM NaCl, 50 mM triethanolamine hydrochloride, pH 7.4.[41] The (γ-B)$_2$ fragments are purified by either high-pressure gel permeation on a TSK G-3000 SW column (Pharmacia LKB, Uppsala, Sweden)[42] or ion-exchange chromatography on a Mono Q HR5/5 column (Pharmacia, Piscataway, NJ). These fragments can also be generated by limited proteolysis of native C$\overline{1}$r by extrinsic proteases such as chymotrypsin, thermolysin, plasmin, or elastase.[42] The proenzyme form of C1r catalytic domains can be obtained by thermolysin cleavage of proenzyme C1r in the presence of NPGB.[43] The monomeric (γ-B) fragment of C$\overline{1}$s is generated by incubation of C$\overline{1}$s with plasmin in the absence of Ca^{2+} ions and isolated by high-pressure hydrophobic interaction chromatography on a TSK-phenyl 5PW column (LKB).[44]

   Interaction (α) Fragments. The α interaction fragments of C$\overline{1}$r and C$\overline{1}$s, derived from the NH$_2$-terminal region of the A chain, are obtained by limited proteolysis of the native proteins with trypsin.[45] Purification is achieved by ion-exchange chromatography on a Mono Q HR5/5 column (Pharmacia) for C$\overline{1}$rα and, in the case of C$\overline{1}$sα, by affinity chromatography on C$\overline{1}$s-Sepharose followed by high-pressure gel permeation on a TSK G-3000 SW column (LKB).[45]

Properties of C1r, C$\overline{1}$r, C1s, and C$\overline{1}$s

Physicochemical Properties

   The proenzyme and activated forms of C1r are noncovalent dimers, dissociating into monomers at acidic pH (5.0 for C1r and 4.0 for C$\overline{1}$r).[46] The dimer has a sedimentation coefficient of 7.1 S, with extreme experimental values of 6.7 and 7.9 S.[16] This value is shifted to 5.0 S on monomerization. From its sedimentation properties, C1r can be described as an elongated molecule with an axial ratio of 10 : 1.[47] The calculated molecular weight

[41] C. L. Villiers, G. J. Arlaud, and M. G. Colomb, *Proc. Natl. Acad. Sci. U.S.A.* **82**, 4477 (1985).

[42] G. J. Arlaud, J. Gagnon, C. L. Villiers, and M. G. Colomb, *Biochemistry* **25**, 5177 (1986).

[43] M. B. Lacroix, C. A. Aude, G. J. Arlaud, and M. G. Colomb, *Biochem. J.* **257**, 885 (1989).

[44] N. M. Thielens, A. Van Dorsselaer, J. Gagnon, and G. J. Arlaud, *Biochemistry* **29**, 3570 (1990).

[45] N. M. Thielens, C. A. Aude, M. B. Lacroix, J. Gagnon, and G. J. Arlaud, *J. Biol. Chem.* **265**, 14469 (1990).

[46] G. J. Arlaud, S. Chesne, C. L. Villiers, and M. G. Colomb, *Biochim. Biophys. Acta* **616**, 105 (1980).

[47] J. Tschopp, W. Villiger, H. Fuchs, E. Kilchherr, and J. Engel, *Proc. Natl. Acad. Sci. U.S.A.* **77**, 7014 (1980).

of C1r is 172,600 (86,300 for the monomer). These values are derived from the amino acid sequence plus 2000 for each carbohydrate moiety.[44] The apparent M_r of C1r is 166,000–188,000[16] under nondenaturing conditions and 83,000–85,000 under denaturing conditions, as determined by gel filtration[11] and SDS polyacrylamide gel electrophoresis.[11]

C1s and C$\overline{1}$s are monomeric proteins and form Ca^{2+}-dependent dimers. The monomer has a sedimentation coefficient of 4.1–4.5 S,[16] with an estimated axial ratio of 6:1,[47] and the dimer sediments at 5.9 S, with extreme experimental values of 5.7 and 6.9 S.[16] The molecular weight of C1s, derived from the amino acid sequence plus 2000 for each carbohydrate moiety, is 78,900. Under nondenaturing conditions, the apparent M_r of C1s is 83,000–85,000 in the presence of EDTA and 170,000–176,000 in the presence of $CaCl_2$. Under denaturing conditions, the M_r of C1s is 83,000–85,000.[16]

Extinction coefficients [ε (1%, 1 cm) at 280 nm] of 12.4 for C$\overline{1}$r and 14.5 for C$\overline{1}$s have been determined experimentally.[44] Partial specific volumes of 0.714–0.722 cm^3/g for C$\overline{1}$r and 0.717–0.724 cm^3/g for C$\overline{1}$s have been calculated from amino acid and carbohydrate compositions.[11,48] In standard electrophoresis C$\overline{1}$s (C1s) migrates as an α_2-globulin and C$\overline{1}$r (C1r) migrates as a β-globulin.[16] Isoelectric points of 4.9 for C$\overline{1}$r[49] and 4.5 for C$\overline{1}$s[13] have been determined.

Structural Properties

The complete primary structures of C1r[50–54] and C1s[55–58] have been determined from either protein or cDNA sequencing. In their proenzyme form, monomeric C1r and C1s are single-chain glycoproteins containing, respectively, 688 and 673 amino acid residues. Activation occurs in each case through cleavage of a single Arg-Ile bond[56,59] that yields active

[48] S. J. Perkins and A. S. Nealis, *Biochem. J.* **263,** 463 (1989).

[49] K. Okamura and S. Fujii, *Biochim. Biophys. Acta* **534,** 258 (1978).

[50] G. J. Arlaud and J. Gagnon, *Biochemistry* **22,** 1758 (1983).

[51] J. Gagnon and G. J. Arlaud, *Biochem. J.* **225,** 135 (1985).

[52] S. P. Leytus, K. Kurachi, K. S. Sakariassen, and E. W. Davie, *Biochemistry* **25,** 4855 (1986).

[53] A. Journet and M. Tosi, *Biochem. J.* **240,** 783 (1986).

[54] G. J. Arlaud, A. C. Willis, and J. Gagnon, *Biochem. J.* **241,** 711 (1987).

[55] P. E. Carter, B. Dunbar, and J. E. Fothergill, *Biochem. J.* **215,** 565 (1983).

[56] S. E. Spycher, H. Nick, and E. E. Rickli, *Eur. J. Biochem.* **156,** 49 (1986).

[57] C. M. Mackinnon, P. E. Carter, S. J. Smyth, B. Dunbar, and J. E. Fothergill, *Eur. J. Biochem.* **169,** 547 (1987).

[58] M. Tosi, C. Duponchel, T. Méo, and C. Julier, *Biochemistry* **26,** 8516 (1987).

[59] G. J. Arlaud and J. Gagnon, *FEBS Lett.* **180,** 234 (1985).

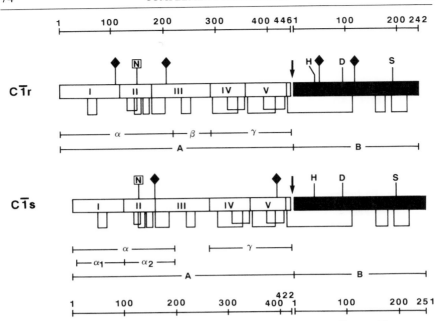

FIG. 2. Schematic representation of the linear structures of human C̄1r and C̄1s, showing the positions of the histidine, aspartic acid, and serine residues (H, D, S) involved in the active site, and of the Asn-linked carbohydrates (◆). N represents the asparagine residues that undergo posttranslational hydroxylation. I–V indicate the structural motifs of the A chains, and α, α₁, α₂, β, and γ indicate the fragments generated on limited proteolysis. The arrows show the Arg-Ile bond cleaved on activation.

proteases comprising two chains linked by a single disulfide bridge (Fig. 2).

The B chains (242 amino acid residues in C̄1r, 251 in C̄1s), derived from the carboxy-terminal part of the proenzymes, are homologous to the catalytic chains of serine proteases such as trypsin, chymotrypsin, and proteolytic blood-clotting factors,[50,55] and contain the triad of amino acids involved in the active site of this family of proteases (His-57, Asp-102, and Ser-195 in chymotrypsinogen). The presence of an Asp residue in the substrate-binding pocket, at positions 185 (C̄1r) and 189 (C̄1s), is indicative of trypsinlike specificities, in agreement with the physiological functions of both enzymes (see below). Compared to other serine proteases, both B chains are poorly disulfide bonded: they contain only two internal disulfide bridges and lack the "histidine loop" disulfide bridge present in all other known mammalian serine proteases.[60]

[60] G. J. Arlaud and J. Gagnon, *Biosci. Rep.* **1,** 779 (1981).

The A chains (446 and 422 amino acid residues in $\overline{C1r}$ and $\overline{C1s}$, respectively), derived from the amino-terminal part of the proenzymes, have a mosaic-like structure. Each can be divided into five structural motifs, including two pairs of internal repeats (I/III and IV/V) and a single copy of motif II (Fig. 2). Motifs I and III are homologous to motifs also found in human bone morphogenetic protein-1,[61] the embryonal protein UVS.2[62] and the A5 antigen[63] from *Xenopus laevis*, and the sea urchin protein uEGF.[64] Motifs II show homology with the epidermal growth factor (EGF) and with the EGF-like domains found in a variety of secreted or membrane-bound proteins, including several proteases of the coagulation system and the low-density lipoprotein receptor.[65] Motif II of Clr contains, at position 135, the single polymorphic site (Ser/Leu) identified in this protein[54] and, at position 150, an *erythro-β*-hydroxyasparagine residue resulting from posttranslational hydroxylation of asparagine.[66] The homologous asparagine residue of Cls (position 134) is only partially (about 50%) hydroxylated.[44] Other cases of EGF-like domains containing a β-hydroxyasparagine or β-hydroxyaspartic acid residue have been documented in blood coagulation factors VII, IX, and X, protein C, protein S, and protein Z,[65] and the presence of such domains can be correlated with a high-affinity Ca^{2+}-binding site. Motifs IV and V of Clr and Cls are tandem repeats of 60–70 amino acid residues, homologous to an ubiquitous sequence element ["short consensus repeat" (SCR) or "complement control protein repeat" (CCP)] found in multiple copies in proteins apparently unrelated to the complement system and various complement proteins sharing the ability to bind fragments C3b and/or C4b.[67]

$\overline{C1r}$ and $\overline{C1s}$ A chains show about 38% sequence homology, including 21 half-cystine residues, of which 20 form 10 internal disulfide bridges, and one is involved in the disulfide bridge to the B chain (Fig. 2). Only a few of these bonds have been formally identified in Clr,[54,68] but the complete

[61] J. M. Wozney, V. Rosen, A. J. Celeste, L. M. Mitsock, M. J. Whitters, R. W. Kriz, R. M. Hewick, and E. A. Wang, *Science* **242**, 1528 (1988).

[62] P. Bork, *FEBS Lett.* **282**, 9 (1991).

[63] S. Takagi, T. Hirata, K. Agata, M. Mochii, G. Eguchi, and H. Fujisawa, *Neuron* **7**, 295 (1991).

[64] M. G. Delgadillo-Reynoso, D. R. Rollo, D. A. Hursh, and R. A. Raff, *J. Mol. Evol.* **29**, 314 (1989).

[65] D. J. G. Rees, I. M. Jones, P. A. Handford, S. J. Walter, M. P. Esnouf, K. J. Smith, and G. G. Brownlee, *EMBO J.* **7**, 2053 (1988).

[66] G. J. Arlaud, A. Van Dorsselaer, A. Bell, M. Mancini, C. A. Aude, and J. Gagnon, *FEBS Lett.* **222**, 129 (1987).

[67] K. B. M. Reid, D. R. Bentley, R. D. Campbell, L. P. Chung, R. B. Sim, T. Kristensen, and B. F. Tack, *Immunol. Today* **7**, 230 (1986).

[68] G. J. Arlaud, M. G. Colomb, and J. Gagnon, *Immunol. Today* **8**, 106 (1987).

disulfide bridge pattern of C1s has been established.[69] Four glycosylation sites are present on C$\overline{1}$r (at positions 108 and 204 of the A chain, and at positions 51 and 118 of the B chain), whereas C$\overline{1}$s contains only two carbohydrates, located at positions 159 and 391 of the A chain (Fig. 2). All carbohydrates are linked to asparagine residues in sequences -Asn-X-Ser/Thr- and, although their structure has not been formally established, their contents in N-acetylglucosamine, mannose, galactose, and sialic acid suggest that they belong to the complex type.[11,54]

Models of the domain structure of C1r and C1s have been elaborated from electron microscopy, limited proteolysis, amino acid sequencing, differential scanning calorimetry, and functional studies.[41,70–73] Native C$\overline{1}$r appears on electron microscopy pictures as a dimeric association of twin monomers interacting through large domains, with smaller domains at each end of the dimer.[41,74] On prolonged incubation at 37°, C$\overline{1}$r undergoes two major autolytic cleavages that remove from the A chain of each monomer fragments α and β, leaving the carboxy-terminal γ fragment disulfide-linked to the catalytic B chain (Fig. 2). The $(\gamma$-B$)_2$ molecule (M_r 100,600) retains a dimeric structure and functional active sites, but lacks the ability to bind Ca^{2+} and to interact with C$\overline{1}$s in the presence of Ca^{2+}.[41,74] Limited proteolysis of C$\overline{1}$r by plasmin, elastase, chymotrypsin, or thermolysin also yields similar $(\gamma$-B$)_2$ fragments, and all cleavages generating the γ fragments occur within a sequence stretch extending from positions 274 to 286 of the A chain.[42] Observation of native C$\overline{1}$s by electron microscopy reveals a two-domain structure comparable to that observed for each monomer of C$\overline{1}$r,[41,70] and limited proteolysis with plasmin also yields a γ-B fragment,[41] which, in this case, is a monomer of M_r 46,000, originating from cleavage of the Lys^{269}-Leu^{270} bond in the A chain.[44] As observed for C$\overline{1}$r, the C$\overline{1}$s γ-B fragment is catalytically active, but lacks the ability to bind C$\overline{1}$r in the presence of Ca^{2+}.[41] The amino-terminal α fragment of C$\overline{1}$r and C$\overline{1}$s can be obtained by trypsin cleavage of the native proteins.[45,71,75] Both fragments C$\overline{1}$rα (M_r 27,600, residues 1–208 of C$\overline{1}$r A chain) and C$\overline{1}$sα (M_r 24,200, residues 1–192 of C$\overline{1}$s A chain) retain the ability to bind Ca^{2+} and to mediate Ca^{2+}-dependent C1r–C1s heteroassociation.

[69] D. Hess, J. Schaller, and E. E. Rickli, *Biochemistry* **30**, 2827 (1991).

[70] V. Weiss, C. Fauser, and J. Engel, *J. Mol. Biol.* **189**, 573 (1986).

[71] T. F. Busby and K. C. Ingham, *Biochemistry* **26**, 5564 (1987).

[72] T. F. Busby and K. C. Ingham, *Biochemistry* **27**, 6127 (1988).

[73] L. V. Medved, T. F. Busby, and K. C. Ingham, *Biochemistry* **28**, 5408 (1989).

[74] C. L. Villiers, G. J. Arlaud, R. H. Painter, and M. G. Colomb, *FEBS Lett.* **117**, 289 (1980).

[75] T. F. Busby and K. C. Ingham, *Biochemistry* **29**, 4613 (1990).

The above observations are consistent with the following scheme of the domain structure of C1r and C1s, in which each monomeric protease comprises two regions[41]: (1) an interaction region (α) comprised of motifs I, II, and part of the motif III of the A chain, responsible for C1r–C1s interaction and Ca^{2+} binding; (2) a catalytic region (γ-B) composed of the B chain, associated with the carboxy-terminal part (motifs IV and V) of the A chain (Fig. 2). Studies by differential scanning calorimetry[73] show that the γ-B region of C1s contains three independently folded domains, corresponding to motifs IV, V, and the B chain. This finding, and neutron diffraction experiments,[76] support the hypothesis that monomeric C1s, and probably also monomeric C1r, comprise at least four distinct domains: one or more in the α region, two in the γ region, and one in the B chain.

The $(\gamma$-B$)_2$ catalytic regions form the core of the C1r–C1r dimer. Assembly of the dimer involves the loose packing of two γ-B monomers,[76] and chemical cross-linking experiments are consistent with a "head to tail" conformation of the dimer, involving heterologous interactions between the γ region of one monomer and the B chain of the other monomer.[42] In contrast, electron microscopy of C1r after negative staining[70] shows "asymmetric X-shaped" molecules compatible with homologous interactions between the γ regions and/or the B chains.[48]

Interaction Properties

A fundamental feature of C1r and C1s is their ability to mediate the heterologous (C1r–C1s) Ca^{2+}-dependent interaction involved in the assembly of the C1s–C1r–C1r–C1s tetramer. This complex has a molecular weight of 330,000 and a sedimentation coefficient of 8.7 S.[26,47] In the absence of C1r, C1s also has the ability to form Ca^{2+}-dependent homodimers. Both the C1r–C1s and C1s–C1s interactions are mediated by the amino-terminal α regions, which can be obtained by limited proteolysis of native C1r and C1s with trypsin (see above). The isolated α fragments retain the ability to mediate Ca^{2+}-dependent associations and contain high-affinity Ca^{2+}-binding sites: each α fragment contains one Ca^{2+}-binding site, with dissociation constants of 38 μM, comparable to those determined for the native proteins.[45] This 1 : 1 stoichiometry is maintained on heterologous C1r–C1s interaction: the C1rα–C1sα dimer binds 2 Ca^{2+} atoms/mol, and the C1s–C1r–C1r–C1s tetramer binds 4 Ca^{2+} atoms/mol. In contrast, the homologous C1s–C1s interaction provides one additional binding site, as the C1sα–C1sα and the C1s–C1s dimers each incorporate

---

[76] G. Zaccai, C. A. Aude, N. M. Thielens, and G. J. Arlaud, *FEBS Lett.* **269,** 19 (1990).

3 Ca^{2+} atoms/mol.[45] The α regions of C1r and C1s exhibit characteristic irreversible low-temperature transitions, with midpoints of 32° for C1r and 37° for C1s, at physiological ionic strength. These transitions are abolished, or shifted to higher temperatures in the presence of Ca^{2+} ions.[71,73] Recent studies based on C1s fragmentation by plasmin[44] and on lactoperoxidase-catalyzed iodination of C1s[77] provide evidence that the structural determinants required for Ca^{2+} binding and Ca^{2+}-dependent protein–protein interactions are contributed by both the amino-terminal $\alpha 1$ and the EGF-like $\alpha 2$ regions (see Fig. 2).

Assembly of the C1 complex occurs through interaction between C1q and the C1s–C1r–C1r–C1s tetramer, and the affinity of the interaction is dependent on the activation state of the tetramer. At 4°, the association constant for the interaction between C1q and the proenzyme tetramer is $(2-7) \times 10^7 \ M^{-1}$, and it decreases to $7 \times 10^6 \ M^{-1}$ when the tetramer is activated.[78,79] The binding of C1s–C1r–C1r–C1s to C1q appears to be a complex process, probably involving multiple binding sites located in both C1r and C1s.[4] The amino-terminal α regions of C1r and C1s not only participate in the assembly of the tetramer, but probably also in the assembly of C1, because a fragment of C1s corresponding to the α region was shown to promote the binding of C1r to C1q.[75]

Enzymatic Properties

Proenzyme C1r and C1s have no detectable esterolytic activity and are not inactivated by DFP or NPGB at the concentrations used during the purification procedures. In the same way, proenzyme C1s has no demonstratable proteolytic activity. In contrast, proenzyme C1r undergoes spontaneous autoactivation on incubation, as detailed below. The active enzymes, C1r and C1s, are very specific serine proteases with trypsinlike specificities. C1r cleaves a single Arg-Ile bond in the sequence Lys-Gln-Arg-Ile-Ile-Gly in proenzyme C1s.[56] Isolated C1r also undergoes autolytic proteolysis involving cleavage of the Arg^{279}-Gly^{280} bond in the sequence Asp-Ser-Arg-Gly-Trp-Lys.[42] C1s cleaves a single Arg-Ala bond in the sequence Leu-Gln-Arg-Ala-Leu-Glu in complement component C4,[80,81] and cleaves a single Arg-Lys bond in the sequence Leu-Gly-Arg-Lys-Ile-Gln in complement component C2.[82] In addition to C4 and C2,

[77] C. Illy, N. M. Thielens, J. Gagnon, and G. J. Arlaud, *Biochemistry* **30**, 7135 (1991).
[78] R. C. Siegel and V. N. Schumaker, *Mol. Immunol.* **20**, 53 (1983).
[79] S. Lakatos, *Biochem. Biophys. Res. Commun.* **149**, 378 (1987).
[80] K. E. Moon, J. P. Gorski, and T. E. Hugli, *J. Biol. Chem.* **256**, 8685 (1981).
[81] E. M. Press and J. Gagnon, *Biochem. J.* **199**, 352 (1981).
[82] M. A. Kerr, *Biochem. J.* **183**, 615 (1979).

its natural substrates, human $\overline{C1s}$ has recently been shown to cleave β_2-microglobulin,[83] type I and II collagen,[84] and the heavy chain of the major histocompatibility complex class I antigens.[85]

$\overline{C1r}$ has a limited esterolytic activity against N-acetyl-L-arginine methyl ester (Ac-Arg-OMe), N-acetylglycine-L-lysine methyl ester (Ac-Gly-Lys-OMe), and several peptide thioesters, including N-carbobenzyloxyglycine-L-arginine thiobenzyl ester (Z-Gly-Arg-S-Bzl).[6,7,16] N^{α}-Carbobenzyloxy-L-lysine p-nitrophenyl ester (Z-Lys-ONp) and N-carbobenzyloxy-L-tyrosine p-nitrophenyl ester (Z-Tyr-ONp) have been described as $\overline{C1r}$ substrates,[6] but a $\overline{C1r}$ preparation was reported not to hydrolyze these esters.[11,16] K_m values range from 0.3×10^{-3} M for Z-Gly-Arg-S-Bzl to 8×10^{-3} M for Ac-Gly-Lys-OMe. It has been shown that the specific esterolytic activity of $\overline{C1r}$ and the number of titratable serine active sites decrease as the $\overline{C1r}$ concentration is increased above 0.5 μM.[6] This unusual effect may result from $\overline{C1r}$ aggregation, with loss of active sites.

In contrast to $\overline{C1r}$, $\overline{C1s}$ hydrolyzes a wide range of synthetic substrates, including peptide 4-nitroanilides, as well as methyl, ethyl, p-nitrophenyl, and thiobenzyl esters of amino acids and peptides.[7,11,12,32,86–88] The spectrum of these substrates is unusual, as they contain either basic or aromatic amino acids.[2] The active site of $\overline{C1s}$ appears to contain a hydrophobic area and an anionic binding site.[12,89] Representative K_m values include N^{α}-benzoyl-L-arginine ethyl ester (Bz-Arg-OEt), 2.5×10^{-3} M[32]; p-tosyl-L-arginine methyl ester (Tos-Arg-OMe), 1.7×10^{-3} M[32]; N-carbobenzyloxy-L-tyrosine p-nitrophenyl ester (Z-Tyr-ONp), $(5.6–7) \times 10^{-5}$ M[11,12]; and N-carbobenzyloxyglycine-L-arginine thiobenzyl ester (Z-Gly-Arg-S-Bzl), 6.9×10^{-5} M.[7]

The esterolytic and proteolytic activities of $\overline{C1r}$ and $\overline{C1s}$ are inhibited irreversibly by $\overline{C1}$ inhibitor, which forms tightly bound complexes containing one molecule of inhibitor and one molecule of protease.[2,16] Although both proteases have a similar affinity for $\overline{C1}$ inhibitor (functional dissociation constant 10^{-7} M), $\overline{C1s}$ reacts four- to fivefold more rapidly

[83] M. H. Nissen, P. Roepstorff, L. Thim, B. Dunbar, and J. E. Fothergill, *Eur. J. Biochem.* **189**, 423 (1990).

[84] K. Yamaguchi, H. Sakiyama, M. Matsumoto, H. Moriya, and S. Sakiyama, *FEBS Lett.* **268**, 206 (1990).

[85] H. Eriksson and M. H. Nissen, *Biochim. Biophys. Acta* **1037**, 209 (1990).

[86] A. L. Haines and I. H. Lepow, *J. Immunol.* **92**, 456 (1964).

[87] J. E. Volanakis, R. E. Schrohenloher, and R. M. Stroud, *J. Immunol.* **119**, 337 (1977).

[88] S. J. Keogh, D. R. K. Harding, and M. J. Hardman, *Biochim. Biophys. Acta* **913**, 39 (1987).

[89] W. J. Canady, S. Westfall, G. H. Wirtz, and D. A. Robinson, *Immunochemistry* **13**, 229 (O976).

than does $C\overline{1}r$.[90] The activation energy for the interaction between $C\overline{1}r$ and $C\overline{1}$ inhibitor (44.3 kcal/mol) is also much higher than that for the $C\overline{1}s$–$C\overline{1}$ inhibitor reaction (11.7 kcal/mol).[90] $C\overline{1}$ inhibitor, a member of the serine protease inhibitor (serpin) family,[91] is the only plasma inhibitor of $C\overline{1}r$ and $C\overline{1}s$.[92] DFP, phenylmethanesulfonyl fluoride (PMSF), and *p*-nitrophenyl-*p'*-guanidinobenzoate (NPGB) also inhibit irreversibly $C\overline{1}r$ and $C\overline{1}s$, but other inhibitors of trypsinlike serine proteases, such as soybean trypsin inhibitor, trasylol, α_1-antitrypsin, and hirudin, are ineffective.[2,16] Leupeptin is a competitive inhibitor of $C\overline{1}r$.[49] A variety of amidines and guanidines inhibit reversibly the enzymatic activities of $C\overline{1}r$ and $C\overline{1}s$.[12,93]

The proteolytic and esterolytic activities of $C\overline{1}r$ show broad pH optima, centered on pH 8.0.[13,16,94] Cleavage of C1s by $C\overline{1}r$ is highly temperature dependent, with an activation energy of 23–32 kcal/mol.[13,16,94] The reaction is slowed down in the presence of Ca^{2+} ions and inhibited at high ionic strength.[13,25,94,95] The esterolytic activity of $C\overline{1}s$ has a broad pH optimum, centered on pH 7.0–8.0,[13,86,89] and is only slightly sensitive to ionic strength and Ca^{2+} ions.[32,86]

Stability and Activation

A fundamental property of isolated proenzyme C1r lies in its ability to activate spontaneously on incubation in buffers containing EDTA. Although there is still some controversy as to whether the observed activation is due to C1r or results from the action of an unrelated trace protease,[26,96,97] several lines of evidence favor this reaction as an intrinsic property of C1r representing the normal C1r activation process as it occurs within the C1 complex (reviewed by Cooper[2]). The activation kinetics are slightly sigmoidal and the reaction, which shows a marked temperature dependence (activation energy, 45 kcal/mol), is independent of C1r con-

[90] R. B. Sim, G. J. Arlaud, and M. G. Colomb, *Biochim. Biophys. Acta* **612**, 433 (1980).

[91] S. C. Bock, K. Skriver, E. Nielsen, H.-C. Thøgersen, B. Wiman, V. H. Donaldson, R. L. Eddy, J. Marrinan, E. Radziejewska, R. Huber, T. B. Shows, and S. Magnusson, *Biochemistry* **25**, 4292 (1986).

[92] R. B. Sim, A. Reboul, G. J. Arlaud, C. L. Villiers, and M. G. Colomb, *FEBS Lett.* **97**, 111 (1979).

[93] S. S. Asghar, K. W. Pondman, and R. H. Cormane, *Biochim. Biophys. Acta* **317**, 539 (1973).

[94] G. B. Naff and O. D. Ratnoff, *J. Exp. Med.* **128**, 571 (1968).

[95] G. J. Arlaud, A. Reboul, and M. G. Colomb, *Biochim. Biophys. Acta* **485**, 227 (1977).

[96] A. W. Dodds, R. B. Sim, R. R. Porter, and M. A. Kerr, *Biochem. J.* **175**, 383 (1978).

[97] R. J. Ziccardi and N. R. Cooper, *J. Immunol.* **116**, 504 (1976).

centration over a wide range, and is inhibited only over a pH range (below 6.5) where dissociation of the C1r–C1r dimer occurs.[26] C1r activation is markedly inhibited or retarded in the presence of Ca^{2+} ions and at high ionic strength.[26,43,97,98] It is also inhibited reversibly by NPGB, partially sensitive to C1̄ inhibitor, but nearly insensitive to DFP.[26,96–99] The proenzyme form of the C1r $(\gamma\text{-B})_2$ catalytic fragments, obtained by limited proteolysis of native C1r with thermolysin, exhibits autoactivation properties comparable to those of the intact molecule.[43] Available experimental data are consistent with a model of C1r activation involving an intramolecular (intradimer) cross-mechanism in which the pro site in each of the paired γB catalytic regions cleaves the susceptible Arg-Ile bond in the adjacent region.[100] Proenzyme C1r isolated by the procedure described above shows about 50% activation after 10–20 min incubation at 37° in EDTA-containing buffers.

Purified activated C1̄r undergoes further autolytic cleavage on prolonged (5–7 hr) incubation at 37° in the presence of EDTA.[26,49] As detailed above, this process removes fragments α and β from the A chain of each C1̄r monomer, yielding dimeric $(\gamma\text{-B})_2$ molecules. The reaction has a pH optimum of 8.5–9.0, a high activation energy (36.8 kcal/mol), and is inhibited by Ca^{2+} ions, NPGB, DFP, and benzamidine.[26] This autolytic degradation does not significantly occur on prolonged storage of C1̄r preparations at 4°.

As measured by differential scanning calorimetry, C1̄r exhibits two irreversible temperature transitions.[71] The low-temperature transition (midpoints 26° at low ionic strength, and 40° in the presence of 0.5 M NaCl) is prevented by Ca^{2+} ions. It results in extensive polymerization of the protein without loss of esterolytic activity. The high-temperature transition (midpoint 53°) is relatively insensitive to Ca^{2+} ions and ionic strength, and is accompanied by a loss of catalytic activity.

C1s exists as a stable proenzyme and C1̄s exists as a stable enzyme. Neither undergoes autolytic cleavage.[25,32] Both forms tend to precipitate on exposure to pH 4.5–5.0, and C1̄s is irreversibly inactivated by exposure to pH values below 5.0 and above 9.5.[86] Three endothermic transitions are observed in C1̄s.[72,73] The first transition (midpoint 37°), which is prevented by Ca^{2+} ions, corresponds to the melting of the amino-terminal $\alpha\beta$ region of the A chain; the second transition (midpoint 49°) corresponds

[98] C. L. Villiers, A.-M. Duplaa, G. J. Arlaud, and M. G. Colomb, *Biochim. Biophys. Acta* **700**, 118 (1982).

[99] C. L. Villiers, G. J. Arlaud, and M. G. Colomb, *Biochem. J.* **215**, 369 (1983).

[100] G. J. Arlaud, M. G. Colomb, and C. L. Villiers, *Biosci. Rep.* **5**, 831 (1985).

to the melting of the catalytic B chain domain, and the third (midpoint 60°) corresponds to the melting of the γ region. The esterolytic activity of $C\overline{1}s$ is decreased by about 90% on heating at 49° for 10 min.[32]

Acknowledgments

We thank Monique B. Lacroix and Catherine A. Aude for helpful discussion, Larry C. Sieker for critical reading of the manuscript, and Florence Dubois for typing the manuscript.

[5] Purification and Properties of Human Factor D

By John E. Volanakis, Scott R. Barnum, and J. Michael Kilpatrick

Introduction

Complement factor D is an essential component of the alternative pathway of complement activation. It is a serine proteinase[1] with low catalytic activity against synthetic substrates[2] and a single known natural substrate, complement factor B. Factor D cleaves the Arg^{233}-Lys^{234} peptide bond of factor B[3] only in the context of a Mg^{2+}-dependent C3bB complex. This cleavage generates two fragments of factor B: Ba, which is released in the fluid phase, and Bb, which remains bound to C3b, completing the assembly of the C3 convertase of the alternative pathway, C3bBb. This extremely restricted substrate specificity is probably important for the regulation of the proteolytic activity of factor D, because no zymogen form of the enzyme exists in the blood.[4]

The serum concentration of factor D, 1.8 ± 0.4 μg/ml,[5] is the lowest of any complement protein. Studies on patients with renal insufficiency[6]

[1] D. T. Fearon, K. F. Austen, and S. Ruddy, *J. Exp. Med.* **139,** 355 (1974).

[2] C.-M. Kam, B. J. McRae, J. W. Harper, M. A. Niemann, J. E. Volanakis, and J. C. Powers, *J. Biol. Chem.* **262,** 3444 (1987).

[3] P. H. Lesavre, T. E. Hugli, A. F. Esser, and H. J. Müller-Eberhard, *J. Immunol.* **123,** 529 (1979).

[4] P. H. Lesavre and H. J. Müller-Eberhard, *J. Exp. Med.* **148,** 1498 (1978).

[5] S. R. Barnum, M. A. Niemann, J. F. Kearney, and J. E. Volanakis, *J. Immunol. Methods* **67,** 303 (1984).

[6] J. E. Volanakis, S. R. Barnum, M. Giddens, and J. H. Galla, *N. Engl. J. Med.* **312,** 395 (1985).

and *in vivo* microperfusion experiments using rat kidneys[7] have indicated that the low serum concentration of factor D is maintained by an extremely rapid catabolic rate. Because of its small size, factor D is filtered through the glomerular membrane and is reabsorbed and catabolized by the proximal tubular epithelial cells. This mechanism results in a fractional catabolic rate of 59.6%/hr.[8] The low serum levels of the enzyme may also contribute to the regulation of its proteolytic activity. Indeed, factor D has been shown to be the limiting enzyme in the activation sequence of the alternative pathway.[4]

Methods for the assay and purification of factor D from human serum have been described previously in this series.[9] In this chapter, we describe a simpler, higher yield method for purifying factor D from urine of patients with tubular dysfunction and from peritoneal dialysis fluid. We also present recent information on the structure and properties of factor D.

Assays

Hemolytic Assays

Several hemolytic assays are available for measuring factor D activity; however, the simplest and most commonly used procedure is that originally described by Lesavre *et al.*[3] This assay involves the use of rabbit erythrocytes and of a serum reagent depleted of factor D (RD). Numerous variations on this method have been developed and one used in our laboratory will be described below in detail. In addition to RD procedures, a number of more quantitative assays using cellular intermediates and purified complement components have been described.[4,10] Earlier versions of these assays required the preparation of EAC43 cells, and the reader is referred to the detailed methods of Rapp and Borsos.[11] A variation used in our laboratory involves the preparation of EC3b cells by direct attachment of C3b to the cell surface by using trypsin as described by Medicus *et al.*[12] Several fluid-phase convertase methods utilizing a cobra venom

[7] P. W. Sanders, J. E. Volanakis, S. G. Rostand, and J. H. Galla, *J. Clin. Invest.* **77,** 1299 (1986).

[8] M. Pascual, G. Steiger, J. Estreicher, K. Macon, J. E. Volanakis, and J. A. Schifferli, *Kidney Int.* **34,** 529 (1988).

[9] K. B. M. Reid, D. M. A. Johnson, J. Gagnon, and R. Prohaska, this series, Vol. 80, p. 134.

[10] D. T. Fearon, K. F. Austen, and S. Ruddy, *J. Exp. Med.* **138,** 1305 (1973).

[11] H. J. Rapp and T. Borsos, "Molecular Basis of Complement Activation." Appleton-Century-Crofts, New York, 1970.

[12] R. G. Medicus, O. Götze, and H. J. Müller-Eberhard, *J. Exp. Med.* **144,** 1076 (1976).

factor complex (CVFB) have also been described, and the reader is referred to Miyata et al.[13] and references therein.

Reagents

VBS: Isotonic veronal-buffered saline containing 150 mM NaCl and 5 mM veronal, pH 7.3

EDTA–GVB: VBS containing 10 mM EDTA and 0.1% gelatin

DGVB–Mg-EGTA: Half-strength VBS containing 2.5% dextrose, 10 mM EGTA, 2.5 mM MgCl$_2$, and 0.1% gelatin

Trypsin–TPCK: L-p-Tosylamino-2-phenylethyl chloromethyl ketone trypsin (Worthington Biochemical Corp., Freehold, NJ)

Soybean trypsin inhibitor (Sigma Chemical Co., St. Louis, MO)

Erythrocytes (E): Sheep and rabbit erythrocytes can be purchased from Colorado Serum Co. (Denver, CO)

EC3b cells: Sheep E are converted to EC3b as described by Ueda et al.[14] Briefly, 5 × 10^9 cells, 800 μg of C3, and 20 μg trypsin–TPCK in 0.5 ml of VBS are incubated for 10 min at room temperature. The reaction is terminated by the addition of a fivefold molar excess of soybean trypsin inhibitor (174 μg) followed by washing the cells three times with 30 ml of DGVB–Mg-EGTA. The cells are resuspended in 0.5 ml DGVB–Mg-EGTA containing 100 μg of factor B and 0.6 μg of factor D and incubated at 37° for 3 min. Following this, 200 μg of C3 is added and the incubation continued for 15 min at 37°. The cells are washed twice in DGVB–Mg-EGTA and the incubation with B, D, and C3 is repeated twice more

D-deficient serum (RD): Normal human serum is depleted of factor D by passing 50 ml of fresh serum over a 2.5 × 30 cm Bio-Rex 70 column (Bio-Rad, Richmond, CA) equilibrated with VBS. The effluent is concentrated to the original serum volume and stored in aliquots at −70°

C3: C3 is prepared according to the method of Gresham et al.[15]

B: Factor B is prepared according to the method of Niemann et al.[16]

P: Properdin is purified according to the method of Reid[17]

Procedures. Hemolytic titrations of factor D using EC3b cells are performed as follows. EC3b cells (1.5 × 10^7) are incubated with 1 μg of

[13] T. Miyata, O. Oda, R. Inagi, S. Sugiyama, A. Miyama, K. Maeda, I. Nakashima, and N. Yamanaka, *Mol. Immunol.* **27**, 637 (1990).

[14] A. Ueda, J. F. Kearney, K. H. Roux, and J. E. Volanakis, *J. Immunol.* **138**, 1143 (1987).

[15] H. D. Gresham, D. F. Matthews, and F. M. Griffin, Jr., *Anal. Biochem.* **154**, 454 (1986).

[16] M. A. Niemann, J. E. Volanakis, and J. E. Mole, *Biochemistry* **19**, 1576 (1980).

[17] K. B. M. Reid, this series, Vol. 80, p. 143.

B, 0.45 μg of P, and dilutions of the factor D source in a total volume of 300 μl of DGVB–Mg-EGTA for 30 min at 30°. The reaction is completed by the addition of 0.5 ml of guinea pig serum diluted 1/25 in EDTA–GVB and further incubation for 60 min at 37°. The reaction is stopped by the addition of 1.0 ml of ice-cold EDTA–GVB. The percentage of lysis is calculated from the absorbance of the supernatants at 413 nm and is used to calculate CH_{63} units.[11]

To measure the hemolytic activity of factor D using a serum reagent depleted of D (RD), a modification of the method of Lesavre *et al.*[3] is used. In this assay 0.25 ml of washed rabbit erythrocytes (1.5×10^8 cells/ ml) is incubated with 0.25 ml of RD diluted 1 : 10 and 0.1 ml of serially diluted sample, at 37° for 60 min. DGVB–Mg-EGTA is used to suspend the cells and to dilute the sera. The reaction is stopped by the addition of 2.0 ml of ice-cold EDTA–GVB. The degree of hemolysis is determined from the absorbance of the supernatant at 413 nm and used to calculate hemolytic activity in CH_{50} units.[11] To screen column fractions for factor D, a 25-μl sample of 1 : 50 dilution of each fraction is substituted into the assay described above. Incubation at 37° is allowed to proceed until visible hemolysis is observed (usually 5–10 min). The reaction is then stopped and percent hemolysis calculated. Similarly, to detect D activity in cell culture supernatants, 0.25 ml of culture supernatant, appropriately diluted, can be substituted into the assay.[18] Appropriate controls should be run when assaying cell culture supernatants because media containing fetal bovine serum (FBS) may cause lysis in the assay due to bovine factor D or a factor D-like enzyme. Pretreatment of FBS with diisopropyl fluoro-phosphate (5 mM, final concentration) inactivates the bovine enzyme and has no effect on cell viability.[18] A solid-phase variation of this assay, described by Martin *et al.*,[19] can detect as little as 2 ng of D when compared to a purified factor D standard or normal human serum of known factor D concentration.

Radioimmunometric Assay

Quantitation of factor D has been facilitated by the development of a solid-phase radioimmunometric assay (RIA) capable of detecting D at concentrations as low as 1–2 ng/ml.[5] The assay employs the IgG fraction of a monospecific rabbit antihuman D serum, isolated by ammonium sulfate precipitation and DEAE chromatography. Successful generation of this

---

[18] S. R. Barnum and J. E. Volanakis, *J. Immunol.* **134,** 1799 (1985).
[19] A. Martin, P. J. Lachmann, L. Halbwachs, and M. J. Hobart, *Immunochemistry* **13,** 317 (1976).

antiserum required that gluteraldehyde-aggregated factor D be used as the immunizing agent,[20] although other investigators have not reported this requirement.[21] The second antibody used in the assay is a mouse monoclonal antihuman D antibody prepared by immunizing BALB/c mice with purified factor D.[20] The antibody, FD-10, is purified from ascites by DEAE chromatography. The FD-10 antibody is radiolabeled by a modification of the chloramine-T method.[5]

Reagents

BSA: Bovine serum albumin, fraction V (Sigma Chemical Co., St. Louis, MO)

TBS: 0.01 M Tris, 0.15 M NaCl, pH 7.2

TBS–BSA: TBS containing 0.1% BSA

Procedures. The factor D RIA is performed as described below. Rabbit IgG anti-D (300 μl; 100 μg/ml or appropriate dilution as determined by the binding of ^{125}I-labeled D to serially diluted antibody) is incubated in microwells (Immulon II Removawell strips, Dynatech Laboratories, Inc., Chantilly, VA) for 1 hr at room temperature. The wells are washed with TBS–BSA and unreacted sites are blocked with 1.0% BSA in TBS for 30–60 min at room temperature. Following an additional wash, 280 μl of purified human D, serially diluted in TBS–BSA or appropriately diluted human serum, culture supernatant sample, or column fractions, is added to the wells and incubated overnight at 4°. The contents of the wells are removed and the wells are washed twice with TBS–BSA. An appropriate dilution (25 μl) of [^{125}I]FD-10 (approximately 100,000 cpm/well) is added, followed by 270 μl of TBS–BSA. After incubation for 4 hr at room temperature, the wells are aspirated, washed twice with TBS–BSA, and counted in a gamma counter. Controls consist of wells coated with BSA instead of rabbit IgG anti-D, without and with added D, and wells coated with rabbit anti-D and no added D. Factor D concentration is calculated from a standard curve, constructed by using purified protein, or a calibrated normal human serum.

Purification

Several methods have been reported for the purification of factor D from human serum.[3,9,20,22] They all use 4–5 liters of serum or plasma,

[20] M. A. Niemann, J. F. Kearney, and J. E. Volanakis, *J. Immunol.* **132,** 809 (1984).

[21] M. Pascual, E. Catana, F. Spertini, K. Macon, J. E. Volanakis, and J. A. Schifferli, *J. Immunol. Methods* **127,** 263 (1990).

[22] A. E. Davis III, C. Zalut, F. S. Rosen, and C. A. Alper, *Biochemistry* **18,** 5082 (1979).

involve several chromatographic steps requiring 2–3 weeks to be carried out, and yield about 1 mg of purified protein. The finding of high concentrations of factor D in the urine of patients with tubular dysfunction, such as patients with Fanconi's syndrome,[6] provided the opportunity to develop a relatively simple, high-yield, and rapid purification procedure.[23] It has been reported that peritoneal dialysis fluid contains factor D and a purification procedure has been described.[24] Protocols for the purification of factor D from urine of patients with Fanconi's syndrome and from peritoneal dialysis fluids are given below.

Purification from Urine

Urine containing high concentrations (5–10 μg/ml) of factor D is collected from patients with Fanconi's syndrome and stored frozen at $-20°$ for up to 1 year. The purification procedure consists of three chromatographic steps: (1) ion-exchange chromatography on Bio-Rex 70, (2) hydroxylapatite HPLC or fast protein liquid chromatography (FPLC), and (3) reversed-phase HPLC. Bio-Rex 70, BioGel HTP, and prepacked 7.8 × 100 mm BioGel HPHT columns are obtained from Bio-Rad Laboratories (Richmond, CA). Prepacked μC_{18} BondClone columns, 3.9 × 300 mm, are purchased from Phenomenex (Torrance, CA) and prepacked C_{18} μBondapak 3.9 × 300 mm columns are purchased from Waters associates (Milford, MA).

Step 1: Ion-Exchange Chromatography. Filter 1 liter of urine through Whatman (Clifton, NJ) No. 1 filter paper and then make the filtrate 2 mM in EDTA by addition of 4 ml of 500 mM EDTA, pH 6.5. The pH of the urine is adjusted to 6.5 with 1 N HCl and the conductivity is adjusted to that of the Bio-Rex 70 starting buffer with distilled water. The urine is applied to a 2.5 × 20 cm column of Bio-Rex 70 equilibrated, at 4°, with 20 mM sodium phosphate, 100 mM NaCl, 2 mM EDTA buffer, pH 6.5. The column is washed with the starting buffer until the absorbance of the effluent at 280 nm is less than 0.1, and the bound factor D is eluted with a 500-ml linear gradient from starting buffer to 500 mM NaCl in starting buffer. Then 5-ml fractions are collected, and their absorbance at 280 nm is measured. The fractions are screened for factor D hemolytic activity using the modified RD assay (Fig. 1). Fractions containing factor D activity are pooled, concentrated by ultrafiltration to 5–10 ml, and dialyzed against 100 mM sodium phosphate, 0.01 mM CaCl$_2$ buffer, pH 6.8.

Step 2: Hydroxylapatite Chromatography. The protein pool from the previous step is subjected to HPLC using a prepacked 7.8 × 100 mm

[23] J. E. Volanakis and K. J. Macon, *Anal. Biochem.* **163**, 242 (1987).
[24] E. Catana and J. A. Schifferli, *J. Immunol. Methods* **138**, 265 (1991).

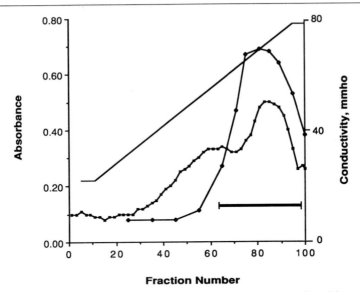

FIG. 1. Ion-exchange chromatography on Bio-Rex 70. Elution profile of factor D from a 2.5 × 20 cm column with a 500-ml gradient of 100 to 500 mM NaCl in 20 mM sodium phosphate, 2 mM EDTA buffer, pH 6.5. Fractions (5 ml) were collected at 4°. Protein concentration (A_{280}, ■), factor D hemolytic activity (A_{416}, ●), and conductivity (—) of the fractions are shown. The horizontal bar indicates the factor D-containing fractions that were pooled for further purification.

BioGel HPHT column and a Waters HPLC system equipped with two M6000 pumps, a Model 680 automated gradient controller, and a U6K universal injector. Absorbance is monitored at 280 nm using a Perkin-Elmer (Norwalk, CT) Model LC 75 detector. The equilibration buffer is 100 mM sodium phosphate, 0.01 mM CaCl$_2$, pH 6.8. Chromatograms are developed using a 30-min linear gradient from equilibration buffer to 200 mM sodium phosphate, 0.01 mM CaCl$_2$, pH 6.8, at a flow rate of 1 ml/min. Usually one-fourth of the sample is applied per run. Protein peaks are pooled and analyzed by electrophoresis on 5–20% polyacrylamide gradient gels containing 0.1% SDS (SDS–PAGE) under reducing conditions (Fig. 2). The factor D-containing peaks (peak 3, Fig. 2) from all runs are pooled and dialyzed against 0.1% trifluoroacetic acid (TFA).

FPLC using a 1.5 × 4.5 cm BioGel HTP column and a Pharmacia FPLC system can substitute for HPLC in this step. The Pharmacia FPLC system is equipped with two P500 pumps, an LCC 500 Plus controller, and a Frac 200 fraction collector. Absorbance is monitored at 280 nm using a UV-M detector. The equilibration buffer is 100 mM sodium phos-

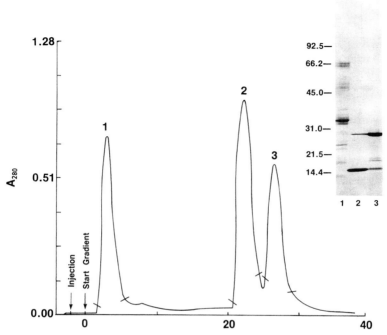

Fig. 2. Hydroxylapatite HPLC using a 7.8 × 100 mm BioGel HPHT column. Elution profile of proteins in the factor D-containing pool from the Bio-Rex 70 chromatography step. The column was eluted using a 30-min linear gradient from 100 to 200 mM sodium phosphate, pH 6.8, containing 0.01 mM $CaCl_2$. SDS–PAGE of the three protein peaks collected is shown in the inset. The position and M_r ($\times 10^{-3}$) of marker proteins are indicated. The abscissa shows time in minutes.

phate, 0.01 mM $CaCl_2$, pH 6.8. Chromatograms are developed with a 27-min linear gradient from 90% equilibration buffer/10% 300 mM sodium phosphate, 0.01 mM $CaCl_2$, pH 6.8, to 40% equilibration buffer/60% 300 mM sodium phosphate, 0.01 mM $CaCl_2$, pH 6.8. The flow rate is maintained at 2 ml/min. A trial run is usually made using 10% of the sample, followed by the remainder of the sample in a single run. Then 2-ml fractions are collected and analyzed by SDS–PAGE, using a 10% polyacrylamide gel and the buffer system of Laemmli.[25] Factor D usually elutes as a single peak at 30–35% of limiting buffer. The factor D-containing fractions are pooled, concentrated by ultrafiltration, and dialyzed against 0.1% TFA.

[25] U. K. Laemmli, *Nature (London)* **227**, 680 (1970).

Step 3: Reversed-Phase HPLC. The factor D-containing pool from the previous step is subjected to reversed-phase HPLC, using either a Waters C_{18} μBondapak column or a Phenomenex $μC_{18}$ column, and the Waters chromatographic system described above. Both columns perform equally well. The column is equilibrated in 35% acetonitrile in 1% TFA. Chromatograms are developed using a 30-min gradient from equilibration buffer to 45% acetonitrile in 1% TFA at a flow rate of 1.5 ml/min. Usually one-fourth of the sample is applied per run. The main protein peak of the chromatogram contains pure factor D (Fig. 3). The factor D-containing peaks from all runs are pooled and dialyzed extensively against water followed by dialysis against 10 m*M* Tris-HCl, 75 m*M* NaCl buffer, pH 7.4.

Table I summarizes the results of the purification procedure, illustrating the yield and purification efficiency of each step. Figure 4 shows an SDS–PAGE of the factor D-containing pools obtained after each step. The Bio-Rex 70 step is the most effective in terms of purification. The

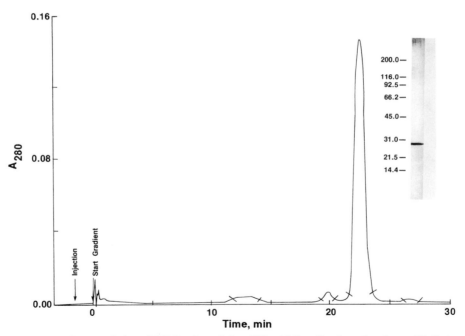

FIG. 3. Reversed-phase HPLC using a 3.9 × 300 m*M* C_{18} μBondapack column. Elution profile of factor D-containing peak from the hydroxylapatite HPLC step. The chromatogram was developed by using a 30-min linear gradient at 1.5 ml/min from 35 to 45% acetonitrile in 1% TFA. SDS–PAGE of the main peak is shown in the inset. The position and M_r ($× 10^{-3}$) of marker proteins are indicated.

TABLE I
PURIFICATION OF FACTOR D FROM URINE OF PATIENT WITH FANCONI'S SYNDROME[a]

Fraction	Volume (ml)	Total protein (A_{280})	Total hemolytic activity $\times 10^{-6}$ (units)	Yield (%)	Specific hemolytic activity $\times 10^{-3}$ (units/A_{280})	Purification factor
Urine	1000	1550	7.1	100	4.6	1
Post-Bio-Rex 70	5.6	49.7	4.9	69	98.6	21
Post-BioGel HPHT	5	9.3	3.3	46	354.8	77
Post-μC$_{18}$	8	5.9[b]	3.3	46	559.3	122

[a] Data from Volanakis and Macon.[23]

[b] Using $\varepsilon_{280}^{1\%} = 16.0$ (arbitrary value corresponding to values determined by amino acid analysis), the final yield was 3.7 mg of factor D.

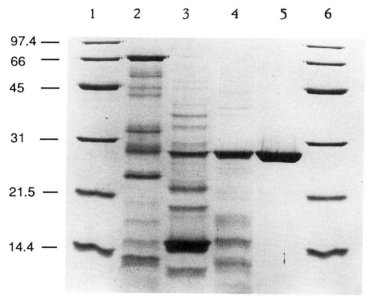

FIG. 4. SDS–PAGE of factor D-containing pools obtained during purification from urine from a patient with Fanconi's syndrome. Lanes 1 and 6, molecular size marker proteins; lane 2, starting urine; lane 3, post-Bio-Rex 70 chromatography pool; lane 4, posthydroxylapatite FPLC pool; lane 5, postreversed-phase HPLC pool. Protein bands were stained with Coomassie brilliant blue.

hydroxylapatite step serves mainly to separate the majority of β_2-microglobulin, the main contaminant following Bio-Rex 70. The remainder of β_2-microglobulin and additional minor contaminants are then removed by the reversed-phase HPLC. No detectable loss in hemolytic activity is noted following the brief exposure of factor D to acetonitrile in the last chromatography step. The entire procedure can be completed in a few days and gives a yield of approximately 3.7 mg of factor D per liter of urine and a recovery of 46% of the original hemolytic activity. Factor D purified by this procedure has been shown to be antigenically, structurally, and functionally identical to factor D purified from human serum.

Purification from Peritoneal Dialysis Fluid

The protocol described above can also be used for the purification of factor D from spent peritoneal fluid from patients undergoing chronic ambulatory peritoneal dialysis (CAPD). Peritoneal fluid is obtained from the Home Dialysis Unit at the University of Alabama at Birmingham. Fluid (2 liters) from each of six different individuals is collected and screened for factor D by the hemolytic assay. In a recent run, four of the six fluids were found to have detectable levels of factor D and were used in the purification procedure. The procedure yielded 1.4 mg of factor D from 8.5 liters of peritoneal dialysis fluid. These results are similar to those reported by Catana and Schifferli,[24] who obtained 2.1 mg of D from 10 liters of peritoneal fluid using a three-step purification protocol consisting of Bio-Rex 70, heparin-Sepharose CL-6B, and Mono S.

The advantage of using peritoneal dialysis fluid as a source of factor D lies in the ready availability of patients undergoing CAPD in most medical centers. In contrast, patients with Fanconi's syndrome are relatively rare. On the other hand, urine from patients with Fanconi's syndrome contains much higher amounts of factor D than does dialysis fluid, making purification of large amounts of the protein more rapid and more efficient.

Structure

More than 90% of the primary structure of human factor D was determined by amino acid sequencing of the intact polypeptide chain and its chemically and enzymatically derived peptide fragments.[26,27] The derived

---

[26] M. A. Niemann, A. S. Bhown, J. C. Bennett, and J. E. Volanakis, *Biochemistry* **23**, 2482 (1984).

[27] D. M. A. Johnson, J. Gagnon, and K. B. M. Reid, *FEBS Lett.* **166**, 347 (1984).

amino acid sequence is in good general agreement with that deduced from the nucleotide sequence of factor D cDNA clones[28] (Fig. 5). Discrepancies between the two sets of data are essentially limited to two positions of the polypeptide chain for which overlapping peptides were not available during amino acid sequencing. The single polypeptide chain of the mature protein consists of 228 amino acid residues with a calculated M_r of 24,376. There are no potential N-glycosylation sites in the amino acid sequence and no carbohydrate is present in the purified protein.[29]

In addition to the mature polypeptide chain, the factor D cDNA[28] encoded a 13-residue-long partial putative leader peptide followed by 5 residues, Arg-Pro-Arg-Gly-Arg-, that could represent an activation peptide. However, factor D purified from serum or urine is devoid of such an NH_2-terminal sequence. In addition, no zymogen form of factor D can be detected in plasma or serum[3] or in supernatants of U937 cells, which synthesize factor D.[18] These combined data suggest that the putative activation peptide is cleaved off either during or shortly after secretion.

The amino acid sequence of factor D shows a high degree of homology to that of other serine proteinases, exhibiting 33–35% amino acid residue identity with pancreatic bovine trypsin, bovine chymotrypsin A, porcine elastase, and human neutrophil elastase.[30] Alignment of the factor D sequence to that of 35 other mammalian serine proteinases[31] indicates that in their majority, conserved and invariable residues of the superfamily, including those important for substrate binding and catalysis, have been retained in factor D. The three residues, His-57, Asp-102, and Ser-197 (chymotrypsinogen numbering), which form the charge-relay system of the catalytic apparatus of all serine proteinases, are found at positions 41, 89, and 183, respectively, in the factor D sequence (Fig. 5). The specificity-determining residue for trypsinlike enzymes, Asp-189 (chymotrypsinogen numbering), is identical with Asp-177 of the factor D sequence. Finally, the eight half-cystine residues of factor D occur at positions analogous to those of other serine proteinases.

Crystals of factor D have been grown by vapor diffusion using polyethylene glycol 6000 and NaCl as precipitants.[32] The crystals are stable in

[28] R. T. White, D. Damm, N. Hancock, B. S. Rosen, B. B. Lowell, P. Usher, J. S. Plier, and B. M. Spiegelman, *J. Biol. Chem.* **267**, 9210 (1992).

[29] M. Tomana, M. Niemann, C. Garner, and J. E. Volanakis, *Mol. Immunol.* **22**, 107 (1985).

[30] M. Carson, S. V. L. Narayana, L. J. DeLucas, O. El-Kabbani, J. M. Kilpatrick, J. E. Volanakis, and C. E. Bugg, *Proc. Conf. Appl. Comput. Visualization Imaging Res. 2nd,* Iowa City, *1991,* 29. (1991).

[31] J. Greer, *Protein Struct. Funct. Genet.* **7**, 317 (1990).

[32] S. V. L. Narayana, J. M. Kilpatrick, O. El-Kabbani, Y. S. Babu, C. E. Bugg, J. E. Volanakis, and L. J. DeLucas, *J. Mol. Biol.* **219**, 1 (1991).

I L G G R E A E A H A R P Y M A S V Q L N G A H L C G G V L V A E Q W V L S A A

H C L E D A A D G K V Q V L L G A H S L S Q P E P S K R L Y D V L R A V P H P D

S Q P D T I D H D L L L Q L S E K A T L G P A V R P L P W Q R V D R D V A P G

T L C D V A G W G I V N H A G R R P D S L Q H V L L P V L D R A T C N R R T H H

D G A I T E R L M C A E S N R R D S C K G D S G G P L V C G G V L E G V V T S G

S R V C G N R K K P G I Y T R V A S Y A A W I D S V L A

FIG. 5. Primary structure of factor D purified from human serum. The one-letter amino acid code has been used. Boxes indicate the putative active center residues, His-41, Asp-89, and Ser-183. The Asp-177 residue, corresponding to the specificity-determining residue of trypsin-like enzymes, is indicated by an asterisk. Cysteine residues are shaded. Data from Refs. 26–28.

the X-ray beam and diffract to beyond 2.5 Å. The factor D crystals are triclinic with a $P1$ space group. The unit cell contains two molecules of the protein related by a noncrystallographic twofold axis. Initial analysis of diffraction data combined with molecular modeling[30] indicate that the overall three-dimensional structure of factor D is similar to that of other serine proteinases of known structure.

Esterolytic Activity

Human factor D isolated from serum of normal individuals or from urine of patients with Fanconi's syndrome exhibits esterolytic activity against peptide thioester substrates.[2] A series of peptide thioesters were used to investigate the specificity and reactivity of factor D. All substrates contained a P_1 Arg residue and the effects of various groups and amino acids in the P_2, P_3, P_4, and P_5 positions were determined using K_{cat}/K_m values as a measure of reactivity. Of 12 dipeptide thioesters tested as substrates, only 3 containing Arg, Val, or Lys in the P_2 position were hydrolyzed at measurable rates. Factor B, the natural substrate of factor D, contains a P_2 Lys at the cleavage site and dipeptide thioesters with P_2 Lys and Arg were among the most reactive substrates. That Z-Val-Arg-SBui was one of the best substrates of factor D, and that Z-Arg-SBzl was also reactive, indicated that a hydrophobic group can also fit into the S_2 subsite of the enzyme.[2] Extension of the peptide thioester to include Gln at the P_3 and P_4 positions, which are also found in the corresponding positions of factor B, resulted in loss of activity. Neither the tri- nor the tetrapeptide thioesters was hydrolyzed. Similarly, tripeptides containing four other amino acid residues at P_3, (Gly, Glu, Lys, Phe) did not react with factor D. Thus, the S_3 subsite of the enzyme is very specific. The results of this study demonstrated that factor D is a specific enzyme of low catalytic efficiency toward thioesters. This conclusion is amply supported by comparing the reactivity of factor D with that of its functional homolog in the classical pathway, C$\overline{1}$s, and with trypsin (Table II). One of the best substrates for factor D, Z-Lys-Arg-SBui, was hydrolyzed 60- and 1000-fold more efficiently by C$\overline{1}$s and trypsin, respectively. In contrast, factor D did not hydrolyze any of the C2- or C4-like peptide thioesters that were excellent substrates for C$\overline{1}$s.

The low esterolytic activity of purified factor D is compatible with the apparent absence of a structural zymogen for the enzyme in blood. That "native" D in blood is in enzymatically active form was demonstrated convincingly by Lesavre and Müller-Eberhard.[4] They showed that distribution of factor D hemolytic activity always overlapped that of antigenically measured factor D protein when plasma or serum was subjected to

TABLE II

K_{cat}/K_m VALUES FOR HYDROLYSIS OF PEPTIDE THIOESTER SUBSTRATES BY FACTOR D, C1s, AND BOVINE TRYPSIN[a]

Homologous cleavage site	Substrate[b]						K_{cat}/K_m ($\times 10^{-3}$)		
	P_5	P_4	P_3	P_2	P_1	P_1'	Factor D	C1s	Trypsin
				Z	-Arg	-SBzl	0.17	ND[d]	3900
Factor B			Z	-Lys	-Arg	-SBui	2.50	160	2500
Factor B		Z	-Gln	-Lys	-Arg	-SBzl	NH[c]	ND	3900
Factor B	Bz	-Gln	-Gln	-Lys	-Arg	-SBzl	NH	ND	1600
C2			Z	-Gly	-Arg	-Sbui	NH	730	4500
C2		Z	-Leu	-Gly	-Arg	-SBzl	NH	1200	ND
C4			Z	-Gln	-Arg	-SBzl	NH	460	ND
C4		Z	-Leu	-Gln	-Arg	-SBzl	NH	460	ND
C4	Z	-Gly	-Leu	-Gln	-Arg	-SBzl	NH	200	ND

[a] Values for factor D from C.-M. Kam, B. J. McRae, J. W. Harper, M. A. Niemann, J. E. Volanakis, and J. C. Powers [*J. Biol. Chem.* **262**, 3444 (1987)]; for C1s from B. J. McRae, T.-Y. Lin, and J. C. Powers [*J. Biol. Chem.* **256**, 12362 (1981)]; and for trypsin from B. J. McRae, K. Burachi, R. L. Heimark, K. Fujikawa, E. W. Davie, and J. C. Powers [*Biochemistry* **20**, 7196 (1981)].

[b] Z, Carbobenzyloxy; Bz, benzoyl; -SBzl, benzyl thioester; -SBui, isobutyl thioester.

[c] NH, No hydrolysis.

[d] ND, Not determined.

various separation procedures. More importantly, the specific hemolytic activity of isolated factor D was identical to that of factor D in serum. In addition, it has been shown that factor D in serum can be inactivated by DFP[1] and also by a series of novel serine proteinase inhibitors derived from isocoumarin.[33] Inhibition of factor D in serum by these inhibitors results in inhibition of the alternative pathway, indicating that no other serum proteinase can substitute for factor D.

Serum Concentration of Factor D in Health and Disease

The concentration of factor D in normal human serum is 1.8 ± 0.4 μg/ ml as measured by RIA in our laboratory.[5] This is in good agreement with values reported from several laboratories.[3,8,34] The serum concentration of factor D in healthy children (range 2–10 years, median 6 years) is approximately 25% lower than in adults.[35] Serum levels of factor D are

[33] C.-M. Kam, T. J. Oglesby, M. K. Pangburn, J. E. Volanakis, and J. C. Powers, *J. Immunol.* **149**, 163 (1992).

[34] L. Truedsson and G. Sturfelt, *J. Immunol. Methods* **63**, 207 (1983).

[35] G. Sturfelt and L. Truedsson, *Acta Pathol. Microbiol. Immunol. Scand.* **91**, 383 (1983).

not significantly changed during the acute-phase response or in immune complex diseases, such a systemic lupus erythrematosus, nor is serum factor D consumed by activation of complement.[5,36] However, because factor D is catabolized by proximal tubular epithelial cells in the kidney,[6,7] it is not surprising that individuals with various diseases associated with renal insufficiency have elevated serum levels of factor D. Thus, individuals with lupus glomerulonephritis, polycystic kidney disease, chronic renal failure, nephrotic syndrome, or on long-term dialysis all have elevated serum levels of D, in some cases reaching amounts 10- to 15-fold higher than normal.[6,8,36] In addition, individuals with nephrotic syndrome, chronic renal failure, and Fanconi's syndrome have detectable levels of D in their urine whereas normal individuals have no factor D in urine.[6,13]

Only one case of complete deficiency of factor D[37] has been described. The individual was reported to suffer from recurrent *Neisseria* infections and to have antigenically and hemolytically undetectable levels of factor D in serum.[37] All other serum complement components were within normal range; however, the total hemolytic complement (CH_{50}) was below normal levels. The mother of the patient is heterozygous deficient for D, but all other family members have normal D levels, suggesting an X-linked mode of inheritance. The chromosomal location of the factor D gene has not been determined.

[36] G. Sturfelt, L. Truedsson, H. Thysell, and L. Björk, *Acta Med. Scand.* **216,** 171 (1984).
[37] P. S. Hiemstra, E. Langeler, B. Coupier, Y. Keepers, P. J. Leijh, M. T. van der Barselaar, D. Overbosch, and M. R. Daha, *J. Clin. Invest.* **84,** 1957 (1989).

[6] C1 Inhibitor

By ALVIN E. DAVIS III, KULWANT S. AULAK, KAMYAR ZAHEDI, JOHN J. BISSLER, and RICHARD A. HARRISON

Introduction

During the late 1950s and early 1960s, a remarkable series of studies by Lepow and colleagues described the isolation and characterization of C1 and its subcomponents: C1q, C1r, and C1s.[1-3] These studies led also to the initial isolation and description of C1 inhibitor (C1 INH) as a heat-

[1] I. H. Lepow, O. D. Ratnoff, F. S. Rosen, and L. Pillemer, *Proc. Soc. Exp. Biol. Med.* **92,** 32 (1956).
[2] O. D. Ratnoff and I. H. Lepow, *J. Exp. Med.* **106,** 327 (1957).
[3] I. H. Lepow, O. D. Ratnoff, and L. R. Levy, *J. Exp. Med.* **107,** 451 (1958).

labile serum protein that inhibited the esterolytic activity of C1[2] and its proteolytic activity against C4 and C2.[4] C1 INH was later shown to be identical to α_2-neuraminoglycoprotein, a protein of unknown function originally identified by Schultze *et al.*[5] *In vitro,* C1 INH subsequently was shown to inactivate several other plasma proteases, including kallikrein, plasmin, plasminogen activator, and factors XIa and XIIa (Hageman factor). C1 INH is the only physiologically important plasma inhibitor of activated C1r and C1s, and it provides a major portion of the plasma inhibitory capacity toward kallikrein and factor XIIa, but probably is not normally an important inactivator of XIa or plasmin. The primary role of C1 INH is as a regulator of the complement and contact (kinin-forming) systems, and perhaps (directly or indirectly) of the fibrinolytic system.[6]

C1 INH is a member of the serine proteinase inhibitor (serpin) superfamily.[7] This family includes many of the other plasma proteinase inhibitors, such as α_1-antitrypsin, antithrombin III, α_2-antiplasmin, and α_1-antichymotrypsin, in addition to several proteins with no known inhibitory function, such as angiotensinogen and thyroxin-binding globulin. Serpins function by providing an exposed region that mimics the natural substrate of target proteases. Proteases recognize and "attempt" to hydrolyze a specific peptide bond (the reactive center) within this region. However, rather than peptide bond cleavage with regeneration of active protease, a tight protease–protease inhibitor complex is formed. Recent information indicates that this complex is a tetrahedral intermediate that, under appropriate conditions, can go on to peptide bond cleavage with formation of an acyl bond with the protease.[8] The amino acids to the amino- and carboxyl-terminal sides of the peptide bond recognized by protease are the P_1 and $P_1{}'$ residues, respectively. Other residues in each direction are numbered consecutively (P_1, P_2, P_3, etc. and $P_1{}'$, $P_2{}'$, $P_3{}'$, etc.). The P_1 residue is a major determinant of protease inhibitory specificity. In C1 inhibitor, the normal P_1 residue is arginine, which is compatible with the specificity of the proteases inactivated by C1 inhibitor.

C1 INH Assays

Functional activity is assayed either by inhibition of the proteolytic action of C1 against C4 and C2,[9] or by inhibition of the esterolytic activity

[4] L. R. Levy and I. H. Lepow, *Proc. Soc. Exp. Biol. Med.* **101,** 608 (1959).
[5] H. E. Schultze, K. Heide, and H. Haupt, *Naturwissenschaften* **49,** 133 (1962).
[6] A. E. Davis III, *Annu. Rev. Immunol.* **6,** 595 (1988).
[7] R. W. Carrell and D. R. Boswell, *in* "Proteinase Inhibitors" (A. Barrett and G. Salvesen, eds.), p. 403. Elsevier, Amsterdam, 1986.
[8] N. R. Mathesen, H. van Halbeck, and J. Travis, *J. Biol. Chem.* **266,** 13489 (1991).
[9] I. Gigli, S. Ruddy, and K. F. Austen, *J. Immunol.* **100,** 1154 (1968).

of C1 or isolated C1s.[2-4,10,11] Inhibition of proteolysis, using a hemolytic assay, is more sensitive than inhibition of esterolysis, but the assay is more complex and requires specific complement reagents. It thus is suitable only for laboratories performing complement hemolytic titrations on a regular basis. In addition to lower sensitivity, assay of C1 INH by inhibition of esterolysis may be complicated by the presence of other proteases in plasma with substrate specificities similar to that of C1s (particularly plasma kallikrein and plasmin). Some of these problems can be minimized through the use of appropriate substrates or specific protease inhibitors. Inhibition of kallikrein, rather than C1s, also has been used to determine C1 INH levels.[12]

Plasma samples should not be stored for prolonged periods at 4° prior to assay as this can lead to complications arising from cold-promoted activation of the contact system. In particular, kallikrein and factor XII activation may result in C1 INH consumption, with artifactual low levels of C1 INH activity.[13] Because EDTA plasmas give lowered titers of C1 INH compared with serum or ACD plasmas (up to 20% loss of activity),[14] heparin or ACD should be used as the anticoagulant in plasma samples. Serum samples rather than plasma samples are preferred if assay by inhibition of C1-mediated hemolysis is intended.

Assay of C1 INH by Inhibition of C1-Mediated Hemolysis[9]

Reagents

Sheep erythrocytes (E); antisheep erythrocytes (EA)
Functionally pure guinea pig or human C1,[15-17] or (preferably) isolated proteolytically active C1s.[18] Functionally pure guinea pig C2[9]
Guinea pig serum: C3-depleted guinea pig serum (R3) prepared by absorption with yeast cell walls or zymosan[16]

[10] B. J. McRae, T.-Y. Lin, and J. C. Powers, *J. Biol. Chem.* **256,** 12362 (1981).

[11] M. Lennick, S. A. Brew, and K. C. Ingham, *Biochemistry* **25,** 3890 (1986).

[12] M. Schapira, L. D. Silver, C. F. Scott, and R. W. Colman, *Blood* **59,** 719 (1982).

[13] E. A. van Rogen, S. Lohman, M. Voss, and K. Pondman, *J. Lab. Clin. Med.* **92,** 152 (1978).

[14] A.-B. Laurell, J. Lindegren, I. Malmros, and H. Martensson, *Scand. J. Clin. Lab. Invest.* **24,** 221 (1969).

[15] G. J. Arlaud, A. Reboul, C. M. Meyer, and M. G. Colomb, *Biochim. Biophys. Acta* **485,** 215 (1977).

[16] R. A. Harrison and P. J. Lachmann, *in* "Handbook of Experimental Immunology" (D. M. Weir, L. A. Herzenberg, C. C. Blackwell, and L. A. Herzenberg, eds.), 4th ed., Chapter 39. Blackwell, Edinburgh, 1985.

[17] W. D. Linscott, *Immunochemistry* **5,** 311 (1968).

[18] D. H. Bing, J. M. Andrews, K. M. Morris, E. Cole, and V. Irish, *Prep. Biochem.* **10,** 269 (1980).

Buffers: $5 \times$ Veronal-buffered saline (VBS; 5.1 g diethyl barbiturate and 41.5 g NaCl per liter, pH 7.4). Gelatin–veronal-buffered saline with Ca^{2+} and Mg^{2+} (GVB^{2+}; VBS, 0.15 mM $CaCl_2$, 0.5 mM $MgCl_2$, 0.1% gelatin). Dextrose–gelatin–veronal-buffered saline with Ca^{2+} and Mg^{2+} [$DGVB^{2+}$; GVB^{2+} with an equal volume of a 5% (w/v) dextrose solution]. Gelatin–veronal-buffered saline with EDTA (GVB–EDTA; VBS, 0.1% gelatin, 0.01 M disodium EDTA, pH 7.4)

Procedure

Preparation of EAC4b. Add 1 volume of antibody, diluted to give 6–10 minimum hemolytic doses (the minimum amount required to give total lysis in the presence of excess quantities of all other components of the system), to 1 volume of 10% E (2.5×10^9 E/ml) and stir gently for 15 min at 4°. Centrifuge, wash twice in at least 50 volumes of GVB^{2+}, and resuspend EA, at 5%, in GVB^{2+}. Test with guinea pig complement: Add 0.25 ml of diluted guinea pig serum (1 : 50 in GVB^{2+}) to 0.02 ml of 5% EA and incubate at 37°. Optimally sensitized EA are lysed within 2 min. Warm 1 ml of 5% EA to 37°, add an appropriate dilution of guinea pig R3 (8 minimum hemolytic doses, generally about 0.25 ml), and stir for 5 min. The cells at this point are EAC142. After 5 min chill rapidly to 0–4°, wash twice with ice-cold GVB–EDTA, then resuspend at 1% (v/v) in GVB–EDTA and incubate at 37° for 90 min. During this incubation C1 and C2b decay off, leaving EAC4b. Finally, wash twice with GVB^{2+} and resuspend in GVB^{2+} at 0.5% (v/v) (1.25×10^8 cells/ml). Alternatively, EAC4b cells can be generated as described by Borsos and Rapp.[19]

Titration of C1, C2, and C-EDTA. The appropriate amounts of C1, C2, and C-EDTA to use in the assay must be determined before incubation of C1 INH-containing samples with C1. First, determine the amount of C2 required to give total lysis of EAC4b in the presence of nonlimiting amounts of C1. Add 50 μl of twofold serial dilutions of C2 to 50 μl of EAC4b (1.25×10^8/ml) and 25 μl of an excess of C1. Incubate for 15 min at 30°, then add 0.1 ml guinea pig serum diluted 1 : 40 in GVB–EDTA. The appropriate amount of C2 to use in assays is a 5- to 10-fold excess over the minimum required to give total lysis of EAC4b. Next, using the amount of C2 determined above, repeat the procedure with twofold serial dilutions of C1. The correct amount of C1 to use should give about 67% lysis. An appropriate dilution of guinea pig complement is usually about 1 : 40 in GVB–EDTA; it may occasionally be necessary to use a higher or lower dilution.

[19] T. Borsos and H. J. Rapp, *J. Immunol.* **99**, 263 (1967).

Preparation of serum samples. Because the assay is highly sensitive and is dependent on inhibition of a small known amount of C1, endogenous C1 should be removed from serum samples before incubation with guinea pig C1. A simple procedure for this is to make an initial 1 : 20 dilution of serum samples in 0.005 M phosphate, pH 6.0, followed by incubation at 0°, either for 2–4 hr or overnight, followed by centrifugation to remove the precipitated C1-containing euglobulin. Further dilutions of the supernatant in GVB^{2+} are required prior to incubation with guinea pig C1. Dilutions in the range of 1 : 50–1 : 1350 are suitable for normal human serum levels of C1 INH. To prevent consumption of C1 INH by plasmin and kallikrein, with consequent anomolously low C1 INH titers, soybean trypsin inhibitor (SBTI; inhibits plasmin and kallikrein, but not C1s) should be added (to 1 mg/ml) to the test serum.

Assay of C1 INH-containing samples. Incubations are set up as indicated in Table I. In addition to the experimental samples, a number of control incubations are required. The "solo" incubation has all of the reagents, but no C1 INH sample, and gives the maximum amount of complement lysis attainable by the system (reagent lysis). The CBC (cells + buffer + complement) incubation contains all the reagents except C1, and gives a measure of background lysis. The 100% lysis and spontaneous lysis (CB) incubations are included as a check on the system. Lysis given by the solo incubation should be about 65–70% of the 100% lysis value, and the CBC value should not be significantly greater than that of the CB incubation.

TABLE I

ASSAY OF C1 INH BY INHIBITION OF C1-MEDIATED HEMOLYSIS

Step	Assay	"Solo"	CBC	CB	100%
Add:					
Sample	0.1 ml	—	—	—	—
Guinea pig C1	0.1 ml	0.1 ml	—	—	—
GVB^{2+}	—	0.1 ml	0.2 ml	0.2 ml	0.2 ml
Incubate 30 min, 30°, then add:					
EAC4b	0.2 ml	0.2 ml	0.2 ml	0.2 ml	0.2 ml
60% sucrose	0.1 ml	0.1 ml	0.1 ml	0.1 ml	—
Incubate 15 min, 30°, then add:					
Guinea pig C2	0.2 ml	0.2 ml	0.2 ml	—	—
GVB^{2+}	—	—	—	0.2 ml	—
Incubate 10 min, 30°, then add:					
C-EDTA	0.6 ml	0.6 ml	0.6 ml	—	—
0.1 M EDTA	—	—	—	0.6 ml	—
H$_2$O	—	—	—	—	0.7 ml
Incubate 60 min, 37°, centrifuge, read supernatant at 412 nm					

Calculation of C1 INH titer. Derivation of the formula given below is discussed in full in Gigli *et al.*[9] The amount of C1-dependent lysis inhibited in each reaction mixture (y) can be represented as follows:

$$y = 1 - \frac{A_{412 \text{ nm}} \text{ sample} - A_{412 \text{ nm}} \text{ reagent lysis}}{A_{412 \text{ nm}} \text{ solo} - A_{412 \text{ nm}} \text{ reagent lysis}}$$

From this equation the average number of effective or functional molecules (z) of C1 INH can be calculated as

$$z = -\ln(1 - y)$$

or

$$z = -\ln \frac{A_{412 \text{ nm}} \text{ sample} - A_{412 \text{ nm}} \text{ reagent lysis}}{A_{412 \text{ nm}} \text{ solo} - A_{412 \text{ nm}} \text{ reagent lysis}}$$

and the titer as

$$z \times \text{dilution} \times 5 \text{ units C1 INH/ml sample}$$

Assay of C1 INH by Inhibition of C1s Esterolytic Activity

Introduction. As with the hemolytic assay, C1 INH activity is assessed indirectly. Samples are incubated with excess activated C1 or C1s and residual C1s activity is determined. The amount of C1 INH present is calculated from the degree of inhibition of C1s. C1s esterolytic activity can be measured using a synthetic ester substrate and either a recording spectrophotometer or, if acetyltyrosine ethyl ester (ATEE) is used as substrate, a pH-stat titrator. The pH-stat procedure is advantageous in that an end point titration is performed and activation of endogenous C1 does not therefore interfere with the assay. However, the assay is cumbersome to perform and is no longer commonly used. Spectrophotometric assays are more convenient. The two "traditional" substrates, ATEE and *N*-carbobenzoxy-L-lysine *p*-nitrophenol ester (ZLNE), have conflicting merits. Assays using ZLNE, although more sensitive than those using ATEE, are less specific for C1s, because ZLNE also is hydrolyzed by plasmin and probably by other less abundant plasma proteases. ZLNE is also unstable at neutral pH. The assay must therefore be performed at pH 6.0, and background hydrolysis corrected by inclusion of a blank cuvette containing sample and substrate but no added C1. In contrast, C1s is the sole plasma protease capable of splitting ATEE, and assays using this substrate are therefore highly specific. However, the wavelength at which hydrolysis is monitored (238 nm) makes it unsuitable for use with assay samples that have a high protein concentration. Visible serum color can also cause interference (ZLNE hydrolysis is monitored at 340 nm), particularly if samples with low C1 INH titers are being analyzed. For most purposes, we have preferred the substrate N-carbobenzyloxy-L-lysine thiobenzyl ester (NZLBz), which combines many of the advantages of ZLNE and ATEE. Although not specific for C1s, it is sensitive, is monitored at 329 nm, and is stable at neutral pH. However, it has not been evaluated carefully for the measurement of C1 INH activity in plasma samples. We have used it primarily for analysis of purified (or partially purified) C1 INH preparations, and in the assay of autoantibodies to C1 INH in patients with acquired angioedema. A number of synthetic tripeptide chromogenic substrates with increased specificity for C1,[20,21] plasmin,[22] or kallikrein[12,23-25] have been described, and these too have been used in spectrophotometric assays of C1 INH activity.

Reagents

Activated C1 or C1s. For most purposes a neutral euglobulin, prepared as described in Linscott,[17] is sufficiently pure, although a purified preparation of activated C1s[18] is preferred for improved sensitivity and specificity. If the euglobulin is used, the final precipitate is redissolved in $2 \times GVB^{2+}$ and stored at $-70°$. This must be activated, by dilution with an equal volume of water and incubation for 1 hr at $37°$, before use in the assay.[16]

ATEE, ZLNE, and NZLBz. A stock ATEE solution (0.075 M in 2-methoxyethanol) can be stored at $4°$ for up to 4 weeks. Alternatively, a stock ATEE solution can be prepared in an aqueous solution (0.01 M in 0.15 M NaCl). Because no organic solvent is involved, this latter stock permits higher final substrate concentrations (up to $8.6 \times 10^{-3} M$) in the assay. A fresh ZLNE stock (0.01 M in 90% acetonitrile) should be prepared each day and held at $4°$. NZLBz is dissolved in methanol (24 mg/50 ml) containing 25 μl of 10 mM HCl/50 ml. For the assay using NZLBz, the chromogen 4,4'-dithiodipyridine (stock solution 4.5 mM) is prepared in 20 mM Tris-HCl, pH 7.6, containing 150 mM NaCl and 9.2% (w/v) dimethyl sulfoxide (DMSO).

Procedure. Direct assay of plasma or serum samples will give a reasonable estimate of C1 INH titer. However, if a high degree of accuracy and sensitivity in C1 INH quantitation is required, endogenous C1 can be removed from serum samples, as described above. SBTI (1 mg/ml) should be added to inhibit plasmin and kallikrein. For normal plasma levels of

[20] T. Nilsson and B. Wiman, *Biochim. Biophys. Acta* **705**, 271 (1982).

[21] B. Wiman and T. Nilsson, *Clin. Chim. Acta* **128**, 359 (1983).

[22] L. F. Kress, J. Catanese, and T. Hirayama, *Biochim. Biophys. Acta* **745**, 113 (1983).

[23] M. J. Gallimore and P. Friberger, *Thomb. Res.* **25**, 293 (1982).

[24] M. J. Gallimore, E. Amundsen, M. Larsbraaten, K. Lyngaas, and E. Fareid, *Thromb. Res.* **16**, 695 (1979).

[25] T. Nilsson, *Thromb. Haemostasis* **49**, 193 (1983).

C1 INH, appropriate dilutions for esterolytic assays range from 1 : 10 to 1 : 50.

Prepare a stock solution of C1 or C1s (about 100 μg/ml) in GVB^{2+}. Next, add 50 μl C1 to 50 μl assay sample and incubate for 15 min at 37°. Finally, add 0.89 ml of assay buffer (0.1 M phosphate/0.1 M NaCl/0.01 M EDTA, pH 6.0, for ZLNE; GVB/EDTA for ATEE), transfer to a 1-ml cuvette, add 10 μl of substrate, mix well, and monitor substrate hydrolysis (at 340 nm for ZLNE, 238 nm for ATEE) using a recording spectrophotometer. For the assay with NZLBz, mix 700 μl of assay buffer (20 mM Tris-HCl, pH 7.6, 150 mM NaCl, 9.2% DMSO) with 150 μl of 4,4'-dithiopyridine and 50 μl of NZLBz. Add C1s, both alone and after incubation with dilutions of assay sample, to a final concentration of approximately 25–50 μg/ml, and monitor hydrolysis at 329 nm. Measure the initial rate of hydrolysis and subtract the rate given by the blank cuvette (sample plus substrate but no added C1) to obtain the corrected value. Although a higher rate of hydrolysis, and therefore increased sensitivity, will be obtained if the assay is performed at 30° or 37°, assay at elevated temperatures does not alter the observed titer of C1 INH.

The C1 INH titer is determined from the degree of inhibition of C1. This can be expressed in arbitrary units (percent inhibition of C1 offered). If activated C1s of known concentration is used, a quantitative estimate of the amount of C1 INH present in the sample can be made.

Determination of C1 INH Antigenic Concentration. The anodal (α_2) mobility of C1 INH in agarose gel (at pH 8.6) makes it an ideal protein for detection and quantitation by electroimmunoassay.[26] Most commercial antisera can be used at approximately 1%, and C1 INH sample loads of 50 ng can be detected by staining the "rockets" with Coomassie brilliant blue.

Isolation of C1 INH

Introduction

A number of isolation procedures for C1 INH have been described.[20,27–30] The first method described here (method 1) gives an excel-

[26] C.-B. Laurell, *Scand. J. Clin. Lab. Invest.* **29,** Suppl. 124, 21 (1972).
[27] P. C. Harpel, this series, Vol. 45, p. 751.
[28] H. Haupt, N. Heimburger, T. Kranz, and H. G. Schwick, *Eur. J. Biochem.* **17,** 254 (1970).
[29] R. B. Sim and A. Reboul, this series, Vol. 80, p. 43.
[30] Y. Pilatte, C. H. Hammer, M. M. Frank, and L. F. Fries, *J. Immunol. Methods* **120,** 37 (1989).

lent yield and purity of C1 INH.[31] As for most plasma proteins, fresh plasma is the starting material of choice, but stored plasma has been used with good results. The procedure described is used routinely for fractionation of between 1 and 2 liters of plasma, but can be scaled up or down. In addition, the method devised by Pilatte *et al.*[30] (method 2), which uses lectin affinity chromatography, is described because of its simplicity and adaptability to isolation from small volumes.

C1 INH Isolation: Method 1

Materials

Chromatographic media. Lysine-Sepharose (Pharmacia, Piscataway, NJ); DEAE-Sephadex A-25 and Sephadex G-150 superfine (Pharmacia); hydroxylapatite (Bio-Rad, Richmond, CA, HTP grade).

Chemicals. Ethylenediaminetetraacetic acid (EDTA); ε-aminocaproic acid (EACA); benzamidine hydrochloride (Sigma, St. Louis, MO); NaOH; KH_2PO_4; KCl; potassium citrate; Na_2SO_4.

Plasma. Blood should be drawn into acid–citrate–dextrose (ACD). If the cells are not required, 0.05 volume of 0.2 M EDTA/0.2 M benzamidine, pH 7.2, is added immediately, the blood chilled to 4°, and the plasma separated by centrifugation (2000 g, 20 min, 4°). If the cells are required, the plasma should first be separated, then EDTA and benzamidine (final concentration of 0.005 M for each) are added to the plasma, before either fractionation or storage at −70°. Prolonged storage of plasma at 4° should be avoided to prevent cold-promoted activation of factor VII, kallikrein, and factor XII, with consequent consumption of C1 INH.[13]

Preparation of columns. Four chromatographic steps are used in the isolation. All columns should be poured and equilibrated before commencing plasma fractionation; unnecessary delays during isolation inevitably lead to decreased yields. All chromatographic steps are performed at 4°.

Lysine-Sepharose. A wide column (200-ml bed volume/1000 ml plasma) with a moderately high flow rate (200–400 ml/hr) should be used. The column is equilibrated in 0.1 M potassium phosphate/0.5 M KCl/0.01 M EDTA/0.005 M benzamidine, pH 7.0. If the column has been used previously plasminogen should first be eluted with equilibration buffer containing 0.2 M EACA, after which the column should be reequilibrated.

DEAE-Sephadex A25. As with the lysine-Sepharose, a wide column (400-ml bed volume/1000 ml plasma) is used. The equilibration buffer is 0.02 M phosphate/0.1 M NaCl/0.005 M EDTA/0.005 M benzamidine, pH 7.0, and a flow rate of 5 ml/cm²/hr is used.

[31] R. A. Harrison, *Biochemistry* **22,** 5001 (1983).

Sephadex G-150 superfine. A 100 × 5 cm column is satisfactory for fractionation of up to 2500 ml of plasma. For larger volumes, a wider column must be used. The column is equilibrated with 0.02 M phosphate/ 0.1 M KCl/0.002 M benzamidine/0.002 M citrate, pH 7.0, at a flow rate of 20 ml/hr.

Hydroxylapatite. A column (250-ml bed volume/1000 ml plasma) with a length : width ratio of greater than 10 : 1 is used. It is equilibrated with 0.02 M phosphate/0.15 M KCl, pH 7.0, at a flow rate of 5 ml/cm^2/hr.

All column buffers are prepared by titration, using sodium hydroxide, of a stock solution of potassium dihydrogen phosphate (1.0 M) to the required pH followed by dilution to the required molarity. Final volume and pH adjustments are made after the addition of protease inhibitors and other salts.

Identification of C1 INH-containing fractions. Either electroimmunoassay ("rockets"), single radial immunodiffusion, or double immunodiffusion analysis can conveniently be used to locate C1 INH-containing column fractions.

Procedure

Precipitation with Na$_2$SO$_4$. Slowly add solid anhydrous Na$_2$SO$_4$ (final concentration 10 g/100 ml) to the stirred plasma and continue to stir gently, at 22°, for 1–2 hr after the final addition. Remove the precipitated protein by centrifugation (10,000 g, 30 min, 22°) and discard the pellet.

Chromatography on lysine-Sepharose. Load the Na$_2$SO$_4$ supernatant onto lysine-Sepharose, wash with equilibration buffer until the A_{280} of the eluate is close to zero, and collect all of the unadsorbed material. It is convenient to collect all of the unbound material in a single vessel rather than to collect fractions.

Chromatography on DEAE-Sephadex A-25. Adjust the pH and conductivity of the above pool to those of the DEAE equilibration buffer. This is most simply done by adjustment of the pH with 1.0 M HCl and dilution with distilled water. Load this C1 INH-containing pool onto DEAE Sephadex, wash the column with 1–2 column volumes of equilibration buffer, and apply a linear gradient, over 10 column volumes, to 0.02 M phosphate/ 0.005 M EDTA/0.005 M benzamidine/0.4 M NaCl, pH 7.0. Pool C1 INH-containing fractions and concentrate to between 20 and 40 ml. Large volumes are most easily concentrated in two stages: to less than 150 ml using a Minitan ultrafiltration cassette (Millipore, Bedford, MA), and then using an ultrafiltration cell (e.g., Amicon, Danvers, MA, with a PM10 membrane) to the required final volume.

Chromatography on Sephadex G-150 superfine. Dialyze the concentrated pool against 0.02 M phosphate/0.1 M KCl/0.002 M benzamidine/ 0.002 M citrate, pH 7.0, then, if necessary, clarify by centrifugation (20,000 *g*, 10 min, 4°). Load the dialyzed pool onto the Sephadex G-150 column and monitor the eluate at both 280 and at 600 nm. Protein-containing fractions eluting before the peak fraction of ceruloplasmin (assessed by absorbance at 600 nm) are assayed for C1 INH, and C1 INH-containing fractions are pooled and concentrated.

Chromatography on hydroxylapatite. Dialyze the concentrated C1 INH pool from the Sephadex G-150 column against 0.02 *M* phosphate/0.15 *M* KCl, pH 7.0, then load onto the hydroxylapatite column. Continue to elute with at least 2 column volumes of equilibration buffer, then apply a linear concentration gradient, 2.5 column volumes on each side, from equilibration conditions to 0.02 *M* phosphate/1.0 *M* KCl, pH 7.0. Locate C1 INH-containing fractions as before and assess the purity of the protein by SDS–PAGE before pooling. Pooled fractions should be concentrated (to about 10 mg/ml), dialyzed against VBS or 0.02 *M* phosphate/0.15 *M* NaCl, pH 7.0 (PBS), and stored frozen at −70°.

The composition of each pool and purity of the isolated C1 INH are illustrated in Fig. 1, and recovery data are given in Table II.

Comments on the Isolation Procedure. The initial steps in the isolation of C1 INH are designed to remove plasminogen and fibrinogen. Although precipitation with Na_2SO_4 removes the bulk of the fibrinogen, it also removes aggregated material and, possibly more importantly, removes coagulation factor XIII. If plasminogen is not removed prior to fractionation of the plasma, activation to plasmin will occur with consequent

TABLE II
C1 ISOLATION: RECOVERY[a]

Step	Total protein (g)[b]	C1 INH (mg)[c]	Purification factor	Yield (%)
Plasma plus protease inhibitors	50.30	228.8	1	100
5% PEG 4000 supernatant	52.75	211.4	0.9	92
Lysine-Sepharose pool	48.25	219.0	1.0	96
DEAE-Sephadex pool	2.23	228.1	22.5	99
Sephadex G-150 pool	0.338	150.9	98.1	66
Hydroxylapatite pool	0.059	164.7	610.0	72

[a] Recovery data taken from a single preparation of C1 INH from 1130 ml of freshly drawn plasma. Reprinted with permission from Ref. 31. Copyright 1983 American Chemical Society.

[b] Total protein was estimated from the absorption at 280 nm. An average extinction coefficient ($A_{1\,cm}^{1\%}$) of 1 was assumed.

[c] C1 INH was estimated by electroimmunoassay. The system was calibrated using purified C1 INH.

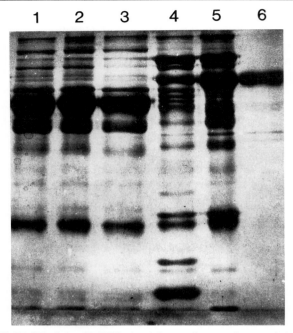

Fig. 1. SDS–PAGE of the C1 INH-containing pools at each step of the fractionation. Samples were prepared in loading buffer containing 1% 2-mercaptoethanol and run on a 10% polyacrylamide gel. Lane 1, plasma; lane 2, 5% PEG 4000 supernatant; lane 3, postlysine-Sepharose pool; lane 4, postDEAE-Sephadex A-50 pool; lane 5, postSephadex G-150 pool; lane 6, posthydroxylapatite pool. Reprinted with permission from Ref. 31. Copyright 1983 American Chemical Society.

proteolysis and consumption of C1 INH. The isolation method, as originally described, used polyethylene glycol (PEG) 4000 (as indicated in Table II), rather than Na_2SO_4, as described above. Na_2SO_4 precipitates fibrinogen and factor XIII as efficiently as does PEG 4000, and allows easier concentration of the supernatant because it can be completely removed by dialysis. PEG is inefficiently removed by dialysis, so that subsequent concentration of a PEG supernatant will induce further protein precipitation.

Resolution of C1 INH from other plasma proteins is achieved by three chromatographic steps: anion exchange, gel filtration, and adsorption onto hydroxylapatite. The anion-exchange resin DEAE-Sephadex is preferred because it gives complete resolution of the complement component C3 and C1 INH. With other anion-exchange resins (e.g., DEAE-Sephacel) there is some overlap of the two peaks. Because the DEAE-Sephadex is only used at high ionic strengths, the problems associated with expansion of the matrix in low-ionic-strength buffers are avoided. The similar elution properties of ceruloplasmin through the first two chromatographic steps make it a good, visible marker for C1 INH and indicate the fractions that should be assayed for the inhibitor. Ceruloplasmin gives the blue-colored fractions and can be monitored at 600 nm.

The conditions used for chromatography on hydroxylapatite are critical. Under the equilibration conditions C1 INH is loosely bound. With a low load volume and a long column it elutes after a small breakthrough peak. This contains a protein that on shorter columns is unrecognized because it has a mobility on SDS–PAGE similar to that of C1 INH. Although it is usually necessary to compromise on the length of column because hydroxylapatite gives a low cross-sectional flow rate, a length : width ratio of less than 10 : 1 should be avoided. Because C1 INH has a low extinction coefficient at 280 nm ($A_{1\ cm}^{1\%} = 3.86$), and because the eluted peak is broad, its location frequently is not obvious on the elution profile. All fractions following the breakthrough peak should therefore be monitored by SDS–PAGE and by immunochemical analysis, as recommended above. Occasionally, some C1 INH elutes with the chloride gradient. While this appears chemically identical to the protein contained in the main peak, its specific activity is low, and it should be pooled separately.

The above procedure has been used to isolate several different dysfunctional C1 INH proteins.[32–35] In all of these, the dysfunctional proteins behaved very much like the normal protein. This was true even for a protein covalently bonded to albumin (due to a reactive center Arg → Cys substitution[36]), where only the elution position from Sephadex G-150 was significantly altered. The corollary of this is that low levels of the normal protein will copurify with dysfunctional proteins. In comparatively few cases can these be readily recognized as being present.[35]

The procedure has many similarities and thus owes much to previously described isolation methods, and the yields achieved compare favorably with those reported by others. Protease inhibitors (e.g., Polybrene, EACA, DFP, and aprotinin) have been recommended by others to prevent contact system activation, and hence to avoid consumption of C1 INH by factor XIIa and kallikrein. In addition, siliconized glassware or plasticware is also often recommended during early fractionation steps in order to prevent contact activation of factor XII by glass surfaces. We

[32] V. H. Donaldson, R. A. Harrison, F. S. Rosen, D. H. Bing, G. Kindness, J. Canar, C. J. Wagner, and S. Awad, *J. Clin. Invest.* **75**, 124 (1985).

[33] R. A. Harrison and F. S. Rosen, *Mol. Immunol.* **19**, 1374 (1982).

[34] K. S. Aulak, P. A. Pemberton, F. S. Rosen, R. W. Carrell, P. J. Lachmann, and R. A. Harrison, *Biochem. J.* **253**, 615 (1988).

[35] R. B. Parad, J. Kramer, R. C. Strunk, F. S. Rosen, and A. E. Davis III, *Proc. Natl. Acad. Sci. U.S.A.* **87**, 6786 (1990).

[36] K. S. Aulak and R. A. Harrison, *Biochem. J.* **271**, 565 (1990).

have found that the addition of benzamidine alone is sufficient to prevent consumption or proteolysis of C1 INH. With the procedure described here, analysis by SDS–PAGE shows only a single band of intact C1 INH (Fig. 1).

C1 INH Isolation: Method 2. Lectin Affinity Chromatography

This protocol is modified only slightly from the method described by Pilatte *et al.*[30] It is included because it is particularly useful for the isolation of small quantities of C1 INH. For example, it can be used for isolation from very small volumes of plasma, or for the isolation of C1 INH expressed in tissue culture.

Materials

Chromatographic media. Jacalin-agarose (Vector Laboratories, Burlingame, CA); Phenyl-Sepharose (Pharmacia).

Chemicals. EDTA; *p*-nitrophenyl-*p*'-guanidobenzoate (NPGB); SBTI; PEG 3350 (J. T. Baker, Phillipsburg, NJ); KH_2PO_4; NaCl; melibiose (Sigma); $(NH_4)_2SO_4$.

Plasma. For collection of blood, EDTA (0.2 M, pH adjusted to 7.4 with NaOH), NPGB (2.5 mM in methanol), and SBTI (500 μM in water) are each added to plastic syringes in quantities to attain final concentrations of 10 mM EDTA, 25 μM NPGB, and 50 μM SBTI. Plasma is separated by centrifugation; all procedures following collection of the blood are performed at 4°.

Preparation of columns. Approximately 1 ml jacalin-agarose/10 ml plasma is sufficient. The jacalin-agarose column is equilibrated in 0.02 M phosphate/0.15 M NaCl, pH 7.0, PBS/10 mM EDTA/25 μM NPGB, pH 7.4. Approximately 1 ml of phenyl-Sepharose/10 ml of starting plasma is sufficient for the isolation. The phenyl-Sepharose column is equilibrated with PBS/0.4 M $(NH_4)_2SO_4$, pH 7.4.

Identification of C1 INH in fractions. The C1 INH elutes from both columns in predictable, sharp peaks in relatively high concentration, and the entire peaks are pooled in each case. It is therefore not essential to test fractions for the purpose of pooling, although we usually monitor fractions by SDS–PAGE in order to visualize the inhibitor (and its state of purity and degradation) during isolation.

Procedure

PEG precipitation. PEG 3350 is added to plasma to achieve a final concentration of 21.4% (w/v), and is incubated with mixing for 1 hr at 4°. The precipitate is removed by centrifugation at 10,000 g for 30 min at 4° and discarded. The final PEG concentration in the plasma is then increased to 45% (w/v), and the pool incubated and centrifuged as above.

Chromatography on jacalin-agarose. The 21.4–45% PEG precipitate is dissolved in PBS/10 mM EDTA/25 μM NPGB, pH 7.4, and applied to the jacalin-agarose column. The column then is washed with the starting buffer containing 0.5 M NaCl until the absorbance at 280 nm is near zero (approximately 10 column volumes). Finally, the column is eluted with 0.125 M melibiose in the same buffer and fractions collected (for a 3-ml column, 1-ml fraction volumes are convenient).

Chromatography on phenyl-Sepharose. The C1 INH-containing pool from the jacalin-agarose column is concentrated approximately 10-fold in an Amicon ultrafiltration cell (PM10 membrane), made 0.4 M in $(NH_4)_2SO_4$ (with a 2 M solution), and applied to the phenyl-Sepharose column. The nonadsorbed protein (containing C1 INH) is then eluted with PBS, pH 7.4, containing 0.4 M $(NH_4)_2SO_4$ (approximately 10 column volumes). The C1 INH-containing pool is again concentrated 5- to 10-fold, as above, and dialyzed into PBS, pH 7.4, or other appropriate buffers.

Comments on Isolation Procedure. This method is equally suitable for the isolation of normal and dysfunctional C1 INH proteins, and has been used even for a mutant with abnormal glycosylation.[35] It is particularly suited to the isolation of C1 INH from small volumes of plasma and, as the quantities of plasma available from patients frequently are limited, can be extremely valuable. It also can be used for recovery of C1 INH from tissue culture supernatants. In addition, we have used the method to isolate C1 INH from several mammalian species. One slight limitation of the method is that it seems to be important to add the protease inhibitors at the time blood is obtained and to perform the PEG precipitation immediately after collection of the plasma in order to prevent contamination of the final product with degraded C1 INH and other contaminants. However, we have used the method, albeit with some variability in success, for the isolation of C1 INH from frozen plasma stored for long periods in the absence of inhibitors. The other major advantage of the method is the speed with which the isolation can be performed. The entire purification procedure easily can be completed in 1 day.

Properties of Human C1 INH and Its Gene

The complete amino acid sequence of C1 INH was determined from a combination of protein and DNA sequencing.[37,38] It consists of a single

---

[37] A. E. Davis III, A. S. Whitehead, R. A. Harrison, A. Dauphinais, G. A. P. Bruns, M. Cicardi, and F. S. Rosen, *Proc. Natl. Acad. Sci. U.S.A.* **83**, 3161 (1986).

[38] S. C. Bock, K. Skriver, E. Nielsen, M. C. Thogersen, B. Wiman, V. H. Donaldson, R. L. Eddy, J. Marrinan, E. Radziejewska, R. Huber, T. Shows, and S. Magnussen, *Biochemistry* **25**, 4292 (1986).

TABLE III
PROPERTIES OF C1 INHIBITOR

Property	Comment
Single polypeptide chain	478 amino acid residues
Apparent M_r ~104,000 on SDS–PAGE	Actual M_r 71,100, polypeptide M_r 52,842
Heavily glycosylated NH$_2$-terminal 100 residues	~30% carbohydrate
Member of serpin superfamily	Reactive center arginine
Target proteases	C1r, C1s, kallikrein, plasmin, factors XIa and XIIa
Single gene copy on chromosome 11	17 kb, 8 exons, introns contain *Alu* repeats

polypeptide chain containing 478 amino acid residues, and is synthesized with an amino-terminal 22-residue signal peptide[38] (Table III). On SDS–PAGE, it has an apparent M_r of 105,000.[31] A similar value (104,000) is found using sedimentation equilibrium analysis.[28] The isolated protein frequently contains variable quantities of C1 INH fragments, particularly at M_r 95,000, although C1 INH-derived peptides of other sizes also are seen. These fragments probably result from combinations of partial deglycosylation, proteolytic cleavage near the amino terminus (the larger product of which retains function), or proteolytic cleavage at or near the reactive center (the products of which are inactive). As a reflection of its low tryptophan content, the protein has a low extinction coefficient; experimental values at 280 nm ($A^{1\%}_{1\,cm}$) of 3.6–5.0 have been reported.[28,31,39] That calculated from the composition of the protein is 3.86. The calculated polypeptide molecular weight of C1 INH is only 52,842. The circulating protein contains approximately 30% carbohydrate (w/w)[40]; this places C1 INH among the most heavily glycosylated plasma proteins. Among the serpins, this degree of glycosylation is unique to C1 INH. Much of the oligosaccharide (7 of 13 units) is contained as galactosamine-based units O-linked to serine or threonine residues. In addition, most of the carbohydrate (10 of the 13 units) is contained within the amino-terminal 100 residues of the protein. Three of the seven O-linked glycosylation units are located within a 35-residue region in which variations of the sequence Gly-Pro-Thr-Thr are repeated seven times (as a result of a 12-nucleotide tandem repeat).[38] The amino-terminal extension has no similarity to any

[39] H. E. Schultze, K. Heide, and H. Haupt, *Naturwissenschaften* **49**, 133 (1962).
[40] S. J. Perkins, K. F. Smith, S. Amatayakul, D. Ashford, T. W. Rademacher, R. A. Dwek, P. J. Lachmann, and R. A. Harrison, *J. Mol. Biol.* **214**, 751 (1990).

of the other serpins, nor does it have any identified homology with other proteins. The region of homology with serpins extends from approximately amino acid 110 to the carboxyl terminus. The functional role of this amino-terminal domain is unknown, although it is not required for protease inhibitory activity.[41,42] Largely because of its high sialic acid content the protein is extremely acidic and migrates in nondenaturing electrophoresis as an α_2-globulin.

As outlined earlier, C1 INH, by both structural and functional criteria, is a member of the serpin superfamily.[7] Comparison of the serpin domain of C1 INH with other serpins reveals amino acid identity of approximately 25%; although low, this is similar to the degree of homology among many of the protease inhibitor members of the family.[37,38] In addition to the similarity in primary sequence, all other available data indicate that C1 INH functions via the same mechanism as other serpins. It forms tight equimolar SDS-stable complexes with target proteases. The P_1 residue is Arg-444, which is consistent with the substrate specificity of its target proteases.[43] Under appropriate conditions, the P_1-P_1' peptide bond may be hydrolyzed with release of the carboxyl-terminal peptide that begins with the P_1' Thr residue. Other properties shared with serpins that indicate similar structure and function include the sensitivity of the region immediately amino terminal to the reactive center to cleavage by nontarget proteases (which implies that this region is exposed on the surface of the molecule),[44] and the characteristic serpin sensitivity to heat denaturation (combined with an increase in heat stability following cleavage at or near the reactive center).[45] The analysis of both naturally occurring and *in vitro*-generated C1 INH mutants, as with other serpins, is now making a significant contribution to our understanding of structure–function relationships among the serpins.

C1 INH is encoded by a gene on chromosome 11 (p11.2-q13).[37,38] The gene is slightly over 17 kb long and contains eight exons.[46] The placement of introns shows no similarity with that of other serpins. The C1 INH gene is unusual in that it contains 17 *Alu* repeat elements within its introns, which constitute nearly a third of the total intronic sequence in the C1 INH gene. In several genes, *Alu* repeats, which are of unknown function

[41] J. O. Minta, *J. Immunol.* **126**, 245 (1981).

[42] M. H. Prandini, A. Reboul, and M. G. Colomb, *Biochem. J.* **237**, 93 (1986).

[43] G. S. Salvesen, J. J. Catanese, L. F. Kress, and J. Travis, *J. Biol. Chem.* **260**, 2432 (1985).

[44] P. A. Pemberton, R. A. Harrison, P. J. Lachmann, and R. W. Carrell, *Biochem. J.* **258**, 193 (1989).

[45] M. Lennick, S. A. Brew, and K. C. Ingham, *Biochemistry* **24**, 2561 (1985).

[46] P. E. Carter, C. Duponchel, M. Tosi, and J. E. Fothergill, *Eur. J. Biochem.* **197**, 301 (1991).

and are structurally related to 7 SL RNA, may predispose to recombination (see below).

Molecular Analysis of C1 Inhibitor Deficiency

The purpose of this discussion is not to provide a detailed description of all described C1 INH mutations, nor to provide a prolonged analysis of C1 INH/serpin function. Rather, the intention is to summarize briefly the emerging contributions of the molecular analysis of C1 INH deficiency to the understanding of both C1 INH function and the mechanisms of mutations in the C1 INH gene. Then we will outline our approach toward analysis of the C1 INH protein and gene in patients with C1 INH deficiency.

Hereditary Angioneurotic Edema

Deficiency of C1 INH results in hereditary angioneurotic edema (HANE), a condition characterized by recurrent episodes of cutaneous or mucosal edema. The disease is inherited in an autosomal dominant fashion, which suggested (and was subsequently proved) that HANE develops in heterozygotes. Type I HANE refers to individuals with decreased functional and antigenic levels of C1 INH in plasma, who express one normal C1 INH allele and who have one "silent" allele. Patients with type II HANE express one normal and one mutant dysfunctional C1 INH molecule. Deficiency of C1 INH results in unregulated activation of both the complement and coagulation systems, with generation of mediators that enhance vascular permeability via one or both systems.

Type I HANE. Even though the data remain somewhat limited, it appears that the distribution of mutations in type I HANE may be unusual. As many as 15–20% of type I mutations result from partial deletions and/or duplications.[47–50] These all occur between the *Alu* repetitive DNA elements in introns in the C1 INH gene. All have occurred between *Alu* repeats oriented in the same direction, and therefore are likely to be a result of homologous recombination and unequal crossing-over. The precise locations of several of these deletions have been defined.[49,50] Deletion and duplication of exon 4, deletion of exon 7, and deletion of exons 4–6 all have been observed, and several involving other exons have been mapped, but not sequenced. It thus is clear that *Alu* repeats are one important element in the development of type I HANE.

---

[47] A. E. Davis III, *Immunodefic. Rev.* **1,** 207 (1989).
[48] M. Tosi, D. Stoppa-Lyonnet, P. E. Carter, and T. Meo, *Behring Inst. Mitt.* **84,** 173 (1989).
[49] D. Stoppa-Lyonnet, P. E. Carter, T. Meo, and M. Tosi, *Proc. Natl. Acad. Sci. U.S.A.* **87,** 1551 (1990).
[50] T. Ariga, P. E. Carter, and A. E. Davis III, *Genomics* **8,** 607 (1990).

	P15					P10					P5				P1	P1'				P5'
AAT	G	T	E	A	A	G	A	M	F	L	E	A	I	P	M	S	I	P	P	E
AT III	G	S	E	A	A	A	S	T	A	V	V	I	A	G	R	S	L	N	P	N
C1 INH	G	V	E	A	A	A	A	S	A	I	A	V	A	-	R	T	L	L	V	F
[a] C1 INH Mo/Ca	-	-	-	-	-	T	-	-	-	-	-	-	-	-	-	-	-	-	-	-
[b] C1 INH Ma	-	-	-	E	-	-	-	-	-	-	-	-	-	-	-	-	-	-	-	-
[c] C1 INH We	-	E	-	-	-	-	-	-	-	-	-	-	-	-	-	-	-	-	-	-
[d] P1 CYS	-	-	-	-	-	-	-	-	-	-	-	-	-	-	C	-	-	-	-	-
[e] P1 HIS	-	-	-	-	-	-	-	-	-	-	-	-	-	-	H	-	-	-	-	-
[f] P1 SER	-	-	-	-	-	-	-	-	-	-	-	-	-	-	S	-	-	-	-	-
[g] P1 LEU	-	-	-	-	-	-	-	-	-	-	-	-	-	-	L	-	-	-	-	-

FIG. 2. Reactive center and hinge region mutations in C1 INH compared with the equivalent regions in α_1-antitrypsin (AAT) and antithrombin III (AT III). References: (a) N. J. Levy et al.[61]; (b) K. Skriver et al.[62]; (c) A. E. Davis III et al.[63a]; (d) K. S. Aulak and R. A. Harrison[36]; (e) K. S. Aulak et al.[34]; (f) K. S. Aulak et al.[56]; (g) D. Frangi et al.[63b].

There are other type I mutations. In addition to deletions, at least one splice junction mutation has been observed,[51] and there are results that indicate that single base mutations may occur more frequently then expected in exon 8.[52–54] The analysis of many more mutations is required, however, to confirm this initial impression.

Type II HANE. At least 65–70% of type II HANE results from reactive center (P1) mutations, mutations replacing the P1 residue (Arg-444).[34,36,55] Most of these result in replacement of the Arg residue with a Cys or His (Fig. 2). Of the other four possible products of point mutation at the Arg-444 codon, Ser[56] and Leu[57] each have been observed only once, and Gly and Pro have not been seen. The most likely explanation for this distribution is the fact that the codon for Arg-444 (CGC) contains a CpG dinucleotide. This dinucleotide is susceptible to mutation via deamidation of 5-methylcytosine in the CpG, resulting in replacement of the C with a T in either the codon or the anticodon. As expected, none of the reactive center mutants have any significant activity against the normal target proteases.

[51] Z. M. Siddique, D. F. Lappin, A. R. McPhaden, and K. Whaley, *Clin. Exp. Immunol.* **86,** Suppl. 1, 12 (1991).

[52] E. Verpy, J. Laurent, P. J. Spath, T. Meo, and M. Tosi, *Complement Inflammation* **8,** 238 (1991).

[53] D. Frangi, M. Cicardi, A. Sica, F. Colotta, A. Agostoni, and A. E. Davis III, *J. Clin. Invest.* **88,** 755 (1991).

[54] Z. M. Siddique, A. R. McPhaden, and K. Whaley, *Clin. Exp. Immunol.* **86,** Suppl. 1, 11 (1991).

[55] K. Skriver, E. Radziejewska, J. A. Siebermann, V. H. Donaldson, and S. C. Bock, *J. Biol. Chem.* **264,** 3066 (1989).

[56] K. S. Aulak, M. Cicardi, and R. A. Harrison, *FEBS Lett.* **266,** 13 (1990).

[57] D. Frangi and A. E. Davis III, unpublished data (1991).

However, analysis of site-directed mutants with a P_1 His replacement revealed the acquisition of inhibitory activity against chymotrypsin.[58] This prompted the detailed analysis of a naturally occurring P_1 His mutant, together with normal human C1 INH and rabbit C1 INH.[59] These studies have shown that, as expected, the P_1 His mutant inhibits chymotrypsin; unexpectedly, both normal human and rabbit C1 INH molecules also inhibited chymotrypsin. All three proteins were shown on SDS–PAGE to form a complex with chymotrypsin. The second-order association rate constant for chymotrypsin with normal human C1 INH was 7.3×10^3 $M^{-1} \sec^{-1}$, and with the P_1 His mutant was $3.4 \times 10^4 M^{-1} \sec^{-1}$. Sequence analysis of the postcomplex C1 INH carboxy-terminal peptide revealed that with the His mutant, this His residue was used as the P_1 residue (cleavage was between the His and the $P_1{}'$ Thr). In contrast with normal C1 INH, chymotrypsin uses the Ala residue at P_2 as the reactive center (cleavage was between the P_2 Ala and P_1 Arg). These data clearly indicate that the P_1–$P_1{}'$ residues in serpins are not restricted to a single site, and that availability within a localized region on the surface of the molecule, as well as absolute position within this reactive center, contribute to P_1 residue definition. However, the higher second-order rate constant for the inhibition reaction with the P_1 His mutant, relative to the normal (P_1 Arg) inhibitor, is likely the conseqence of His being in the preferred P_1 position rather than His being a better substrate than Ala. A similar observation also has been made with α_2-antiplasmin.[60]

The reactive center region of serpins extends from approximately P_{15} to $P_5{}'$, and the hinge region encompasses P_{10} through P_{17}. Three different C1 INH hinge region mutants have been described: a P_{10} Ala $\rightarrow$ Thr, a P_{12} Ala $\rightarrow$ Glu, and a P_{14} Val $\rightarrow$ Glu (Fig. 2).[61–63b] These, like similar mutations in antithrombin III, have lost inhibitory activity. Most serpin

[58] E. Eldering, C. C. M. Huijbregts, and C. E. Hack, *Complement Inflammation* **6**, 333 (1989).

[59] K. S. Aulak, A. E. Davis III, V. H. Donaldson, and R. A. Harrison, *Protein Sci.* **2**, 727 (1993).

[60] J. Potempa, B. H. Shieh, and T. J. Travis, *Science* **241**, 699 (1988).

[61] N. J. Levy, N. Ramesh, M. Cicardi, R. A. Harrison, and A. E. Davis III, *Proc. Natl. Acad. Sci. U.S.A.* **87**, 265 (1990).

[62] K. Skriver, W. R. Wikoff, P. A. Pattson, R. Tausk, M. Schapira, A. P. Kaplan, and S. C. Bock, *J. Biol. Chem.* **266**, 9216 (1991).

[63] A. E. Davis III, K. S. Aulak, R. B. Parad, H. P. Stecklein, E. Eldering, C. E. Hack, J. Kramer, R. C. Strunk, and F. S. Rosen, *Complement Inflammation* **8**, 138 (1991).

[63a] A. E. Davis III, K. S. Aulak, R. B. Parad, H. P. Stecklein, E. Eldering, C. E. Hack, J. Kramer, R. C. Strunk, J. Bissler, and F. S. Rosen, *Nature Gene* **1**, 354 (1992).

[63b] D. Frangi, K. S. Aulak, M. Cicardi, R. A. Harrison, and A. E. Davis III, *FEBS Lett.* **301**, 34 (1992).

hinge region mutants are converted from inhibitors to substrates and are efficiently cleaved by target proteases; both the P_{12} and P_{14} mutants fit this category. The P_{10} Ala → Thr mutant, however, neither complexes with, nor is cleaved appreciably by, C1s.

All of these data are consistent with the hypothesis that partial insertion of the hinge region (as strand A4) into β sheet A is required for serpin function. Mutations that either restrict this movement or block insertion interfere with function. In cleaved α_1-antitrypsin and α_1-antichymotrypsin, the P_{10}, P_{12}, and P_{14} amino acid side chains all face the core of the molecule. All the above mutations are to bulkier, more hydrophilic side chains than those in the normal molecule. This, however, does not explain why the P_{12} and P_{14} mutants are converted to substrates, while the P_{10} Ala → Thr mutant is not. This mutant has been shown to polymerize via the insertion of the reactive center loop of one molecule into the A β sheet of another,[64] as also occurs with $Z_{\alpha 1}$-antitrypsin,[65] and with antitrypsin cleaved between P_{14} and P_{10}.[66] It has not been proved, however, that the lack of conversion to substrate is secondary to polymerization.

Definition of Mutations in HANE

Ideally, the characterization of mutations requires molecular genetic techniques for analysis of type I HANE, and a combination of molecular genetic and protein chemical analyses for analysis of mutations resulting in dysfunctional C1 INH proteins (type II). The recommended first steps differ for types I and II, but subsequent analysis is the same for both.

Type I, Step 1. As discussed, type I HANE frequently results from relatively large intragenic partial deletions (or insertions) that result from unequal crossing over. Accordingly, the most straightforward method to identify these is by Southern blot analysis of genomic DNA. This is conveniently accomplished using the restriction endonuclease, *Bcl*I, which releases a 21-kb fragment that contains the entire coding sequence of the C1 INH gene.[55] Blots are then hybridized with a full-length C1 INH cDNA probe. Because patients with HANE are all heterozygous, two fragments will be visualized, one of which is 21 kb and a second which is either smaller (deletion) or larger (insertion). If more detailed delineation of deletions is desired, hybridizations with polymerase chain reaction (PCR)-generated probes for each individual exon may be performed.

[64] K. S. Aulak, E. Eldering, C. E. Hack, Y. P. T. Lubbers, R. A. Harrison, A. Mast, M. Cicardi, and A. E. Davis III, *J. Biol. Chem.,* in press.

[65] D. A. Lomas, D. Ll. Evans, J. Finch, and R. W. Carrell, *Nature (London)* **357,** 605 (1992).

[66] A. E. Mast, J. J. Enghild, and G. Salvesen, *Biochemistry* **31,** 2720 (1992).

Type II, Step 1. The purpose of these analyses is to determine whether the mutant results from a reactive center mutation, because these account for the majority of dysfunctional C1 INH proteins. Reactive center Arg → Cys mutants may be identified in whole plasma by immunoelectrophoresis, because most of the dysfunctional protein exists as a covalent complex with albumin and is of more anodal mobility than is free C1 INH. Further analysis for reactive center mutants is performed essentially as described by Aulak and Harrison.[36] Test plasma (1 volume, usually 0.4–1.0 ml) is incubated with anti-C1 INH-Sepharose (1 volume) for 2 hr at 4°. The Sepharose is pelleted by centrifugation, and washed with PBS, pH 7.4, followed by PBS, 1 M NaCl, pH 7.4, and then again with PBS. Washed resin (25 μl) is then incubated in TPCK-trypsin (Sigma) (20 μl, 0.43 μM, in 50 mM Tris-HCl, 100 mM NaCl, pH 8.0) at 37° for 40 min. Nonreducing SDS–PAGE sample buffer is then added, the sample is incubated for 2 min at 100°, and subjected to SDS–PAGE [10% (w/v) polyacrylamide]. A control sample that has not been digested with trypsin is also analyzed by SDS–PAGE. C1 INH with a normal reactive center Arg residue is cleaved primarily to an 85-kDa fragment that results from cleavage at the reactive center and at a site near the amino terminus. Reactive center mutants, because of the absence of the basic residue at P1, are reduced to 96 kDa due to removal of the amino-terminal fragment only (Fig. 3). This provides a relatively simple, rapid approach toward the identification of a P1 mutant. If desired, the precise mutation may be determined by digestion of immunoprecipitated protein with *Pseudomonas aeruginosa* elastase, which hydrolyzes the P4–P3 peptide bond, followed by isolation of the released carboxy-terminal peptide by reversed-phase HPLC, and sequence analysis of the isolated peptide.

Jacalin-agarose also may be used (as described previously) to obtain C1 INH from very small volumes of plasma for these analyses. The immunochemical method is more convenient to screen samples for susceptibility to trypsin, whereas the jacalin-agarose technique may be preferable to obtain peptide for sequence determination.

Types I and II, Step 2. The limited data currently available suggest that the number of mutations in exon 8 in type I HANE as well as in type II may exceed the number expected if mutations were randomly distributed through the gene.[52–54] Therefore, for all type I and type II samples for which a mutation has not been identified using the techniques described in step 1, exon 8 is amplified from genomic DNA by PCR. This amplified DNA is sequenced in order to identify the potential mutations. Thus far, as might be expected, most mutations have been single base changes (or single base insertions or deletions) that create premature stop codons. In addition, one example of a short deletion and one example of a short

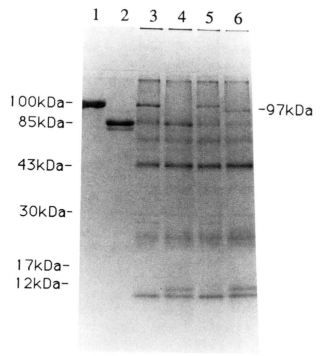

FIG. 3. SDS–PAGE of trypsin-digested C1 INH immobilized on anti-C1 INH Sepharose. Samples were analyzed on a 10–20% polyacrylamide gradient gel in the absence of reducing agent. Lanes 1 and 2, isolated normal C1 INH; lanes 3 and 4, C1 INH immunoprecipitated from normal human plasma; lanes 5 and 6, protein from the plasma of a patient with a C1 INH P_1 Arg → His substitution. Samples in lanes 1, 3, and 5 are undigested; samples in lanes 2, 4, and 6 were digested with trypsin.

duplication within exon 8 have been identified (J. Bissler and A. Davis, unpublished data, 1993). The DNA sequence analysis is most conveniently accomplished by direct sequence analysis using *Taq* polymerase of PCR-amplified DNA (using the kit from BRL, Gaithersburg, MD). Eliminating subcloning of amplified DNA avoids the problems of PCR-associated errors, and is also particularly valuable in a heterozygous disorder, such as HANE, because both the normal and the mutant nucleotide are detected on the same sequence.

Types I and II, Step 3. The screening procedure we have used for approximate localization of mutations within individual exons is the method of single-stranded conformation polymorphism[64,65] of PCR-amplified DNA. Oligonucleotide primer pairs have been constructed that span each exon and include the sequences through the intron–exon junctions.

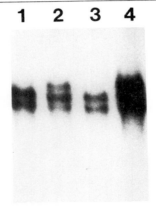

FIG. 4. Single-stranded conformation polymorphism of PCR-amplified DNA from exon 8 of the C1 INH gene. Lane 1, normal; lane 2, insertion of an A after nucleotide 1304[51]; lane 3, C → T replacement at nucleotide 1431[55]; lane 4, T deletion at nucleotide 1298.[51] Each sample has one band in common (the lower, more anodal band), which presumably represents the normal allele. The other bands differ in mobility from either band in the normal sample. In most instances of single base substitutions, as opposed to single base insertions or deletions, the mobility difference is very slight, but reproducible.

Standard PCR reactions (30 cycles with predetermined optimum annealing temperature) with genomic DNA are set up with the dCTP in the reaction mixture replaced with [α-^{32}P]dCTP. PCR product (1 μl) is mixed with formamide (9 μl) and denatured by boiling for 2–5 min. The denatured product (2 μl) is then loaded on a 5% polyacrylamide gel (0.5 mm) (in 0.09 M Tris–borate, 0.002 M EDTA, pH 8.0) and subjected to electrophoresis (10 W, overnight). The gel is then dried and exposed for 18 hr to Kodak (Rochester, NY) XAR-5 film. Several manipulations may be used to enhance the ability to visualize polymorphisms. These include the addition of 10% glycerol to the polyacrylamide gel, electrophoresis at 4° rather than room temperature, and the use of either a more dilute or more concentrated running buffer. An example of single-stranded conformation polymorphism of known C1 INH mutants is shown in Fig. 4. After identification of polymorphic exons using this technique, the appropriate amplified exons are sequenced as described above.

[67] M. Orita, H. Iwahana, H. Kanazawa, K. Hayashi, and T. Sekiya, *Proc. Natl. Acad. Sci. U.S.A.* **86**, 2766 (1989).

[68] M. Dean, M. B. White, J. Amos, B. Gerrard, C. Stewart, K.-T. Khaw, and M. Leppert, *Cell (Cambridge, Mass.)* **61**, 863 (1990).

[7] α-Macroglobulins: Detection and Characterization

By Guy Salvesen *and* Jan J. Enghild

Introduction

α-Macroglobulins (αMs) are large proteins that bind proteinases, thereby decreasing their activity on protein substrates. Proteins with the special proteinase-binding characteristics of αMs are well represented in advanced animal phyla, wherein they are located primarily in the circulatory fluids, often at very high concentrations. Apart from being fascinating examples of molecular evolution, αMs are of more pragmatic interest to those of us investigating biological and chemical aspects of proteolysis. This is due to their lack of specificity; almost all proteinases, unless they are extremely selective or very large, are bound and inhibited by αMs. This is often a problem for investigators who want to isolate proteinases from animals or tissue culture, and ways have been developed to eliminate or inactivate αMs from these sources. On the other hand, the very nonspecificity of αMs makes them ideal tools for characterizing aspects of proteolysis in defined or undefined systems. In this chapter we describe procedures appropriate for the identification, isolation, characterization, elimination, and use of αMs. Human $α_2$-macroglobulin, the best characterized αM, was previously featured in this series in 1982.[1] Most of the methods identified therein are still appropriate, though we recommend a different purification procedure and slightly different conditions for polyacrylamide gel electrophoresis conditions. More detailed reviews can be found for the structure and biology of αMs.[2,3]

Mechanism

All known αMs are monomers, dimers, or tetramers having identical subunits of about 180 kDa. In the middle of each subunit is a stretch of amino acid residues that is highly susceptible to proteolysis. Cleavage of any of the peptide bonds in this region, the "bait region," initiates a series of changes that often result in compacting of the subunits. The inhibitory capacity of the αMs results from a conformational change following pro-

---

[1] A. J. Barrett, this series, Vol. 80, p. 737.

[2] R. C. Roberts, *Rev. Hematol.* **2**, 129 (1986).

[3] L. Sottrup-Jensen, *in* "The Plasma Proteins" (F. W. Putman, ed.), Vol. 5, p. 191. Academic Press, Orlando, FL, 1987.

METHODS IN ENZYMOLOGY, VOL. 223

teinase reaction that physically entraps the reacting proteinase molecule(s).[4] αMs are thus mechanistically distinct from the active site-directed proteinase inhibitors that directly block the active sites of proteinases.[5] αM-bound proteinases retain activity against small substrates such as peptides, but are almost completely inactive against large ones (proteins larger than 30 kDa). This unusual reactivity results from steric hindrance of proteinases by the large αM molecule. The retention of activity by αM-proteinase complexes can complicate analysis of proteolytic activities. For example, Dahlmann et al.[6] realized that a high-molecular-weight proteinase reported by several authors to be present in rat muscle was really a complex of rat α_1-macroglobulin with small lysosomal proteinases, possibly resulting from isolation artifacts.

Unique to αMs is an intrachain thiol ester[7] located two-thirds of the distance from the N terminus of each subunit (Fig. 1). This group is usually reactive with small amines in native αMs, and becomes highly reactive with a broader range of nucleophiles during the conformational change caused by reaction with proteinases. Often, proteinase molecules become covalently linked to the thiol ester, though this seems to be essential only for inhibition by the monomeric αMs, such as rat α_1-inhibitor 3[8] and the dimeric human pregnancy zone protein.[9] Although complement components C3 and C4 probably originated from the common αM ancestor, based on sequence criteria and the possession of a thiol ester, there is no evidence that they inhibit proteinases, and they will not be considered in this chapter.

Purification of Human α_2-Macroglobulin

In dealing with human α_2-macroglobulin it is important to remember that the protein can exist in two forms. The electrophoretically "slow" (S) form is fully active and may bind up to two molecules of some proteinases. On reaction with proteinases, or on reaction with some small primary amines (and in the absence of any modifying reagents), the S form is irreversibly converted to the electrophoretically "fast" (F) form. The F form is not able to inhibit proteinases.

The starting material for purification of human α_2-macroglobulin (200

[4] A. J. Barrett and P. M. Starkey, Biochem. J. 133, 709 (1973).

[5] M. Laskowski, Jr. and I. Kato, Annu. Rev. Biochem. 49, 593 (1980).

[6] B. Dahlmann, L. Kuehn, P. C. Heinrich, H. Kirschke, and B. Wiederanders, Biochim. Biophys. Acta 991, 253 (1989).

[7] J. B. Howard, Proc. Natl. Acad. Sci. U.S.A. 78, 2235 (1981).

[8] J. J. Enghild, G. Salvesen, I. B. Thøgersen, and S. V. Pizzo, J. Biol. Chem. 264, 11428 (1989).

[9] U. Christensen, M. Simonsen, N. Harrit, and L. Sottrup-Jensen, Biochemistry 28, 9324 (1989).

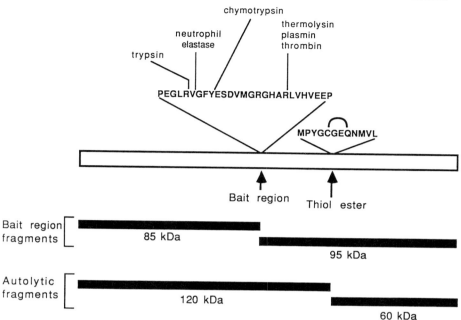

Fig. 1. The subunit of human α_2-macroglobulin. Human α_2-macroglobulin is a tetramer composed of identical subunits assembled from pairs of disulfide-linked dimers. The bait region of each subunit comprises a stretch of amino acids whose side chains fit the specificity of a large variety of proteinases. Cleavage within this region initiates a series of conformational changes that result in entrapment of attacking proteinases. The primary recognition sites of some proteinases [R. C. Roberts, *Rev. Hematol.* **2**, 129 (1986); L. Sottrup-Jensen, *in* "The Plasma Proteins" (F. W. Putman, ed.), Vol. 5, p. 191. Academic Press, Orlando, FL, 1987] are identified in the figure. Secondary cleavages occur, sometimes outside the bait region, but these later events are probably not directly involved in proteinase inhibition. The site of cleavage by human neutrophil elastase reported earlier [G. D. Virca, G. S. Salvesen, and J. Travis, *Hoppe-Seyler's Z. Physiol. Chem.* **364**, 1297 (1983)] is a secondary cleavage site. The site identified above is the primary cleavage site (J. J. Enghild and G. Salvesen, unpublished results, 1990). Cleavage of the bait region results in the characteristic bait region fragments seen in reduced SDS–PAGE. The autolytic fragments of 60 and 120 kDa seen in reduced SDS–PAGE are due to nonenzymatic cleavage of the subunit chain, most likely mediated by attack of the G–E peptide bond nitrogen on the carbonyl of the thiol ester.

ml) is usually outdated citrated or EDTA-treated plasma or plasmapheresis plasma. There is no need to use haptoglobin-selected plasma. Serum should not be used because it contains F form α_2-macroglobulin. If the plasma is obtained frozen it must be thawed at 37° in a water bath. The plasma should not be thawed at room temperature or at 4° because these procedures may cause precipitation of proteins and activation of the intrinsic pathway of coagulation. All subsequent steps of the procedure, based on that of Kurecki et al.,[10] are carried out at 4° except the zinc–chelate chromatography, which is performed at room temperature.

Step 1: Dialysis against Water. The plasma is transferred to a dialysis membrane with a 15-kDa or higher molecular weight cutoff. During the dialysis the plasma triples in volume and it is important to take that into account when the dialysis membrane is sized. The plasma is dialyzed twice against 12 liters of deionized water at 4°; most conveniently the dialysis is started in the morning and changed at the end of the work day so that the second dialysis continues overnight. We use a Nalgene (Nalge, Rochester, NY) pipette washer as a container for the dialysis step. The following day the precipitate is removed by centrifugation at 5000 g for 20 min.

Step 2: Zinc–Chelate Chromatography. A chromatography column containing a 100-ml bed volume of chelating Sepharose FF (Pharmacia, Piscataway, NJ) is prepared by washing the medium first with 500 ml of 0.1 M sodium phosphate buffer, pH 5.0, then with 500 ml of water, and then with 80 ml of $ZnCl_2$ (0.3 g/100 ml). All steps are carried out at a flow rate of 500 ml/hr. The sample supernatant is applied to the equilibrated column. Following sample application the column is washed with 0.1 M sodium phosphate buffer, pH 7.0, until the absorbance is below 0.01 at 280 nm. The column is then washed with 0.1 M sodium phosphate buffer, pH 6.0, again until the absorbance is below 0.01 at 280 nm. The bound human α_2-macroglobulin is eluted with 0.1 M sodium phosphate buffer, pH 5, and 5-ml fractions are collected. Only one peak of protein elutes from the column and contains 80% pure human α_2-macroglobulin. The pH of the collected sample is raised to pH 7.0 by titrating with 1 M sodium phosphate buffer, pH 7, and the protein is concentrated by ultrafiltration to 40 ml, using a 100-kDa membrane.

Step 3: Gel Filtration. Gel filtration is performed on a 2.5 × 125 cm column of Sephacryl S-300 HR (Pharmacia) equilibrated with 50 mM sodium phosphate buffer, 150 mM NaCl, pH 7.0. Three samples of 14 ml each are loaded sequentially onto the gel-filtration column, which is run at 20 ml/hr at 4°. The purified human α_2-macroglobulin is concentrated by ultrafiltration as described above and stored in aliquots at $-20°$ or $-80°$.

The yield is about 250 mg and should be in the S form when analyzed by pore-limit PAGE. Time is of the essence in this procedure because prolonged manipulations can cause production of the F form. Increasing

[10] T. Kurecki, L. F. Kress, and M. Laskowski, Sr., *Anal. Biochem.* **99**, 415 (1979).

the volume of the starting plasma should be avoided because this leads to inefficient removal of contaminants in the initial step.

Nonhuman α-Macroglobulins

Examples of αMs are found in all chordates and arthropods examined to date, and our recent work[11] indicates that the αM lineage extends as far as the octopus, a member of the molluscs. We do not yet know if αMs are represented in animals that lack a closed circulatory system, though it is clear that the proteins are present at very high concentrations in the hemolymph of arthropods and molluscs. Given the lack of an advanced immune system in these phyla, and the rather nonspecific nature of proteinase binding by αMs, it is likely that ancestral αMs evolved to control invasion mediated by pathogenic proteinases. An additional defensive function of ancestral αMs is suggested by the ability of limac, the αM from the horseshoe crab *Limulus polyphemus*, to participate in a lytic system for foreign cells.[12]

It is not known whether αMs regulate the coagulation proteinases of modern arthropods and molluscs, nor is it clear whether they regulate the coagulation, fibrinolytic, or complement proteinases of advanced animals. It is now generally accepted that the most intensively studied αM, human α_2-macroglobulin, does not usually participate in regulating these three proteolytic systems. Rather, this is accomplished by faster-acting, more specific proteinase inhibitors such as antithrombin III, α_2-antiplasmin, and C1 inhibitor.

Purification of αMs is accomplished by a combination of standard chromatographic techniques, often incorporating a gel-filtration step to take advantage of the large size of the proteins. We usually assay fractions using the metalloproteinase thermolysin, because most biological fluids contain no other inhibitors of this proteinase. For those cases when a particular αM does not inhibit thermolysin, for example, rat α_1-inhibitor 3, fractions are assayed by using the serine proteinase pig pancreatic elastase. Those wishing to purify αMs are directed to the following procedures: Human pregnancy zone protein,[13] bovine α_2-macroglobulin,[14] rat

[11] I. B. Thøgersen, G. Salvesen, F. H. Brucato, S. V. Pizzo, and J. J. Enghild, *Biochem. J.* **285**, 521 (1992).

[12] J. J. Enghild, I. B. Thøgersen, G. Salvesen, G. H. Fey, N. L. Figler, S. L. Gonias, and S. V. Pizzo, *Biochemistry* **29**, 10070 (1990).

[13] O. Sand, J. Folkersen, J. G. Westergaard, and L. Sottrup-Jensen, *J. Biol. Chem.* **260**, 15723 (1985).

[14] J. J. Enghild, I. B. Thøgersen, P. A. Roche, and S. V. Pizzo, *Biochemistry* **28**, 1406 (1989).

α_1-macroglobulin,[15] rat α_2-macroglobulin,[15] rat α_1-inhibitor 3,[8] chicken plasma α-macroglobulin,[16] chicken ovastatin,[16] and *L. polyphemus* α-macroglobulin.[12]

Storage of Purified αMs

Slow hydrolysis of the thiol ester of many αMs, leading to inactivation and thiol-disulfide multimerization, occurs during prolonged storage. The rate of hydrolysis is increased with pH, so we usually store αMs at 5–10 mg/ml in 50 mM sodium phosphate buffer, pH 7, 150 mM NaCl. A previously uncharacterized αM should be stored at 0° until test portions have been frozen and thawed without loss of activity. Repeated freeze thawing inactivates αMs, so samples are stored in small portions (0.1–1 ml) at −20°, in a non-frost-free freezer if available. Most αMs retain full activity for several months under these conditions, but it is advisable to check activity when a new portion is thawed for use.

Polyacrylamide Gel Electrophoresis of α-Macroglobulins

Two types of PAGE are frequently used to analyze αMs: nondenaturing pore-limit PAGE and sodium dodecyl sulfate (SDS) PAGE. Both techniques give information about the purity, subunit composition, and proteinase-binding characteristics of αMs, and several publications have provided detailed descriptions of their uses.[8,12,17]

Pore-Limit PAGE

This type of PAGE separates proteins according to their radius of gyration and is especially useful for large globular proteins. The gel system we use is based on that described by Manwell[18] and consists of a continuously recirculating Tris/boric acid/EDTA buffer with pH 9.1. Most αMs have an isoelectric point well below this pH and move toward the anode in the system. To prepare 500 ml of a 10× concentrated stock solution of Tris/boric acid/EDTA (TBE) buffer, dissolve 54 g of Tris base and 27.5 g of boric acid in 480 ml and add 20 ml of a 0.5 M EDTA solution,

[15] K. Lonberg-Holm, D. L. Reed, R. C. Roberts, R. R. Hebert, M. C. Hillman, and R. M. Kutney, *J. Biol. Chem.* **262**, 438 (1987).
[16] H. Nagase and E. D. Harris, *J. Biol. Chem.* **258**, 7490 (1983).
[17] A. J. Barrett, M. A. Brown, and C. A. Sayers, *Biochem. J.* **181**, 401 (1979).
[18] C. Manwell, *Biochem. J.* **165**, 487 (1977).

pH 8 (EDTA titrated to pH 8 with NaOH). The final pH of the buffer is not adjusted.

We cast eight gels (1.5 mm × 10 cm × 10 cm) in a casette using a commercially available double-syringe gradient former and casting tank. The gels are cast as a linear gradient of 4–20% in acrylamide, with a methylenebisacrylamide : acrylamide ratio of 1 : 40, and stored in closed containers containing water-saturated tissue paper for up to 1 month at 4°. The gels are run with continuous buffer recirculation from the lower to upper reservoirs for 16–24 hr at 100 V constant voltage, at which time the proteins have reached an acrylamide concentration beyond which they cannot penetrate further.

Preparation of Samples. Pore-limit PAGE is the most convenient way of observing S-to-F transitions on proteinase reaction or methylamine incorporation. It is also used to determine subunit compositions because monomeric, dimeric, and tetrameric molecules are readily visualized (see Fig. 2). The samples may contain 4 M urea or 10 mM dithiothreitol without any major loss in resolution. Pore-limit PAGE is a measure of the size, not mass, of globular proteins, so care should be exercised when interpreting results. Reduced SDS–PAGE is a much more reliable way of determining the mass of individual subunits and subunit derivatives.

SDS–PAGE

Most likely any discontinuous stacking SDS–PAGE gel system is acceptable for analyzing αMs. We use the 2-amino-2-methyl-1,3-propanediol/HCl/glycine system described by Bury[19] because it resolves the bait region fragments better than Tris-based systems. Gels containing a gradient of acrylamide are particularly useful when analyzing αM–proteinase interactions because they allow the relatively large αM subunits and the relatively small proteinase chains to be visualized on the same gel.

Linear gradient gels containing 5% acrylamide at the top and 15% acrylamide at the bottom are cast as described above, using appropriate buffer and acrylamide stock components. Gels are stored as described above. A well-forming, or stacking, gel is added just before the sample is applied.

Preparation of Samples. αMs that contain a thiol ester are susceptible to autolytic cleavage, the rate of which increases as the pH of the SDS sample buffer and the temperature of sample preparation increase. Nor-

[19] A. F. Bury, *J. Chromatogr.* **213**, 419 (1981).

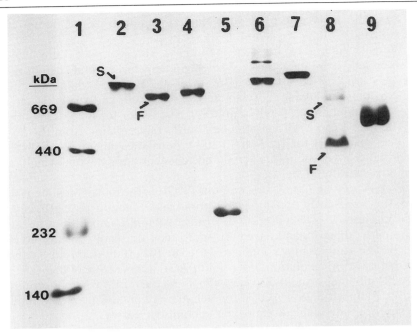

FIG. 2. Pore-limit PAGE of various αMs. A 4–20% pore-limit gel was cast as described in the text and used to analyze 10–15 μg of various αMs. We have included the gel to give an idea of the heterogeneity in subunit composition and electrophoretic characteristics of some αMs that we have worked on. The gel lanes contain the following components: (1) A standard mixture of thyroglobulin (669 kDa), ferritin (440 kDa), catalase (232 kDa), and lactate dehydrogenase (140 kDa); (2) human α_2-macroglobulin; (3) human α_2-macroglobulin saturated with pig pancreatic elastase; (4) rat α_1-macroglobulin; (5) rat α_1-inhibitor 3; (6) chicken plasma α-macroglobulin; (7) ovostatin; (8) limac; (9) *Octopus vulgaris* α-macroglobulin. Lanes 2–4, 6, and 7 contain tetrameric αMs, lanes 8 and 9 contain dimeric ones, and lane 5 contains a monomeric one. Note the S-to-F transition of human α_2-macroglobulin following saturation with a proteinase (lanes 2 and 3). The preparation of limac contained a significant proportion of F form, and note the large change in migration distance of the S and F forms, compared with human α_2-macroglobulin. The large aggregates of chicken plasma α-macroglobulin result from prolonged storage (about 2 years at −20° in this sample), whereas the smeary band of *O. vulgaris* α-macroglobulin results from extensive carbohydrate heterogeneity.

mally, αM samples are boiled for 5 min in sample buffer containing 1% SDS and 10 mM dithiothreitol. If the generation of autolytic fragments (see Fig. 1) is a problem, this may be overcome by treating the samples for 30 min at 37° in a neutral buffer containing 1% SDS and 10 mM dithiothreitol. Sample buffer is then added to increase the pH to the desired value.

TABLE I
TREATMENT OF SAMPLES TO PREVENT PROTEOLYTIC ARTIFACTS

Proteinase class to be inhibited	Inhibitor[a]	Stock solution[b]
Serine	3,4-Dichloroisocoumarin	2 mM in dimethyl sulfoxide
Serine	Phenylmethylsulfonyl fluoride	20 M in 2-propanol
Metalloproteinase	EDTA	0.1 M in water
Cysteine	E-64	0.4 mM in water

[a] Reagents (available from Sigma or Boehringer, among others) are added when a particular proteinase has been allowed to react with a sample of αM, or when αM-associated proteolysis is a problem. The stocks are diluted 20-fold into the sample and allowed to inhibit proteolytic activity for 15 min before addition of SDS. It is very important to ensure that the concentration of organic solvent does not exceed 10% of the sample volume because the solvent will interfere with SDS binding and resolution of protein chains in the gel.

[b] The stock solutions are stable for 6 months at $-20°$, except for the EDTA, which is stable at room temperature indefinitely. The frozen stocks should not be refrozen once thawed.

Unfortunately, the 37° treatment results in more proteolytic artifacts than does the high-temperature treatment, so it is important to pretreat samples with proteinase inhibitors (small enough to gain access to trapped proteinases) before adding SDS.[20] Table I outlines methods for decreasing artifactual proteolysis before analyzing samples by SDS–PAGE.

Determination of Concentration of Purified αMs

Purified human α_2-macroglobulin is quantitated based on spectrophotometric analysis using $A_{1\,cm,\,1\%,\,280\,nm} = 9$. This gives the total concentration of the protein. Determination of the active concentration (S form) of human α_2-macroglobulin, or αMs whose extinction coefficient is unknown, can be carried out by titration with trypsin and observing inhibition using hide powder azure as substrate. Alternatively, the active concentration of several αMs can be determined by saturation with trypsin and observing the appearance of thiol groups. The latter procedure relies on the cleavage of internal thiol esters during the S-to-F change (see section on thiol ester).

[20] G. Salvesen, and H. Nagase, in "Proteolytic Enzymes: A Practical Approach" (R. Beynon and J. Bond, eds.), p. 83. IRL Press, Oxford, UK, 1989.

Titration with Trypsin

Reagents

Assay buffer: 0.1 M Tris/HCl, pH 8, 10 mM CaCl$_2$, 0.05% Triton X-100

Trypsin solution: TPCK-treated bovine trypsin (Sigma) is active site titrated[21] and dissolved to a concentration of 10^{-5} M (about 0.3 mg/ml) in 1 mM HCl

Substrate suspension: Hide powder azure is an insoluble particulate protein mixture that has been dyed blue. It is available from several sources, though we find that material from Sigma gives consistent results. It is suspended at 12.5 mg/ml in assay buffer containing 0.6 M sucrose. Store at 4° and use within 2 days

Stopping reagent: 3 M sodium formate/formic acid buffer, pH 3

Procedure. Dissolve αM to a concentration of approximately 10^{-7} M

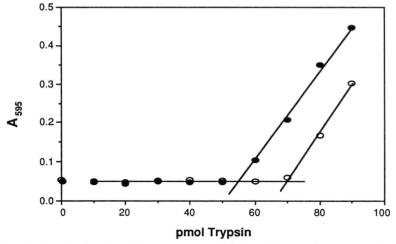

FIG. 3. Trypsin titration of human α_2-macroglobulin. The procedure was carried out exactly as described in the text (●), or by a variation in which trypsin was added in a single portion to the human α_2-macroglobulin sample (○). The absorbance at 595 nm (y axis) was plotted against the amount of trypsin added (x axis); the intersection of the lines gives the amount of trypsin required to saturate the inhibitory capacity of the 50 pmol sample of αM. When trypsin was added slowly, by the normal procedure, the value was 54 pmol trypsin, giving a calculated activity for the αM of 104%. When trypsin was added rapidly, the value was 70 pmol trypsin, giving a calculated activity of 140%. The latter value reflects the ability of human α_2-macroglobulin to bind up to two molecules of trypsin per molecule of αM. We use this example to emphasize the importance of adding the trypsin slowly when accurate calculations of the relative activity of an αM samples are required.

in assay buffer. Place 0.5-ml portions (50 pmol) in each of 10 1.5-ml microfuge tubes. Dilute trypsin in assay buffer to give equal increments in the range 10^{-7} to 9×10^{-7} M. Over a period of 1 min, with continuous mixing, add 100 μl from each trypsin solution to each portion of αM, as follows. Take up the 100 μl into a pipette tip and allow a drop (about 15 μl) to fall into the αM solution, cap the tube and vortex briefly, add another drop, and repeat until all the trypsin is added. Repeat this procedure for each portion of αM. Slow addition of trypsin favors the formation of equimolar complexes (Fig. 3). Add 100 μl of assay buffer to the last tube to serve as a control blank. Incubate at room temperature for 15 min.

Add 0.4 ml of substrate suspension using a blue pipette tip with 5 mm cut from the end. Cap the tubes and incubate with sufficient agitation to keep the substrate in suspension. We use a test tube rack taped to a rotating table, with tubes in a horizontal position. Terminate the reaction after the solution in the tube containing the most trypsin is visibly blue (30–60 min at 23°), by adding 0.2 ml of stopping reagent.

Pellet the suspensions by centrifugation for 5 min at 12,000–15,000 rpm in a microcentrifuge. Determine the $A_{595\,nm}$ of the supernatants and plot these values against trypsin concentration to determine the equivalence point, enabling calculation of the active αM concentration (Fig. 3).

Thiol Determination

Reagents

Assay buffer: 0.1 M Tris/HCl, pH 8

Trypsin solution: TPCK-treated bovine trypsin (Sigma) is dissolved to a concentration of 10^{-4} M (about 3 mg/ml) in 1 mM HCl. The trypsin need not be active site titrated because it will only be used to saturate the αM

Substrate stock: DTNB is dissolved to 10 mM in dimethylformamide

Procedure. About 1 nmol of αM in 0.9 ml of assay buffer is added to a 1-ml, 1-cm light path spectrophotometer cuvette. DTNB (100 μl of stock) is added and mixed by inverting the cuvette four times. The spectrophotometer is zeroed on this solution at 410 nm. Trypsin (50 μl of stock) is added to the cuvette and mixed by inverting the cuvette four times. The observed change in absorbance at 410 nm, caused by cleavage of thiol esters, is noted once a stable readout is achieved. This is normally within seconds of trypsin addition.

The concentration of DTNB-reactive thiols is calculated by dividing the absorbance at 410 nm by 14,150 (the molar extinction coefficient of

TNB^{2-}).[22] The calculated value is equivalent to the molar concentration of subunits of active αM in the cuvette. The value is divided by four (for tetrameric αMs) or by two (for dimeric αMs) to give the concentration of active molecules.

Identification in Crude Mixtures

The presence of αMs in samples of interest can be documented by exploiting their ability to bind proteinases in such a way that activity against small substrates is maintained. A sample is incubated with a proteinase after which an active site-directed proteinase inhibitor is added in severalfold excess over the proteinase. Finally, a small synthetic substrate is added to measure residual proteinase activity. The proteinase is usually trypsin and the inhibitor is usually soybean trypsin inhibitor or human α_1-proteinase inhibitor. Because soybean trypsin inhibitor (21 kDa) gains access to proteinases trapped by several αMs, we prefer to use α_1-proteinase inhibitor (52 kDa) to inhibit excess trypsin. The presence of αM activity in the sample is demonstrated by the retention of tryptic activity in the presence of the active site-directed inhibitor, which results from the ability of most αMs to shield trypsin from reaction with large proteins.[23] This method is useful for demonstrating the presence of αMs in samples, but the variability in the degree of shielding of αM-bound trypsin from the inhibitors limits its appropriateness in αM quantitations.

Crude samples that contain αMs often contain inhibitors of serine and cysteine proteinases. Most αMs inhibit the metalloproteinase thermolysin, but we know of no other inhibitors of this enzyme in animals. Consequently, the first choice in quantitating the active concentration of total αMs in biological fluids is to measure their inhibition of thermolysin. This procedure will not measure rat α_1-inhibitor 3 because this αM does not inhibit thermolysin.

Titration with Thermolysin

Reagents

Assay buffer: 0.1 M Tris/HCl, pH 8, 10 mM CaCl$_2$, 0.05% Triton X-100

Thermolysin solution: Dissolve thermolysin (protease type X, Sigma) in assay buffer and dilute to a working concentration of 10^{-7} M (about 3 μg/ml). The solution is stable for several hours at 4°

---

[21] T. Chase, Jr. and E. Shaw, *Biochem. Biophys. Res. Commun.* **29**, 508 (1967).

[22] P. W. Riddles, R. L. Blakeley, and B. Zerner, this series, Vol. 91, p. 49.

[23] P. O. Ganrot, *Clin. Chim. Acta* **14**, 493 (1966).

Substrate suspension: Hide powder azure as previously described; it is
    suspended at 12.5 mg/ml in assay buffer containing 0.6 M sucrose.
    Store at 4° and use within 2 days
Stopping reagent: 0.5 M EDTA, pH 8
    Procedure. Mix 0.3-ml portions of thermolysin solution with 0.3-ml
portions of twofold serial dilutions of sample containing αM. Incubate for
15 min at 23°. Incubate with substrate suspension as described previously
in "titration with trypsin."
    Pellet the suspensions by centrifugation for 5 min at 12,000–15,000
rpm in a microcentrifuge. Determine the $A_{595\,nm}$ of the supernatants
and plot these values against thermolysin concentration to determine
the equivalence point (Fig. 4), enabling calculation of the active αM concentration.

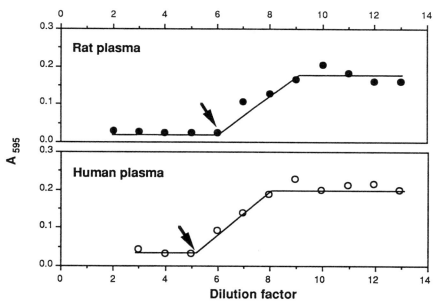

FIG. 4. Thermolysin titration of crude samples. In this example, samples of human
and rat blood plasma were titrated with thermolysin exactly as described in the text. The
absorbance at 595 nm (y axis) was plotted against the plasma dilution factor (x axis); the
point of inflection (arrows) gives the amount of total thermolysin inhibitory activity in each
sample according to the relationship: [thermolysin inhibitor] = [thermolysin] $\times$ $2^{(dilution\ factor)}$.
This gives $10^{-7} \times 2^{5.2} = 3.7\ \mu M$ for the human sample and $10^{-7} \times 2^{6} = 6.4\ \mu M$ for the rat
sample. On the assumption that the only inhibitors of thermolysin in animals are αMs, these
concentrations equate to total αM inhibitory activity in the samples. Unfortunately the αM
with the highest concentration in normal rat plasma, the monomeric α_1-inhibitor 3, does not
inhibit thermolysin, so it is not measurable by this technique.

Trypsin-Shielding Assay (Ganrot Assay)

The trypsin-shielding assay is sometimes known as the Ganrot assay after its originator.[23]

Reagents

Assay buffer: 0.1 M Tris/HCl, pH 8, 0.05% Triton X-100

Trypsin solution: TPCK-treated bovine trypsin (Sigma) is dissolved to a concentration of 10^{-4} M (about 3 mg/ml) in assay buffer. The trypsin need not be active site titrated because it will only be used to saturate the αM. The solution should be stored on ice and used within 1 day

α_1-Proteinase inhibitor stock solution (α_1PI): A stock solution of α_1PI is prepared at 5×10^{-4} M (about 25 mg/ml) in assay buffer. It is stable for several months when stored in portions at $-20°$, but it should not be thawed and refrozen

Substrate stock solution: N^α-Benzoyl-DL-arginine p-nitroanilide (Sigma) is dissolved in dimethyl sulfoxide to a concentration of 10 mM. The substrate stock is stable at 4° for 1 month

Procedure. Relatively large quantities of trypsin are used to ensure saturation of all trypsin inhibitors. Mix 0.4 ml of trypsin solution with 0.4 ml of sample containing αM in a 1-ml, 1-cm light path spectrophotometer cuvette. Incubate for 15 min at 23°. Add 0.4 ml of α_1PI solution and incubate for 5 min at 23°. Add 0.1 ml of substrate stock solution and determine the change in absorbance at 405 nm using a recording spectrophotometer. If a recording spectrophotometer is not available the reaction may be carried out in microcentrifuge tubes and stopped by adding 0.3 ml of 3 M formic acid/sodium formate buffer, pH 3, when the contents of the tube develop a definite yellow color.

Although we do not consider this to be a quantitative technique, it is possible to obtain a crude estimate of αM inhibitory capacity by comparing the absorbance reading to a standard curve constructed with serial trypsin dilutions.

Characterization of Thiol Esters

A putative intrachain β-Cys-Gly-Glu-γ-Glu thiol ester is thought to be present in most αMs (Fig. 1); the only well-characterized αM without a thiol ester is ovostatin from chicken egg white.[16] Evidence for the presence of a thiol ester is indirect but explains three characteristic reactions that αMs undergo, all of which result in the appearance of a β-cysteinyl SH group and a γ-substituted Glx residue. The three reactions are (1) direct reaction with small amine nucleophiles such as methylamine, (2) activation of the thiol ester by proteolytic cleavage of the bait region and subsequent reaction of the thiol ester with water or ε-amino groups of Lys residues on the proteinase surface, and (3) attack of the carbonyl of the ester by the peptide bond nitrogen, resulting in autolytic cleavage of the peptide chain on the N-terminal side of the Glx residue of the ester. All three reactions may be exploited in detection of the thiol ester.

Covalent Incorporation of [³H]Methylamine

Procedure. The purified αM (1 mg), or a crude mixture such as plasma, synovial fluid, or cell culture medium, is reacted with 10 μCi [³H]methylamine at pH 8.0; the pH of the αM-containing buffer may be adjusted by adding an appropriate volume of a 2 M Tris/HCl buffer, pH 8.3. The reaction is continued overnight at 23°. The following day the reaction is quenched by adding nonradioactive methylamine to 0.2 M and incubation is continued for 2 hr. Covalent incorporation of the radioactive methylamine is detected by examining the sample by reduced SDS–PAGE. Following electrophoresis the gel is stained, destained, and equilibrated with gentle shaking in water for 1 hr with two changes, followed by 1 hr of soaking in 1 M salicylic acid (Sigma). The gel is dried at 60° until just dry. Overdrying the gel, or drying above 60°, results in sublimation of the salicylate. The dried gel is allowed to expose X-ray film overnight at −70°. Though we prefer the salicylate fluorography procedure of Chamberlain[24] because of its convenience, other methods should work equally well. Methylamine will incorporate into αM subunits, so a radioactive band migrating with a size of about 180 kDa in reduced SDS–PAGE is indicative of a thiol ester-containing αM.

Covalent Incorporation of Pig Pancreatic Elastase

Procedure. Thiol ester-containing αMs usually, but not always,[12] covalently incorporate pig pancreatic elastase (Sigma). This enzyme is preferred to trypsin or chymotrypsin because it is much more readily inhibited prior to SDS–PAGE, and because it is a single chain proteinase, simplifying interpretation of results. The proteinase is labeled with ¹²⁵I to a specific activity of about 10⁶ counts per minute (cpm)/μg; most radiolabeling procedures based on reaction with tyrosine side chains are acceptable, but procedures that result in modification of lysines (Bolton–Hunter) should be avoided. The purified αM (10 μg) is mixed with 0.01–0.1 μg of ¹²⁵I-labeled elastase for 5 min at 37° and is stopped by adding 3,4-dichloroisocoumarin[20] to a final concentration of 50 μM and incubating

[24] J. P. Chamberlain, *Anal. Biochem.* **98,** 132 (1979).

for 15 min at 37° before the SDS-containing sample buffer is added. Because pig pancreatic elastase is more resistant to denaturation than αMs, it is important to inactivate it with 3,4-dichloroisocoumarin before adding the SDS sample buffer. Proteinases will incorporate into αM subunits only when they have been cleaved in the bait region, so a radioactive band migrating with a size of about 120 kDa in reduced SDS–PAGE is indicative of a thiol ester-containing αM. In crude samples that may contain other inhibitors of elastase, ^{125}I-labeled thermolysin is substituted for elastase. EDTA is added to 50 mM final concentration to inhibit the proteinase before addition of SDS sample buffer. Commercial thermolysin displays several bands in SDS page, in the range 20–50 kDa, so control samples with no αM sample should be included. Otherwise, interpretation is as for elastase.

Autolytic Fragmentation

Procedure. The pH of the purified αM solution is raised to pH 9 by titration with a 1.0 M solution of 2-amino-2-methyl-1,3-propanediol. The sample is made 5 mM in dithiothreitol (DTT), 1% in SDS, and boiled for 10 min. A portion (10 μg) is analyzed by reduced SDS–PAGE. Make sure that the final salt concentration is not above 0.2 M before the SDS–PAGE analysis. Three bands should appear on the gel: a band at approximately 180, representing the intact subunit, and the characteristic 120- and 60-kDa autolytic cleavage fragments. In the case of αMs that contain posttranslational proteolytic modifications, the size of the autolytic fragments may vary, but should add up to the size of the parent molecule. The autolytic cleavage, because it depends on the presence of an intact thiol ester, is usually prevented by pretreatment of αMs with methylamine, which may serve as a control.

Inhibition of Covalent Linking

The importance of the covalent linking reaction of αMs in inhibiting proteinases can be determined by allowing reaction in the presence of the α-effect nucleophile β-aminopropionitrile. This compound has no short-term effect on the activity of αMs, but competes efficiently with proteinases for nucleophilic substitution into activated thiol esters.[25] The procedure is split into two experiments: the first measures the decrease in covalent linking using radioiodinated proteinases, the second measures any change in inhibition of unlabeled proteinases.

[25] G. S. Salvesen, C. A. Sayers, and A. J. Barrett, *Biochem. J.* **195**, 453 (1981).

Reagents. A stock solution of β-aminopropionitrile is prepared by dissolving the monofumarate salt (Sigma) to a concentration of 100 mM in water. The pH is raised to pH 8.0 by addition of NaOH. Proteinases are radiolabeled with ^{125}I as described earlier.

Procedure. A sample of αM is mixed with an equal volume of β-aminopropionitrile stock, a radioiodinated proteinase is added, and incubation is allowed to proceed for 30 min at 25°. We use pig pancreatic elastase, bovine trypsin, and thermolysin; the αM must be in at least twofold molar excess over the proteinase. Samples allowed to react in the presence of β-aminopropionitrile, and control samples reacted in the absence of the nucleophile, are treated with 3,4-dichloroisocoumarin or EDTA and are analyzed by SDS–PAGE as described earlier. The gel should also contain samples of 3,4-dichloroisocoumarin- or EDTA-treated proteinases so that the migration of proteinases or contaminants within the proteinase preparations can be checked. Gels are stained, destained, dried, and autoradiographed. The radioactive bands are cut from the gel and counted in a gamma counter. The degree of covalent linking is determined by dividing the αM-associated radioactive counts by the total counts in each gel lane. β-Aminopropionitrile at a concentration of 50 mM does not abolish covalent linking, but does decrease it significantly to a degree dependent on the αM used.

The effect of decreasing the covalent linkage on inhibition of proteinases is assessed by performing hide powder azure assays following proteinase titration in the presence of 50 mM β-aminopropionitrile. This technique has enabled us to show that inhibition of proteinases by tetrameric human $α_2$-macroglobulin is not dependent on covalent linking,[23] but that inhibition by monomeric rat $α_1$-inhibitor 3 is totally dependent on the linking reaction.[8]

Inactivation of α-Macroglobulins

The presence of αMs in samples often interferes with the detection and analysis of proteinases; the following method is thus used when a proteinase must be isolated or measured from a solution containing αMs. The method is based on the ability of methylamine to react with thiol esters and cause a loss of proteinase-binding potential. The method does not inactivate bovine $α_2$-macroglobulin or chicken ovostatin, but is particularly useful when used as a first step during the isolation of active proteinases from animal plasmas.

Procedure. Add solid methylamine hydrochloride to the sample to a final concentration of 0.2 M, and 2 M Tris/HCl buffer, pH 8.5 to raise the pH to 8.3. For human $α_2$-macroglobulin incubate for 4 hr at 23°. Other

PIR (protein)

Pir1:Mahu	α2-macroglobulin precursor - Human 1,474aa
Pir3:Ps0078	α2-macroglobulin pseudogene product - Human (fragment) 150aa
Pir3:S09107	α2-macroglobulin - Human (fragment) 75aa
Pir2:A26122	α2-macroglobulin precursor - Rat 1,472aa
Pir2:A05278	α2-macroglobulin - Rat (fragments) 157aa
Pir2:A22614	α2-macroglobulin - Rat (fragment) 114aa
Pir2:S03431	α2-macroglobulin - Rat (fragment) 164aa
Pir2:A26124	α1-macroglobulin - Rat (fragment) 20aa
Pir2:B20872	α2-macroglobulin - Chicken (fragment) 14aa
Pir2:A20872	Ovostatin - Chicken (fragment) 13aa
Pir3:A33715	Ovostatin - Chicken (fragment) 34aa
Pir2:S00121	Ovostatin - Duck (fragment) 32aa
Pir3:S00150	Ovostatin - Duck (fragment) 14aa
Pir2:A29481	α2-macroglobulin homolog - American lobster (fragments) 38aa
Pir3:A32977	α2-macroglobulin - Crayfish (fragment) 34aa
Pir3:A36260	αmacroglobulin complement homolog - Atlantic horseshoe crab (fragment) 57aa

GenBank (nucleic acid)

Gb_pr:Huma2m	M11313 Human α2-macroglobulin mRNA, complete cds. 6/89 4,577bp
Gb_pr:Huma2mgl	M36501 Human α2-macroglobulin mRNA, 3' end. 9/90 2,041bp
Gb_pr:Hummga2ps	M24415 Human α2-macroglobulin pseudogene, partial cds. 6/90 3,003bp

Gb_pr:Humxt00239	M62177 Human expressed sequence tag (EST00239 to α_2-macroglobulin). 7/91 337bp
Gb_ro:Rata2m	J02635 Rat liver α_2-macroglobulin mRNA, cds.3/90 4,595bp
Gb_ro:Rata2m1	M11792 Rat α_2-macroglobulin (α_2M) mRNA near 5' end, partial cds. 9/88 150bp
Gb_ro:Rata2m2	M11793 Rat α_2-macroglobulin (α_2M) mRNA at HindIII site, partial cds. 9/88 321bp
Gb_ro:Rata2mac1	M23567 Rat α_2-macroglobulin gene, 5' end. 9/89 7,763bp
Gb_ro:Rata2mac2	M23566 Rat α_2-macroglobulin gene, 3' end. 9/89 250bp
Gb_ro:Rata2mg1	X13983 Rat α_2-macroglobulin gene exon 1. 9/90 1,525bp
Gb_ro:Rata2mg2	X13984 Rat α_2-macroglobulin gene exon 2. 9/90 1,056bp
Gb_ro:Rata2mg3	X13985 Rat α_2-macroglobulin gene exon 3 and 4. 9/90 774bp
Gb_ro:Ratmgaa21	M22667 Rat α_2-macroglobulin gene, exon 1. 3/90 1,270bp
Gb_ro:Ratmgaa22	M22668 Rat α_2-macroglobulin gene, exon 2. 3/90 277bp
Gb_ro:Ratmgaa23	M22669 Rat α_2-macroglobulin gene, exons 3 and 4. 3/90427bp
Gb_ro:Ratmgaa24	M22670 Rat α_2-macroglobulin gene, exons 5 and 6. 3/90842bp
Gb_ro:Ratmgaa2a	M29481 Rat α_2-macroglobulin mRNA, partial cds. 3/90 348bp
Gb_ro:Rata1i4	X52984 Rat mRNA for α_1-inhibitor 3, variant 7/91 4,620bp
Gb_ro:Ratinh3a1a	M28297 Rat negative acute-phase protein α_1-inhibitor 3 mRNA, 3' end. 4/91 801bp
Gb_ro:Ratinh3a1b	M28298 Rat negative acute-phase protein α_1-inhibitor 3 mRNA, 3' end. 4/91 339bp
Gb_ro:Hamalinh3	M73991 Mesocricetus auratus α_1-inhibitor 3 mRNA. 7/911,963bp

Fig. 5. Entry codes and accession numbers for various αMs.

αMs may require prolonged incubation periods because they are less sensitive to inactivation by methylamine.[12]

Uses of α-Macroglobulins as Reagents for Proteinases

We do not often use αMs as additives to prevent adventitious proteolysis in sensitive samples for the following reasons: (1) proteinases in complex with αMs may retain significant activity against small or disordered proteins and (2) it is not possible to rule out the presence of important contaminants in αM preparations. We recommend other procedures for decreasing adventitious proteolysis.[20] Nevertheless, there are certain situations in which αMs, especially well-characterized ones with broad specificities, such as human α_2-macroglobulin, are useful in analyzing certain aspects of proteolysis.

Removal of Contaminating Proteinases. Immobilized bovine α_2-macroglobulin (Boehringer, Mannheim, Indianapolis, IN) is sometimes used to remove proteinases, but care should be taken that the ligand (which sometimes binds cationic proteins nonspecifically[1]) does not remove the activity of interest.

Identification of Proteinases in Mixtures. Human α_2-macroglobulin can be used to distinguish proteinases from their inactive zymogens: the following example was kindly provided by Dr. Hideaki Nagase, University of Kansas Medical Center, Kansas City, Kansas. Members of the matrix metalloproteinase superfamily undergo a stepwise series of proteolytic cleavages during their activation, with generation of intermediates that are not proteolytically active.[26] To discriminate between the various species, human α_2-macroglobulin is mixed with a partially activated sample, the mixture is analyzed by SDS–PAGE, and is compared to an equivalent sample lacking the αM. Because only active endopeptidases are trapped by human α_2-macroglobulin, the metalloproteinase intermediate bands that are markedly decreased in the αM-containing sample must be the fully active species, whereas bands that do not diminish are inactive intermediates.

This procedure can be adapted to identify proteinases present in other complex mixtures. For example, rat α_1-macroglobulin has been used to identify and allow sequence analysis of a poliovirus proteinase from crude infected cell extracts.[27]

---

[26] H. Nagase, Y. Ogata, K. Suzuki, J. J. Enghild, and G. Salvesen, *Biochem. Soc. Trans.* **19,** 715 (1991).

[27] B. D. Korant and K. Londberg-Holm, *in* "Proteinase Action" (P. Elodi, ed.), p. 435. Akadémiai Kiadó, Budapest, 1984.

Many active site-directed proteinase inhibitors are used as affinity adsorbents for proteinase purification. However, the lack of selectivity of αMs and the variable amount of covalent linking[28] of proteinases argue against their use as affinity adsorbents. Methods have yet to be developed to overcome these problems, although the use of chicken ovostatin (which does not covalently link proteinases) and the preincubation of samples with β-aminopropionitrile before addition of human α_2-macroglobulin (see section on "inhibition of covalent linking") are avenues worth pursuing.

Concluding Comments

Figure 5 presents entry codes and accession numbers for various α-macroglobulins. Release 29.0 of the National Biomedical Research Foundation Protein Identification Resource (PIR) and release 69.0 of GenBank nucleic acid sequence data base were scanned for αMs.[29] Figure 5 contains an edited version of the results of the search, and is presented as an aid to those investigators interested in protein and nucleic acid sequence information of αMs.

---

[28] G. S. Salvesen and A. J. Barrett, *Biochem. J.* **187,** 695 (1980).
[29] J. Devereux, P. Haeberli, and O. Smithies, *Nucleic Acids Res.* **12,** 387 (1984).

Section II

Fibrinolysis

[8] Streptokinase–Plasmin(ogen) Activator Assays

By James T. Radek, Donald J. Davidson,
and Francis J. Castellino

Introduction

Streptokinase (SK), a catabolic protein produced by the Lancefield group C strains of the β-hemolytic streptococci, forms a 1 : 1 (mole : mole) stoichiometric complex with either human plasminogen (HPg) or plasmin (HPm). These complexes are capable of activating plasminogen to plasmin.[1–3] At early reaction times, at least two species of streptokinase–plasminogen activator (designated SK–HPg* and SK–HPg') exist with differing sensitivities for inhibition by chloride ions and enhancement by fibrinogen.[4–6] These species appear sequentially, with SK–HPg* forming before SK–HPg'. The SK–HPg' is then converted to another activator complex consisting of equimolar levels of SK and HPm (SK–HPm).

To investigate the enzymatic properties of these complexes, methods are described to preform each of the activator species using differing incubation times and temperatures. Streptokinase activator species were formed with native plasminogen as well as fully functional recombinant wild type, and site-directed mutant plasminogens.[6,7] These mutants are activation cleavage-site-resistant variants of HPg that cannot be converted to HPm. Activator species are assayed for steady-state amidolytic and activator activities under distinct assay conditions. In addition, the activator species are tested for susceptibility to irreversible inhibition by the active-site titrants diisopropyl fluorophosphate (DFP) and N^{α}-p-tosyl-L-lysine chloromethyl ketone (TLCK).

Proteins

Native human plasminogen with glutamic acid at its amino-terminal ([Glu^1]HPg, affinity chromatography variant I) is purified from fresh

[1] G. Markus and W. C. Werkheiser, *J. Biol. Chem.* **239**, 2637 (1964).
[2] L. A. Schick and F. J. Castellino, *Biochemistry* **12**, 4315 (1973).
[3] F. F. Buck, B. C. W. Hummel, and E. C. DeRenzo, *J. Biol. Chem.* **243**, 3648 (1968).
[4] B. A. K. Chibber and F. J. Castellino, *J. Biol. Chem.* **261**, 5289 (1986).
[5] B. A. K. Chibber, J. P. Morris, and F. J. Castellino, *Biochemistry* **24**, 3429 (1985).
[6] B. A. K. Chibber, J. T. Radek, J. P. Morris, and F. J. Castellino, *Proc. Natl. Acad. Sci. U.S.A.* **83**, 1237 (1986).
[7] D. J. Davidson, D. L. Higgins, and F. J. Castellino, *Biochemistry* **29**, 3585 (1990).

plasma by a modification[8] of the procedure of Deutsch and Mertz.[9] Citrated human plasma (~1 liter) is cooled to 4° and made 10 mM in DFP and 5 mM in benzamidine. The plasma is passed through a glass funnel containing Whatman (Clifton, NJ) #1 filter paper to remove lipids and is mixed with 150 ml of Sepharose-4B-L-lysine[9] for 2 hr at 4°. The resin is washed, at room temperature, on a course-fritted glass funnel with 0.3 M sodium phosphate and 5 mM benzamidine (pH 8.0) at 175 ml/hr until the absorbancy of the eluate at 280 nm is less than 0.05. The resin is transferred to a glass column (2.5 × 60 cm) and the bound plasminogen is eluted, at 4°, by applying a buffer of 0.1 M sodium phosphate, 5 mM benzamidine, and 15 mM ε-aminocaproic acid, pH 8.0. The fractions containing plasminogen, as measured by absorbancy at 280 nm, are pooled and dialyzed against 2 × 2 liters of 0.1 M sodium phosphate, 5 mM benzamidine, pH 8.0 (4°). The dialyzed plasminogen is applied to a Sepharose-4B-L-lysine column (2.5 × 30 cm) and washed at 4° with 0.1 M sodium phosphate, 5 mM benzamidine, pH 8.0, until the absorbance at 280 nm of the eluate is less than 0.05. The two carbohydrate variants of plasminogen are released sequentially from the column by the application of a 500-ml linear gradient of 3–12 mM ε-aminocaproic acid in 0.1 M sodium phosphate, 5 mM benzamidine, pH 8.0 (75 ml/hr, 5-ml fractions). Affinity chromatography variant I plasminogen is pooled and dialyzed against 4 × 4 liters of distilled water without benzamidine, lyophilized, and stored at −20°. Bovine plasminogen (BPg) is purified from fresh bovine plasma in the same manner. Recombinant (r) wild-type (wt) [Glu^1]HPg and R^{561}E-r[Glu^1]HPg insect cell (i) expressed are purified as above from the culture media of *Spodoptera frugiperda* cells, which have been infected with the recombinant baculovirus, pAV6 containing the wt-ir[Glu^1]HPg or R^{561}E-ir[Glu^1]HPg cDNA.[10] The R^{561}S-[Glu^1]HPg (R^{561}S-crHPg) is similarly purified from the culture supernate of CHO (c) cells transfected in a similar manner to that previously described.[11] Site-directed mutagenesis of the cDNA for HPg is accomplished as described.[7] Deglycosylation of the R^{561}S-crHPg is accomplished in 10 mM sodium phosphate, pH 7.4, with treatment by glycopeptidase F (Boehringer Mannheim, Indianapolis, IN) at a concentration of 0.4 unit of enzyme/μg of R^{561}S-crHPg. The reaction is allowed to proceed for 24 hr at 37°. The protein is then separated from the liberated oligosaccarides by centrifugation in molecular weight cut-off 10,000 (Centricon-

[8] W. J. Brockway and F. J. Castellino, *Arch. Biochem. Biophys.* **151**, 194 (1972).

[9] D. G. Deutsch and E. T. Mertz, *Science* **170**, 1095 (1970).

[10] J. Whitefleet-Smith, E. Rosen, J. McLinden, V. A. Ploplis, M. J. Fraser, J. E. Tomlinson, J. W. McLean, and F. J. Castellino, *Arch. Biochem. Biophys.* **271**, 390 (1989).

[11] G. Urlaub and L. A. Chasin, *Proc. Natl. Acad. Sci. U.S.A.* **77**, 4216 (1980).

10) microconcentrator tubes (Amicon, Danvers, MA).[7] Streptokinase is purified from Kabikinase (AB Kabi, Stockholm, Sweden) as previously described by Castellino et al.[12]

Protein concentrations are determined spectrophotometrically using the following molecular sizes and extinction coefficients (1%, 1 cm, 280 nm): HPg, 94 kDa per mole and 17.0[13]; SK, 47 kDa per mole and 9.5.[14]

Expression Systems

Supercoiled plasmid DNAs used in the insect and mammalian cell transfections are purified using ultracentrifugation in a CsCl/ethidium bromide gradient (1.57 g/ml CsCl in 17 ml containing 10 mg of ethidium bromide). Samples are centrifuged for 7 hr at 15° in an L5 65 ultracentrifuge using a VTi 65.1 vertical rotor (Beckman, Palo Alto, CA). The DNA–ethidium bromide complexes are removed with a syringe and the ethidium bromide is extracted from the plasmid DNAs with an equal volume of CsCl-saturated 2-propanol. Plasmid DNAs are dialyzed against 1.0 mM Tris-HCl, 0.1 mM EDTA, pH 7.1, for 12 hr and stored at −70°.

A baculovirus vector-infected insect system is used to express wt-ir[Glu¹]HPg and R⁵⁶¹E-ir[Glu¹]HPg cDNAs.[15] The cDNA from ir[Glu¹]HPg or R⁵⁶¹E-ir[Glu¹]HPg is inserted adjacent to the polyhedrin promoter in the genome of the baculovirus, *Autographa californica* nuclear polyhedrosis virus, which is used to infect cultured cells of the farm armyworm, *Spodoptera frugiperda*. A *Bam*HI/*Nae*I fragment containing the entire HPg cDNA is excised from p199PN127.6 and inserted between the *Bam*HI and *Sma*I sites of the viral transfer vector, pAV6. The resulting plasmids contain the AcMNPV polyhedrin (PH) promoter linked to the appropriate wild-type or mutant HPg signal and mature coding sequence. *Spodoptera frugiperda* cells (3 × 10⁶), in Hink's medium (GIBCO, Grand Island, NY) plus 10% fetal bovine serum, are placed on a 60-mm culture dish. After 2 hr, the medium is removed and the cells are incubated in a 0.5-ml suspension containing wild-type DNA from *Autographa californica* (0.1 μg) plus the pAV6HPg transfer plasmid DNA (1 μg) containing the appropriate HPg cDNA coding sequence in 0.14 M NaCl, 5 mM KCl, 0.5 mM Na₂HPO₄, 0.1% dextrose, and 20 mM HEPES–NaOH, pH 7.2. After incubation overnight, the DNA solutions are removed and the cells are incubated in Hink's medium plus 10% fetal bovine serum for 5 days. Under

---

[12] F. J. Castellino, J. M. Sodetz, W. J. Brockway, and G. E. Siefring, Jr., this series, Vol. 45, p. 244.

[13] B. N. Violand and F. J. Castellino, *J. Biol. Chem.* **251**, 3906 (1976).

[14] W. J. Brockway and F. J. Castellino, *Biochemistry* **13**, 2063 (1974).

[15] B. G. Cosaro and M. J. Fraser, *J. Tissue Cult. Methods* **12**, 7 (1988).

the conditions of cell growth employed, recombinant HPg is secreted into the medium after 24 hr postinfection, at which point no antigen remains inside the cells. Beyond 48 hr, the protein undergoes substantial proteolysis.

DG-44 cells (CHO), which do not contain the dihydrofolate reductase gene (dhfr$^-$), are plated at 3×10^5 cells per 60-mm dish overnight in Ham's F12 medium (GIBCO).[11] The cDNA of the R^{561}S-rHPg is inserted into the pSVL$_{neo}$ plasmid (Promega, Madison, WI) by ligation of the KnpI- and NaeI-cleaved R^{561}S-rHPg fragment with the KnpI and SmaI cleavage sites of the plasmid. Twenty micrograms of the pSVLneoR^{561}S-rHPg plasmid and 100 ng of a dihydrofolate reductase minigene are diluted in 1.0 ml of HEPES-buffered saline (21 mM HEPES, 0.14 M NaCl, 5 mM KCl, 0.7 mM Na$_2$HPO$_4$, 6 mM dextrose, pH 7.1). Filter-sterilized 2 M CaCl$_2$ is added to a final concentration of 0.125 M and the solution is incubated at room temperature for 20–30 min. The growth medium is removed from the DG-44 cells and 0.5 ml of the DNA suspension is added for 20 min at room temperature. The DNA solution is removed and 5 ml of Dulbecco's modified Eagle's medium (DMEM)/Ham's F12/10% fetal bovine serum is added to the plates and incubated for 4–5 hr at 37°. Cells are then washed once with normal medium and a solution of 20% glycerol in PBS (v/v) is added and incubated for 5 min. The glycerol solution is removed and the cells are washed twice in normal medium and then grown for 3 days in normal medium. Cells are then passaged (1 : 10) and placed in α-MEM (containing no added nucleosides; Sigma, St. Louis, MO) with 10% dialyzed fetal bovine serum. Confluent, transfected DG-44 cells are split 1 : 15 into α-MEM medium containing 10% dialyzed fetal calf serum. Cells are fed after 5 days and incubated for 5 days. Colonies are then cloned into 96-well plates containing 0, 0.04, and 0.08 μM of the folate antagonist methotrexate. Medium from confluent wells is assayed for the presence of the mutant plasminogen by enzyme-linked immunosorbent assay (ELISA). This step amplification/cloning procedure is repeated up to 80 μM methotrexate. Positive clones, at 80 μM methotrexate, are split into 6-well plates and on reaching confluency are further split into 25 cm^2 flasks.

Amidolytic Assays of Stoichiometric Complexes of SK–HPg* and SK–HPg'

Preincubations of SK and HPg are carried out at 4° in a low-ionic-strength buffer containing 10 mM HEPES–NaOH, pH 7.4. Stock solutions of each protein, dissolved to about 1 mg/ml under the above conditions, are used. Equimolar amounts of SK and HPg (final concentration, 100 nM) are added in a total volume of 1 ml. Aliquots are removed at 2.5 min for SK–HPg* and 30 min for SK–HPg' and assayed for their levels of amidase activity. The activator species are shown to be plasmin free by reduced sodium dodecyl sulfate polyacrylamide gel electrophoresis (SDS–PAGE).[6,16]

The chromogenic substrate H-D-Val-L-Leu-L-Lys-p-nitroanilide (S2251; Helena Laboratories, Beaumont, TX) is used to determine the amidolytic steady-state kinetic constants of the SK–HPg complexes. All assays are performed at 4° in a buffer containing 10 mM HEPES–NaOH, 0.15 M sodium acetate, pH 7.4. Each assay mixture has measured volumes of substrate, buffer, added salts, and appropriate enzyme (added last), such that the final volume is 0.8 ml. Substrate concentrations are 0.0625 to 0.5 mM. Hydrolysis of S2251 is monitored continuously at 405 nm. Reaction rates are calculated from the data using an extinction coefficient for the concentration of p-nitroanilide (1 M, 1 cm, 405 nm) of 10,000.[17] The K_m and k_{cat} values are then determined from Lineweaver–Burk plots. Where possible, the enzyme active site concentration is measured by titration with p-nitrophenyl p'-guanidinobenzoate.[18]

The K_i values for chloride ions (Cl⁻) are determined for SK–HPg* and SK–HPg' using amidolytic assays to which are added various concentrations (0–0.15 M) of NaCl. In all cases, the total ionic strength due to salt is kept constant at 0.15 by addition of the appropriate amount of sodium acetate, the anion component that does not inhibit the enzymes assayed.[4] The inhibition constants are determined from a Dixon plot of the data.

Assay for Irreversible Inhibition of SK–HPg* and SK–HPg' Activator Species by DFP and TLCK

The streptokinase activator species are preformed by incubating equimolar concentrations (100 nM) of SK and HPg for 2.5 min (SK–HPg*) or 30 min (SK–HPg') in 10 mM HEPES–NaOH, pH 7.4, at 4°. Enzyme solutions are adjusted to 0.1 M in sodium acetate by addition of the appropriate amount of a stock solution of 100 mM HEPES–NaOH, 1 M sodium acetate, pH 7.4. This mixture is then assayed for its amidolytic activity by placing 0.08 ml in a spectrophotometer cuvette containing 10 mM HEPES–NaOH, 0.5 mM S2251, pH 7.4, at 4° (total buffer volume, 0.8 ml). The rates of amidolytic activity of the formed enzymes are measured at 405 nm wavelength. The inhibitors (DFP or TLCK) are then

[16] U. K. Laemmli, *Nature (London)* **227**, 680 (1970).
[17] R. C. Wohl, L. Summaria, and K. C. Robbins, *J. Biol. Chem.* **255**, 2005 (1980).
[18] T. Chase, Jr., and E. Shaw, *Biochemistry* **8**, 2212 (1969).

added in varying concentrations (0–30 mM for DFP and 0–40 mM for TLCK) and incubated at 4°. At varying time points, 0.08 ml of the inhibitor–enzyme solutions is removed and assayed in the same buffer, and the rate of residual amidolytic activity is determined. The pH of each inhibitor–enzyme sample is checked periodically to ensure that no large change in pH occurs as a result of nonspecific hydrolysis of inhibitors. In addition, a separate sample is set up in which the pH of an unbuffered solution containing sodium acetate is monitored to determine the rate of nonspecific hydrolysis of DFP and TLCK. The treatment of the kinetic data obtained from the titration of an enzyme with an inhibitor that forms a dead-end acyl-enzyme has been discussed previously.[19,20]

Amidolytic Assays of Stoichiometric Complexes of SK–HPg and SK–HPm

The SK–HPg complex forms as a result of preincubation of 1 : 1 (mole : mole) stoichiometric amounts of SK and HPg for 30 sec at 25°. The SK–HPm complexes are generated by incubation under identical conditions for 3 min. The complexes are analyzed for their plasmin content by SDS–PAGE under reducing conditions.[16]

A quantity of 0.2 ml of a buffer consisting of 100 mM HEPES–NaOH, 10 mM ε-aminocaproic acid,[21] pH 7.4, is placed in a spectrophotometer cuvette maintained at 25°. The desired concentration of S2251 is added followed by the required amount of H_2O. The preformed activator species is added at this time (final concentration: 6–10 nM). The rate of hydrolysis of the substrate is recorded for 2–5 min at 405 nm. The absorbancies are converted to initial rates of amidolysis using the extinction coefficient for p-nitroanilide (1 M, 1 cm, 405 nm) of 10,000. The data are analyzed on Lineweaver–Burk plots.

Plasminogen Activator Assays of SK–HPg and SK–HPm Complexes

The SK–HPg and SK–HPm activator complexes are preformed under the identical incubation conditions described above for the amidolytic assays.

---

[19] M. L. Bender, M. L. Begué-Cantón, R. L. Blakely, L. J. Brubacher, J. Feder, C. R. Gunter, F. J. Kézdy, J. V. Killheffer, Jr., T. H. Marshall, C. G. Miller, R. W. Roeske, and J. K. Stoops, J. Am. Chem. Soc. **88**, 5890 (1966).

[20] M. L. Bender and J. K. Stoops, J. Am. Chem. Soc. **87**, 1622 (1965).

[21] The ε-aminocaproic acid was present in the buffers in order to provide for maximal activation rates and eliminate any differences in those rates due to variable amounts of [Glu¹]HPg and [Lys⁷⁸]HPg, a proteolytically derived form of HPg, in the complexes of the assay mixtures.

TABLE I
STEADY-STATE KINETIC CONSTANTS FOR AMIDOLYTIC ACTIVITY TOWARD
S2251 OF SK–HPg* AND SK–HPg' ACTIVATOR SPECIES

Activator species	K_m (mM)	k_{cat} (min^{-1})	k_{cat}/K_m (mM min^{-1})
SK–HPg*[b]	0.52 ± 0.06	90 ± 11	173
SK–HPg'[c]	0.23 ± 0.02	130 ± 9	565

Inhibitor of SK–HPg'	K_i (mM)	k_2 (min^{-1})	k_2/K_i (mM min^{-1})
DFP[d]	11 ± 2	0.036 ± 0.005	3.3×10^{-3}
TLCK[e]	100 ± 12	0.040 ± 0.007	4.0×10^{-4}

[a] At 4°; in sodium acetate buffer.

[b] Human plasma plasminogen. The SK–HPg* species was formed by preincubation of SK and HPg for 2.5 min at 4°.

[c] Human plasma plasminogen. The SK–HPg' species was formed by preincubation of SK and HPg for 30 min at 4°.

[d] The SK–HPg' species was formed by preincubation of equimolar concentrations of SK and HPg for 30 min at 4°. Inhibition was carried out at varying concentrations of DFP in 10 mM HEPES–NaOH, 0.1 M sodium acetate, pH 7.4, at 4°. Aliquots (0.08 ml) were removed with time and assayed for their remaining amidolytic activity.

[e] The SK–HPg' species was formed by preincubation of equimolar concentrations of SK and HPg for 30 min at 4°. Inhibition was carried out at varying concentrations of TLCK in 10 mM HEPES–NaOH, 0.1 M sodium acetate, pH 7.4, at 4°. Aliquots (0.08 ml) were removed with time and assayed for their remaining amidolytic activity.

An amount of 0.2 ml of 100 mM HEPES–NaOH, 10 mM ε-aminocaproic acid, pH 7.4, is added to a spectrophotometer cuvette, maintained at 25°. After this, 0.08 ml of S2251 (final concentration, 0.5 mM) is added, followed by various concentrations of BPg (0.2–10 μM), and, last, the SK–HPg or SK–HPm activator complex (final concentration, 0.2 nM). The rate of BPg activation is monitored in continuous assay by recording the increasing rate of release of p-nitroanilide from the S2251 by the bovine plasmin generated as a result of activation. The data are analyzed by Lineweaver–Burk plots as previously published by Urano et al.[22] BPg has been shown to be insensitive to activation by SK alone.

Results

Kinetic constants for the amidase activities of the SK–HPg* and SK–HPg' activator species are listed in Table I. Each activator species

[22] T. Urano, V. Sator de Serrano, B. A. K. Chibber, and F. J. Castellino, *J. Biol. Chem.* **262**, 15959 (1987).

possesses distinct kinetic constants under the conditions employed. The SK–HPG′ species is a considerably more effective amidase than is SK–HPg*, primarily as a result of a decrease in the K_m value. The k_i for Cl⁻ with SK–HPg* is much lower than the same K_i with SK–HPg′ (8.5 ± 1.0 versus 250 ± 20 mM). Thus, at a Cl⁻ concentration of 0.15 M, the activity of SK–HPg* is approximately 6% of the maximal and that of SK–HPg′ is approximately 40% of maximal.

Figure 1 contains the first-order plots for the inhibition of SK–HPg* by 30 mM DFP and 40 mM TLCK. The rate of residual activity for the SK–HPg* activator species was not constant. A distinct recovery of activity was found at each subsequent time point where residual activity was measured. This phenomenon was found at all concentrations of DFP and TLCK. The enzyme was irreversibly inhibited only minimally (<20%) over the course of over 100 min even at the highest concentrations of inhibitors. Because the extent of irreversible inhibition was not easily measured under the experimental conditions, kinetic constants for SK–HPg* could not be determined. Rates of residual activity with inhibition of the SK–HPg′ species by DFP and TLCK, however, were constant

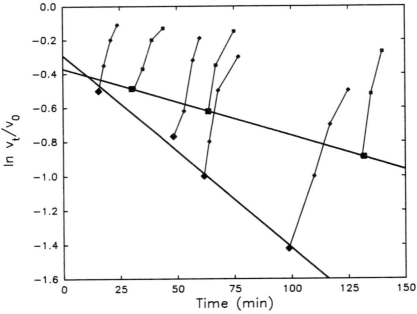

FIG. 1. First-order plots of the inhibition of the SK–HPg* activator species by 30 mM DFP (◆) and 40 mM TLCK (■) in 10 mM HEPES–NaOH, 0.1 M sodium acetate, pH 7.4. Smaller diamonds (DFP) and squares (TLCK) represent the recovery of activity with time.

TABLE II
STEADY-STATE KINETIC CONSTANTS FOR AMIDOLYTIC ACTIVITY TOWARD
S2251 OF EQUIMOLAR COMPLEXES OF SK WITH PLASMINOGENS AND PLASMINS

Activator species	K_m (mM)	k_{cat} (min^{-1})	k_{cat}/K_m (mM min^{-1})
SK–HPm[b]	0.35 ± 0.05	312 ± 16	891
SK–wt-irHPm	0.33 ± 0.04	312 ± 12	945
SK–R^{561}E-irHPg	0.17 ± 0.05	272 ± 18	1600
SK–R^{561}S-crHPg	0.28 ± 0.06	368 ± 20	1314
SK–R^{561}S-crHPgΔCHO[c]	0.50 ± 0.06	377 ± 18	754

[a] At 25°.
[b] Human plasma plasmin.
[c] HPg variant deglycosylated at Asn − 289.

with time and showed no recovery. Thus, inhibition constants were determined and are also listed in Table I.

The steady-state amidolytic activities of complexes of SK with wild-type and variant plasmin(ogen)s are listed in Table II and compared with that of SK and native human plasmin(ogen). In the case of the plasmin-generated enzymes, SDS–PAGE analysis showed that they were composed of SK and [Lys^{78}]HPm.[23] The data show that the SK–HPm complexes, containing either human plasma plasmin or wt-irHPm, have virtually identical steady-state kinetic constants toward S2251. Similarly, the complexes containing the variant plasminogens displayed very similar kinetic constants. Comparison of the kinetic values between the SK–HPg and SK–HPm complexes suggests that only small differences exist in the amidolytic steady-state properties.

Table III provides steady-state parameters that reflect the abilities of the various SK–HPg and SK–HPm complexes to activate plasminogen. BPg was chosen as the activator substrate due to its insensitivity to activation by SK alone. The complex containing wt-irHPm is considerably more active than that containing plasma HPm, primarily due to an increase in the k_{cat}. It is equally apparent that all complexes consisting primarily of HPg were significantly more effective activator enzymes than were their HPm counterparts. This was due mostly to decreases in the K_m values for the SK–HPg activator complexes.

The mammalian cell-expressed variant HPg (R^{561}S-crHPg) demonstrated similar, but not identical, steady-state kinetic parameters toward activation of BPg as the insect cell-expressed variant HPg. The major

[23] [Lys^{78}]HPm arises from the proteolytic cleavage of the activation site Arg^{561}-Val^{562}, and also at Lys^{77}-Lys^{78}.

TABLE III

STEADY-STATE KINETIC CONSTANTS FOR PLASMINOGEN ACTIVATOR ACTIVITY
TOWARD BOVINE PLASMINOGEN OF EQUIMOLAR COMPLEXES OF SK WITH
PLASMINOGENS AND PLASMINS

Activator species	K_m (μM)	k_{cat} (min^{-1})	k_{cat}/K_m (μM min^{-1})
SK–HPg[b]	1.72 ± 0.06	1.79 ± 0.07	1.04
SK–HPm[c]	7.00 ± 1.14	0.61 ± 0.09	0.09
SK–wt-irHPg[d]	0.98 ± 0.12	4.14 ± 0.52	4.22
SK–wt-irHPm[e]	9.80 ± 1.06	2.28 ± 0.23	0.32
SK–R^{561}E-irHPg	0.72 ± 0.07	7.00 ± 0.84	9.72
SK–R^{561}S-crHPg	0.98 ± 0.10	3.54 ± 0.36	3.61
SK–R^{561}S-crHPgΔCHO[f]	0.49 ± 0.11	8.52 ± 1.00	17.4

[a] At 25°.

[b] Human plasma plasminogen. This complex was formed from the incubation
of SK and HPg for 30 sec. Estimates from reduced SDS–PAGE show that
the relative percentage of complexes containing HPg is 80% with the remaining 20% composed of HPm.

[c] Human plasma plasmin. This complex was formed from the incubation of
SK and HPg for 3 min.

[d] Insect cell-expressed wild-type human plasminogen. This complex was
formed as a result of preincubation of SK with wt-irHPg for 30 sec. Estimates
from SDS–PAGE demonstrated that 80% of the complexes had wt-irHPg
and 20% had wt-irHPm.

[e] Insect cell-expressed wild-type human plasmin. This complex was formed
from the preincubation of SK and wt-irHPg for 3 min. Reduced SDS–PAGE
showed that 100% of the complexes contained wt-irHPm.

[f] Equimolar complex of SK and R^{561}S-crHPg, deglycosylated at Asn-289.

difference was a twofold higher k_{cat} with the insect-expressed variant
in the complex as compared to the mammalian cell-expressed variant.
Deglycosylating the insect cell-expressed rHPg produced a protein that,
when complexed to SK, was nearly five times more efficient (2.4-fold
lower K_m and 2-fold higher k_{cat}) as a BPg activator (Table III). These
differences were not revealed in the amidolytic activities of these enzyme
complexes (Table II).

Discussion

The methods described for the preformations of the activator species
are useful because they (1) are simple to perform, (2) include all the
currently known activator species, and (3) can be used in conjunction
with other analyses not described here.

The studies on the earliest forming activator species demonstrate that
there exist at least two forms of the activator (SK–HPg* and SK–HPg′)

that have significantly distinct amidolytic activities and inhibition profiles to classic serine protease active-site titrants (Fig. 1 and Table I). In addition, the amidolytic activity of SK–HPg* is inhibited by Cl^- to a much greater extent than is SK–HPg'. This phenomenon suggests that some interesting reciprocal points of control might exist for the activation of HPg by SK. Rapid active site generation in plasma would be inhibited by the presence of physiological levels of sodium chloride. However, this situation appears to be reversed when fibrinogen is present.[4,5] Thus, fibrinogen and Cl^- may play opposite roles in regulating activation of HPg by SK.

The studies performed on the early (SK–HPg) and the late-forming species (SK–HPm) clearly suggest significantly greater potency of the former over the latter for activating BPg. This is verified by studying mutant forms of HPg that cannot be converted to HPm in the activator complex.

Variations in glycosylation, resulting from expression of cDNA in different cell systems, have also been implicated to play a role in influencing the kinetic properties of activation. Those plasminogen variants expressed in the nonmammalian system and the insect cell-expressed variant lacking its sugar residues showed even greater activator activity than did native and mammalian cell-expressed variants.

The contribution that each of the earlier forming activator species (SK–HPg* and SK–HPg') makes to the kinetic properties of the SK–HPg complex is not known. This is in part due to the difficulty in studying SK–HPg* and SK–HPg' at temperatures higher than 4°.[6] However, the stabilized SK–HPg complex performs its activator function significantly more efficiently than does SK–HPm. These results may be of great utility in the study of structure–function relationships with HPg and may be of use in designing better reagents for thrombolytic therapy.

[9] Activation of Human Plasminogen by Recombinant Staphylokinase

By THANG TRIEU,* DETLEV BEHNKE, DIETER GERLACH, and JORDAN TANG

Introduction

Staphylokinase is an extracellular protein produced by certain strains of *Staphylococcus aureus*. It has been known since 1948 that staphylokinase is an activator of human plasminogen and can cause the lysis of blood clots.[1] Although many laboratories have studied this protein, only limited information is available about its properties and its mode of human plasminogen activation. Staphylokinase has been purified by several investigators[1–4] and studied for molecular properties. However, the molecular structure of this protein was greatly clarified only after the cloning and sequencing of staphylokinase genes.[5–7] From these works it is clear that staphylokinase contains 136 amino acids and is 15.4 kDa in size.

Evidence indicates that staphylokinase is not a protease, but it activates human plasminogen by forming a complex to plasminogen, as in the case of plasminogen activation by streptokinase.[8] It is well known that streptokinase forms a stoichiometric tightly binding, but noncovalent, complex with human plasminogen. This complex is a plasminogen activator that converts excess plasminogen to plasmin. The hypothesis that staphylokinase activates human plasminogen by a similar mechanism is implied by the fact that it possesses no measurable proteolytic activity.[1] The treatment of staphylokinase with serine protease inhibitors, such as diisopropyl fluorophosphate (DFP), phenylmethylsulfonyl fluoride (PMSF), or *p*-nitrophenyl *p*'-guanidinobenzoate (NPGB), did not diminish its activity for plasminogen activation.[4] Additionally, the kinetics of sta-

* Deceased.

[1] C. H. Lack and K. L. A. Glanville, this series, Vol. 19, p. 706.
[2] M. Nick, J. Bruckler, W. Schaeg, K. D. Hasche, and H. Blobel, *Zentralbl. Bakteriol., Parasitenkd., Infektionskr. Hyg., Abt. I: Orig., Reihe A* **237**, 160 (1977).
[3] S. Arvidson, R. Ericksson, T. Holme, R. Mollby, T. Wadstron, and O. Vesterberg, *Contrib. Microbiol. Immunol.* **1**, 406 (1973).
[4] K. W. Jackson, N. Esmon, and J. Tang, this series, Vol. 80, p. 387.
[5] T. Sako and N. Tsuchida, *Nucleic Acids Res.* **11**, 7679 (1983).
[6] T. Sako, S. Sawaki, T. Sakurai, S. Ito, Y. Yoshizawa, and I. Kondo, *Mol. Gen. Genet.* **190**, 271 (1983).
[7] D. Behnke and D. Gerlach, *Mol. Gen. Genet.* **210**, 528 (1987).
[8] F. J. Castellino, *Trends Biochem. Sci.* (Pers. Ed.) **4**, 1 (1979).

phylokinase activation of plasminogen is consistent with the mechanism of binding activation.[9] However, there is no direct evidence that demonstrates the binding of staphylokinase to plasminogen. Moreover, the detailed mechanism of Glu-plasminogen conversion to plasmin by staphylokinase is not well understood. The absence of detailed information on staphylokinase is mainly due to the lack of purified protein for such studies. However, the successful expression of the staphylokinase gene in other bacteria, especially *Bacillus subtilis*,[7] and purification in reasonable quantity have enabled us to carry out detailed experiments in staphylokinase activation of and interaction with plasminogen. These results are described in this chapter.

Experimental Procedures

Materials

Staphylokinase is purified from the supernatant of a late log-phase culture of *Bacillus subtilis* GB500 carrying the recombinant plasmid pDB15 according to the method previously described.[7,10] Human Glu-plasminogen, a gift from Dr. Kenneth G. Jackson, is purified from human plasma using lysine-sepharose chromatography as described.[11] Human plasmin was a gift from Dr. Fletcher B. Taylor, Oklahoma Medical Research Foundation. Plasmin chromogenic substrate, D-Val-Leu-Lys-*p*-itroanilide (Kabi S2251), is obtained from Kabi (Stockholm, Sweden). Other reagents used are the highest grade obtained commercially.

The following stock solutions and reagents are prepared and stored at $-20°$ for use in all experiments: Kabi S2251, 30 mg/ml of 1.0 M NaCl; Glu-plasminogen, 2 mg/ml in 0.04 M Tris-HCl, pH 9.0, 0.08 M NaCl, 0.003 M EDTA, and 25% glycerol; human plasmin, 0.456 mg/ml in 0.001 M HCl and 50% glycerol; diluted staphylokinase and staphylokinase-10, stored in 50 mM Tris-HCl, pH 7.5, 1% bovine serum albumin (BSA). Purified staphylokinase and staphylokinase-10 are stored in 20 mM Tris, pH 8.0, as concentrated solutions.

Assay for Human Plasminogen Activation

Human plasminogen activation is assayed using the two-stage assay procedure of Radcliff[12] with minor modifications. In the first stage, each

---

[9] B. Kowalska-Loth and K. Zakrzewski, *Acta Biochim. Pol.* **22**, 327 (1975).
[10] D. Gerlach, R. Kraft, and D. Behnke, *Zentralbl. Bakteriol., Mikrobiol. Hyg., Ser. A* **269**, 314 (1988).
[11] K. C. Robins and L. Summaria, this series, Vol. 45, p. 257.
[12] R. Radcliff and T. Heinze, *in* "The Regulation of Coagulation" (K. G. Mann and F. B. Taylor, eds.), p. 551. Elsevier, New York, 1980.

assay sample consists of 10 μg of plasminogen, 0.5 mg of bovine serum albumin, and an appropriate amount of activator in 50 μl of 0.05 M Tris-HCl, pH 7.5. (The amount of activator is usually in the range of 1 to 20 ng. The activation amount outside of this range can be assayed by adjusting the incubation time.) After incubation at 37° for 5 min, 28 μl of 1 M NaCl is added to stop the further activation of plasminogen. For the second stage, 60 μg of Kabi S2251 is added in 2 μl of water, and the solution is incubated at 37° for 10 min. At the end of incubation, the tube is moved to an ice bath and 0.9 ml of 5% (v/v) acetic acid is immediately added. The solutions in the tubes are transferred to a cuvette and the absorbance at 405 nm is determined. A separate blank without activator is included in each assay.

Polyacrylamide Gel Electrophoresis and Agarose–Casein Overlay Method for Locating Staphylokinase and Its Complex in Gel

In a series of tubes placed in an ice bath, plasminogen is mixed with different amounts of staphylokinase in 10 μl of 50 mM Tris-Cl, pH 7.5, with 1% bovine serum albumin. The final concentration of plasminogen is 1.48×10^{-11} M and the staphylokinase concentrations range from 1.06×10^{-11} M to 2.64×10^{-11} M. After incubating at 37° for 10 min, the tubes are again placed in ice bath. Electrophoresis sample buffer[13] (10 μl) is added to each solution. Alkaline native polyacrylamide gel (4%) electrophoresis is carried out as described.[13] The samples of staphylokinase or its complexes are applied to gel electrophoresis without mercaptoethanol and heating. After the electrophoresis, one of the glass plates is removed and the gel attached to the other plate is placed in a closed humidified container that is preequilibrated with water vapor at 50°. A solution consisting of 1% agarose, 1% skim milk, 50 mM Tris-HCl, pH 7.5, 0.15 N NaCl is prepared, boiled to completely dissolve agarose, and cooled to 50°. Glu-plasminogen (100 μg to every 30 ml of this solution) is then added followed by thorough mixing. This agarose solution is then poured over the polyacrylamide gel in the humidified container, and the container with the gel is incubated at 37°. The cleared zones, which result from the digestion of casein by plasmin, appear at different time of incubation (ranging from a few minutes to overnight), depending on the amount of activator present in the gel.

[13] F. M. Ausubel, R. Brent, R. E. Kingston, D. D. Moore, J. G. Seidman, J. A. Smith, and K. Struhl, eds., "Current Protocols in Molecular Biology," p. 10.2.5. Wiley, New York, 1987.

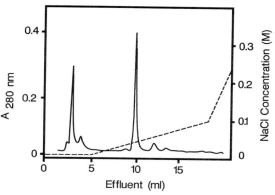

FIG. 1. Ion-exchange chromatographic separation of two structural versions of staphylokinase. Partial purified recombinant staphylokinase was chromatographed on a Mono Q column
in an FPLC instrument. The material eluted in the first peak was staphylokinase and that
in the second peak was staphylokinase-10 (staphylokinase without the amino-terminal 10
residues; see text). The solid line represents protein concentration and the dashed line
represents NaCl concentration.

Chromatographic Separation of Staphylokinase and Staphylokinase-10

A lyophilized preparation of partially purified recombinant staphylokinase[7] is dissolved in 20 mM Tris-HCl, pH 8.0, to a concentration of 1
mg/ml. An aliquot of 500 μl is chromatographed on a Mono Q ion-exchange
column in a Pharmacia (Piscataway, NJ) fast protein liquid chromatography (FPLC) instrument. The proteins are eluted with linear gradient, from
0 to 0.1 M of NaCl, in 20 mM Tris-HCl, pH 8.0. Staphylokinase is eluted
at the breakthrough position and staphylokinase-10 is eluted at 50 mM
NaCl (Fig. 1). Eluent from each of the staphylokinase peaks is collected,
concentrated with a YM10 (25 mm) membrane in an Amicon (Danvers,
MA) ultrafiltration apparatus, and dialyzed against 20 mM Tris-HCl, pH
8.0. The protein solutions are stored at −20° until use.

Determination of Protein Concentrations

The concentrations of human plasminogen, plasmin, staphylokinase,
and staphylokinase-10 used in the specific activity and stoichiometry studies are determined with a Bio-Rad (Richmond, CA) Protein Assay
Kit. The experimental conditions are those described by the manufacturer.

Amino-Terminal Sequence Determination

The amino-terminal sequences of two staphylokinase fractions obtained from HPLC are determined by automated Edman degradations in a Gas-phase Sequencer Applied Biosystem (Foster City, CA) Model 470A with an on-line HPLC Model 120A for PTH-amino acid identification.

Preparation of Human Lys-Plasminogen and Human Plasmin

Lys-plasminogen is prepared from human Glu-plasminogen essentially as described by Castellino and Powell.[14] Glu-plasminogen, 2 mg/ml of storage buffer, is incubated in 0.04 M L-lysine at 37° for 1 hr, then human plasmin is added at a weight ratio of 100 to 1 for Glu-plasminogen to plasmin. The solution is incubated at 37° and the conversion of Glu-plasminogen to Lys-plasminogen is monitored by a 7.5% SDS–polyacrylamide electrophoresis. The conversion is completed after 5 hr of incubation.

Results

Separation of Recombinant Staphylokinase Species

When the partially purified recombinant staphylokinase was subjected to amino-terminal sequence analysis, a mixture of two sequences was observed; these could be assigned to be SSSFDKGK and KGDDASYF on the basis of the known gene structure. The former was derived from the known amino-terminal sequence of staphylokinase,[5,7] and the latter was apparently generated from the staphylokinase without the amino-terminal 10 residues (staphylokinase-10). The protein was subjected to FPLC separation on an ion-exchange Mono Q column from which two forms of staphylokinase were separated (Fig. 1). The amino-terminal sequences from material obtained in these peaks confirmed the identities of the two staphylokinases. To assess the purity of the isolated material, the staphylokinases from FPLC were subjected to polyacrylamide electrophoresis. As shown in Fig. 2, each protein produced a single band in SDS–polyacrylamide electrophoresis and the activity of the gel bands was demonstrated directly by its plasminogen activation and caseinolysis in an overlay assay.

---

[14] F. J. Castellino and J. R. Powell, this series, Vol. 80, p. 365.

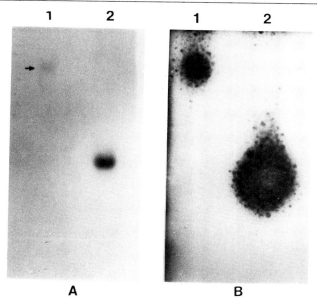

FIG. 2. Polyacrylamide gel electrophoresis of two structural versions of staphylokinase obtained from ion-exchange chromatography of Fig. 1. (A) Results of electrophoresis stained with Coomassie blue; (B) result of casein–agarose overlay assay on a separate electrophoresis. The dark zones resulted from the casein hydrolysis due to plasminogen activation by staphylokinases; lane 1, peak 1 of chromatography in Fig. 1 (staphylokinase); lane 2, peak 2 of chromatography in Fig. 1 (staphylokinase-10).

Activation of Human Plasminogen by Staphylokinase and Staphylokinase-10

Both staphylokinase and staphylokinase-10 activated human Glu-plasminogen with a distinct lag period (Fig. 3). The addition of free lysine to the system, at the beginning of the activation, prolonged the lag periods. Preincubation of free lysine with Glu-plasminogen produced the longest lag period (Fig. 3). When Glu-plasminogen was first converted to Lys-plasminogen then subjected to the activation of staphylokinases, a lag period was not observed for either staphylokinase (Fig. 4). These observations support the view that the conversion of Glu-plasminogen to Lys-plasminogen is an obligatory pathway for the staphylokinase activation and this conversion is retarded by the binding of free lysine to Glu-plasminogen.

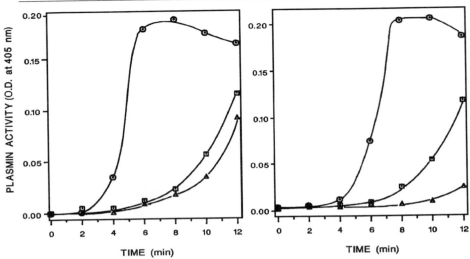

FIG. 3. Time course of human Glu-plasminogen activation by staphylokinase (left) and staphylokinase-10 (right). ○, No addition; □, 40 m*M* of L-lysine added at 0 time; △, Glu-plasminogen was preincubated in 40 m*M* L-lysine for 1 hr at 37° prior to the addition of activator.

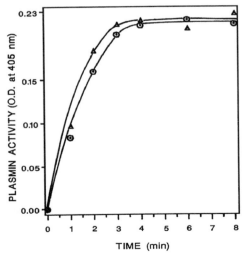

FIG. 4. Time course of human Lys-plasminogen activation by staphylokinase (△) and staphylokinase-10 (○). The activity caused by the small amount of plasmin added for conversion of Glu-plasminogen to Lys-plasminogen was subtracted. In order to achieve a better assay, the concentration of Kabi substrate was increased to twice that described in the text.

Specific Activities of Staphylokinase and Staphylokinase-10

The specific activities of staphylokinase and staphylokinase-10 were analyzed using both Glu-plasminogen and Lys-plasminogen as substrates. For Glu-plasminogen, the time–activity curves of human Glu-plasminogen activation by both staphylokinases were determined at different activator concentrations. Using the assay conditions described previously, 11.88 ng of staphylokinase and 3.78 ng of staphylokinase-10 produced nearly the same activation curves. This indicates that staphylokinase-10 (217.43 μg of plasmin/min/μg staphylokinase) has a specific activity about three times of that of staphylokinase (72.96 μg of plasminogen/min/μg staphylokinase-10). The specific activities of staphylokinases toward Lys-plasminogen were calculated from the initial velocity of time–activity curves of the two-stage assay because Lys-plasminogen activation by either staphylokinase produced no lag period. The specific activity of staphylokinase was 55.7 μg of plasmin/min/μg activator, whereas that for staphylokinase-10 was 144.0 μg plasmin/min/μg activator. The ratio of specific activities of staphylokinase-10 to staphylokinase was thus 2.6. We have determined the specific activities of staphylokinase-10 over the concentration range of 0.76 to 3.62 ng of activator per assay tube. The specific activities in this concentration range were essentially the same as reported above for both Glu-plasminogen and Lys-plasminogen.

Stoichiometry of Staphylokinase Binding to Human Plasmin and
  Human Glu-plasminogen

Previous kinetic studies have established that the activation of plasminogen by staphylokinase is accomplished by the formation of a staphylokinase–plasminogen complex,[9] an activation mechanism similar to that of human plasminogen activation by streptokinase.[8] Streptokinase forms a 1 : 1 complex with either plasminogen or plasmin. Because the molecular mass of staphylokinase (15,446 Da as calculated from a sequence) is considerably smaller than that of streptokinase (47,000 Da), the molar ratio of the proteins in the complex needs to be determined. Also, the binding between staphylokinase and plasminogen has not yet been directly demonstrated. We therefore devised experiments to determine whether staphylokinase binds to plasminogen and plasmin, and to determine their binding stoichiometry. Samples with a constant amount of plasmin (or plasminogen) and increasing amounts of staphylokinase-10 (14,298 Da) were subjected to polyacrylamide gel electrophoresis followed by casein–agarose overlay to reveal the positions of staphylokinase-10 and its complex. Figure 5 (lane 1) shows the free staphylokinase-10 position in electrophoresis. Human plasmin (Fig. 5, lane 2) is essentially devoid of plasminogen

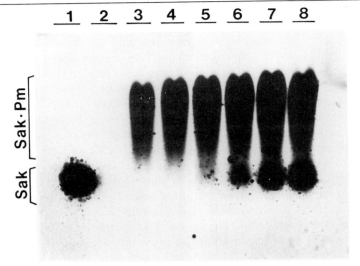

FIG. 5. Casein-agarose gel overlay assay of gel electrophoresis of the complex between staphylokinase-10 and human plasmin. In the samples, different amounts of staphylokinase-10 were mixed with 1.2 μg of human plasmin and the samples were incubated at 37° for 10 min to allow binding and subjected to electrophoresis in a 4% native alkaline polyacrylamide gel. Casein–agarose gel overlay was used to reveal the location of staphylokinase-10 (Sak) and its complex with plasmin (Sak.Pm). Lane 1, Staphylokinase-10; lanes 2–8, the molar ratios of staphylokinase-10 to plasmin were 0, 0.74, 0.93, 1.11, 1.30, 1.49, and 1.86, respectively.

activation activity and was not visible. Increasing amounts of staphylokinase-10 were added to the same amount of plasmin in lanes 3–8. An increase in the amounts of plasmin–staphylokinase-10 complex from lane 3 to lane 5 was visible. Lanes 6–8 contained increasing excess amounts of staphylokinase-10. Thus, the saturation point for plasmin binding to staphylokinase-10 was about lane 5, which had a molar ratio of staphylokinase-10 to human plasmin of 1.1 : 1.

Similar experiments were carried out using Glu-plasminogen instead of plasmin. A typical result is shown in Fig. 6, in which free staphylokinase-10 spots were clearly visible (lanes 1, 7, and 8). However, the complex of Glu-plasminogen and staphylokinase-10 was not clearly visible (Fig. 6, lanes 3–6). Only after overnight incubation of the casein–agarose overlay at 37° did the complex in lanes 3–6 appear as weak streaking bands at positions with slightly faster electrophoretic mobility than that of staphylokinase-10 (results not seen in Fig. 6). Why the Glu-plasminogen–staphylokinase-10 complex was relatively slow and insensitive in casein–gel overlay assay is uncertain. However, the results in Fig. 6 clearly indicated

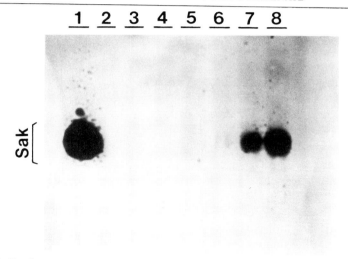

FIG. 6. Casein–agarose gel overlay assay of gel electrophoresis of the complex between staphylokinase-10 (Sak) and human Glu-plasminogen. The experimental conditions were as given in Fig. 5 except that Glu-plasminogen was used instead of plasmin. Lane 1, Staphyloki-nase-10; lanes 2–8, the molar ratios of staphylokinase-10 to Glu-plasminogen were 0, 0.15, 0.29, 0.57, 0.86, 1.15, and 1.40.

that the saturation point for Glu-plasminogen was between lanes 6 and 7, which had ratios of Glu-plasminogen to staphylokinase-10 of 0.86 : 1 and 1.15 : 1, respectively.

The results described above indicate that staphylokinase-10 forms a tightly binding complex with either human plasmin or Glu-plasminogen that remains associated during polyacrylamide gel electrophoresis. The molar stoichiometry of these complexes is 1 : 1 for both plasminogen and plasmin complexes. Both complexes are plasminogen activators based on their ability to activate the plasminogen in the casein–gel overlay, resulting in caseinolysis.

Discussion

Staphylokinase and staphylokinase-10 (without their amino-terminal 10 residues) are both capable of the activation of human plasminogen. From the point of view of structure and function, it is clear that the missing 10 residues at the amino terminus of staphylokinase are not essential for human plasminogen activation. The time course of human Glu-plasmino-gen activation by either of the staphylokinases showed a lag period in the appearance of plasmin activity (Fig. 3). This lag period disappeared when

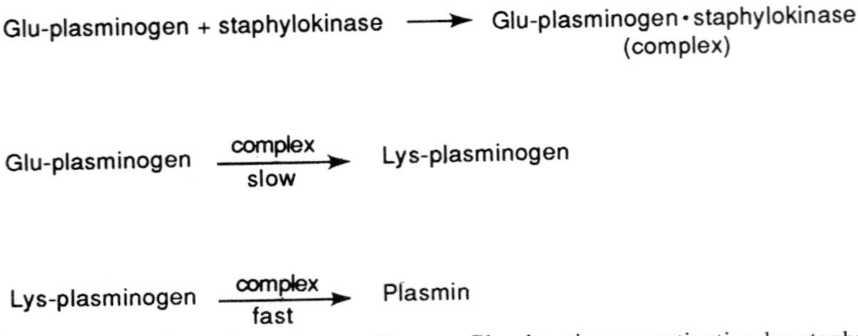

Glu-plasminogen + staphylokinase $\longrightarrow$ Glu-plasminogen · staphylokinase
(complex)

Glu-plasminogen $\xrightarrow[\text{slow}]{\text{complex}}$ Lys-plasminogen

Lys-plasminogen $\xrightarrow[\text{fast}]{\text{complex}}$ Plasmin

FIG. 7. Proposed reaction scheme of human Glu-plasminogen activation by staphylokinase or staphylokinase-10.

Glu-plasminogen was converted to Lys-plasminogen prior to activation by either of the staphylokinases (Fig. 4). The current results agree with our previous finding that native staphylokinase isolated from *Staphylococcus aureus* activated human Glu-plasminogen with a lag.[15] These observations further suggest that in the activation of Glu-plasminogen by either staphylokinase, the conversion from Glu-plasminogen to Lys-plasminogen is an obligatory path and is also the rate-limiting step (Fig. 7). This latter aspect was also confirmed by the specific activity calculations, respectively, for the activation of Glu-plasminogen and of Lys-plasminogen. The presence of lag periods can be explained as follows. During the activation of Glu-plasminogen by the staphylokinases, the initial plasmin production was limited by the slow conversion of Glu-plasminogen to Lys-plasminogen (Fig. 7). However, the plasmin produced was capable of converting Glu-plasminogen to Lys-plasminogen, which appears to be the main reason for the rate acceleration. Because free lysine increased the lag period, it is assumed that lysine retarded the conversion of Glu-plasminogen to Lys-plasminogen. The formation of Lys-plasminogen as an obligatory intermediate of staphylokinase activation of Glu-plasminogen distinguishes this mechanism from the Glu-plasminogen activation by either streptokinase or urokinase. In these cases, Lys-plasminogen is not an obligatory intermediate.[8]

Interestingly, staphylokinase-10 has about threefold greater specific activity than staphylokinase in the activation of both Glu- and Lys-plasminogens. These observations imply that the catalytic efficiency of Glu-plasminogen–staphylokinase-10 complex is about three times higher than that of Glu-plasminogen–staphylokinase complex. The higher specific activity of staphylokinase-10 provokes the possibility that the native staphy-

[15] K. W. Jackson, N. Esmon, and J. Tang, *in* "The Regulation of Coagulation" (K. G. Mann and F. B. Taylor, eds.), p. 515. Elsevier, New York, 1980.

lokinase is a precursor (pro) form that can be "activated" to a smaller activator, such as staphylokinase-10. However, at present there is no evidence to support such a hypothesis and it is more likely that the shorter staphylokinase results from proteolysis by extracellular proteases of *B. subtilis*.

The mechanism by which staphylokinase activates human plasminogen was studied by Kowalska-Loth and Zakrzewski,[9] who concluded that the two proteins formed a complex that was responsible for plasminogen activation. This is also the activation mechanism of streptokinase for human plasminogen.[8] The molar binding stoichiometries of staphylokinase-10 to Glu-plasminogen and to plasmin are clearly shown to be 1 : 1 (Figs. 5 and 6). Because the staphylokinase–plasmin complex remained intact in gel electrophoresis, this binding appears to be high affinity. We have also demonstrated that the complex that appeared on gel electrophoresis possessed activity for plasminogen activation and thus confirmed that the complexes are plasminogen activators. Due to the limitation in purified material, the stoichiometry of staphylokinase binding to plasminogen was not tested. However, because the kinetics of activation of two staphylokinases were very similar, it seems safe to assume that staphylokinase also forms a 1 : 1 molar complex with plasminogen or plasmin. It is interesting to note that the amino acid sequences of streptokinase[16] and staphylokinase[5] reveal no apparent sequence similarity. The similr activation mechanism of these two proteins would suggest that they share some similarity in either the tertiary structure or, at the minimum, the plasminogen-binding sites.

Acknowledgment

This work was supported by NIH Research Grants HL-32128 and HL-34367.

[16] K. W. Jackson and J. Tang, *Biochemistry* **21**, 6620 (1982).

[10] Expression of Human Plasminogen cDNA in Lepidopteran Insect Cells and Analysis of Asparagine-Linked Glycosylation Patterns of Recombinant Plasminogens

By Francis J. Castellino, Donald J. Davidson, Elliot Rosen, and James McLinden

Introduction

There has been an increasing number of reports detailing the use of lepidopteran insect cells to express heterologous cDNAs that have been inserted by homologous recombination into the genome of an infective baculovirus. The reasons for the popularity of this approach include (1) the high levels of expression that are possible when heterologous cDNA is linked to the late polyhedrin (PH) promoter,[1] (2) the ease of screening recombinant viruses[2] when insertion is such that the viral PH gene is strategically interrupted by the heterologous cDNA, and (3) the fact that many post- and cotranslational reactions, typical of higher order eukaryotic processing events that lead to mature proteins, have been found to occur in lepidopteran-based expression systems. These reactions include signal peptide cleavage, proteolysis, nuclear translocation, fatty acid acylation, phosphorylation, and N- and O-linked glycosylation.[3]

Our use of this system was governed by an inability to express the human plasminogen (HPg) cDNA in mammalian systems due to the nearly ubiquitous presence of plasminogen activators in such cells. These latter substances catalyze conversion of the recombinant (r) HPg to plasmin, which undergoes proteolysis. The invertebrate cells that we employ do not possess plasminogen activators, and good yields of r-HPg can be easily obtained using these invertebrate-based expression systems.[4] However, our continued and intensive use of these expression systems results from the highly interesting observations that we have made regarding the nature of N-linked glycosylation of r-HPg produced by certain lepidopteran insect cells, namely, *Spodoptera frugiperda* IPLB-SF-21AE ovary cells[5] and a

[1] Y. Matsura, R. D. Possee, H. J. Overton, and D. H. L. Bishop, *J. Gen. Virol.* **68**, 1233 (1987).

[2] M. J. Fraser and W. F. Hink, *J. Invertebr. Pathol.* **40**, 89 (1982).

[3] V. A. Luckow and M. D. Summers, *Bio/Technology* **6**, 47 (1988).

[4] J. Whitefleet-Smith, E. Rosen, J. McLinden, V. A. Ploplis, M. J. Fraser, J. E. Tomlinson, J. W. McLean, and F. J. Castellino, *Arch. Biochem. Biophys.* **271**, 390 (1989).

[5] J. L. Vaughn, R. H. Goodwin, G. J. Tompkins, and P. McCawley, *In Vitro* **13**, 213 (1977).

subclone (SF-9) of this cell line, *Manduca sexta* CM-1 embryonic tissue cells,[6] and *Mamestra brassicae* IZD-MBO503 cells.[7] The results of some of these investigations are summarized herein.

Construction of Transfer Vector

A vector (pAV6) is constructed that includes DNA *Autographa californica* nuclear polyhedrosis virus (AcMNPV) DNA sequences surrounding the PH gene. This plasmid contains AcMNPV sequences from an *Xho*I site 1.8 kb 5' of the PH gene, to a position 8 nucleotides 5' of the ATG initiation codon for PH. This plasmid also includes a 1.5-kb fragment extending from a *Kpn*I site within the PH gene to a *Bam*HI site 3' of this same gene. A multiple cloning site, containing recognition sequences for restriction endonucleases, *Bam*HI, *Sma*I, *Bst*EII, and *Kpn*I, has been inserted between viral sequences flanking the 5' and 3' regions of the PH gene.

The vector, pAV6, is constructed according to the following protocol. A 7.2-kb *Eco*RI fragment (p*Eco*RI-I) of *A. californica* DNA containing the PH gene and flanking sequences[8,9] is cut with the restriction endonucleases *Xho*I/*Bam*HI, and the *Xho*I/*Bam*HI fragment is ligated into the *Sal*I/*Bam*HI sites of plasmid mp19 (Bethesda Research Laboratories, Gaithersburg, MD), generating the new plasmid, mp19*Xho–Bam*. This latter plasmid is cut with *Eco*RV/*Kpn*I, and ligated to the synthetic oligonucleotide shown in Scheme I. This nucleotide replaces the AcMNPV sequence from an *Eco*RV site 5' of the PH gene to the *Bam*HI site within the gene. It also includes the AcMNPV sequences from the *Eco*RV site to the putative CAP site, and contains the indicated multiple cloning site. As a result of the ligation, a new plasmid, mp19AID, is produced.

Next, pUC12 (Bethesda Research Laboratories) is cut with *Hin*dIII/*Sst*I. Plasmid mp19AID is cut with *Hin*dIII/*Kpn*I, and a 1.8-kb fragment containing sequences flanking the 5' terminus of the PH gene is isolated. Additionally, p*Eco*I-I is cut with *Bam*HI/*Kpn*I, and a 1.5-kb fragment extending from the *Kpn*I site within the PH gene to a *Bam*HI site in the 3' flanking region of PH is isolated. These three fragments (pUC12, 1.8 kb, and 1.5 kb) are ligated along with the following oligonucleotide, 5'-GATCAGCT, which adapts a *Bam*HI site to an *Sst*I site, producing plasmid pAV1. pAV1 is excised with *Bam*HI/*Kpn*I, and the resulting fragment

[6] P. E. Eide, J. M. Caldwell, and E. P. Marks, *In Vitro* **11**, 395 (1975).
[7] S. A. Weiss, C. G. Smith, S. S. Kalter, and J. L. Vaughn, *In Vitro* **17**, (1981).
[8] G. E. Smith, J. M. Vlak, and M. D. Summers, *J. Virol.* **45**, 215 (1983).
[9] G. E. Smith, M. J. Fraser, and M. D. Summers, *J. Virol.* **46**, 584 (1983).

transcription initiation
—

5'-<u>GATATC</u>ATGGAGATAATTAAAAATGATAACCATCTCGCAAAGGATCCGAATTC

3'-<u>CTATAG</u>TACCTCTATTAATTTTACTATTGGTAGAGCGTTT<u>CCTAGGCTTAAG</u>

    EcoRV   *BamHI, EcoRI,*

<u>GTCGACGGTAC</u>-3'

<u>CAGCTGC</u>-5'

*SaI, Kpn*I

SCHEME I

       *BgI*I   *Sma*I

  *Xba*I   *Sst*I *Nru*I   *Bam*HI   *Bst*EII   *Kpn*I

5'-GATCTAGATCTGAGCTCGCGATGGATCCCGGGTAACCGGTAC-3'

3'-ATCATGACTCGAGCGCTACCTAGGGCCCATTGGC-5'

SCHEME II

is ligated to a synthetic oligonucleotide (Scheme II), resulting in plasmid pAV2.

A plasmid, pDS, is constructed by treating pBR322 (Bethesda Research Laboratories) with *Hin*dIII/*Sa*lI, filling with Klenow fragment, and ligating. This plasmid lacks the sequences between the *Hin*dIII and *Sa*lI restriction sites, has lost the *Sa*lI site, and retains the *Hin*dIII site. Plasmid pAV2 is cut with *Hin*dIII/*Bam*HI, and the 1.8-kb fragment, containing the 5' flanking sequence of PH is isolated. Another preparation of pAV2 is then cut with *Bam*HI/*Eco*RI and the 1.5-kb fragment, extending from the *Bam*HI site in the multiple cloning region to the *Eco*RI site adjacent to the 3' PH flanking sequence, is isolated. Plasmid pDS was treated with *Hin*dIII/*Eco*RI and ligated to the two fragments generated from pAV2, generating plasmid pAV3.

Next, a *Bam*HI/*Sa*lI fragment from vector pAC373[10] which contains AcMNPV DNA from a *Sa*lI site situated approximately 0.8 kb 5' of the PH gene to a *Bam*HI site (in its multiple cloning region) inserted at nucleotide −8 from the ATG initiation sequence of PH, is ligated to the *Bam*HI/*Sa*lI-cut pAV3. This provides plasmid pAV4.

[10] G. E. Smith, M. D. Summers, and M. F. Fraser, *Mol. Cell. Biol.* **3,** 2156 (1983).

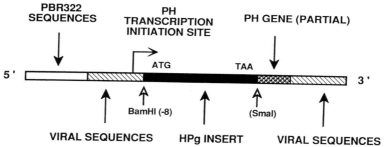

Fig. 1. Schematic diagram of pAV6/HPg. The vector contains the HPg gene inserted into the multiple cloning site of pAV6. The vector contains 1.8 kb of AcMNPV sequences 5' of the PH gene from a *Xho*I site (eliminated in the construction) to position −8 of the PH promoter (position 1 begins with the ATG initiation codon). The vector also includes 1.5 kb of AcMNPV sequences 3' of the PH gene from a *Kpn*I site in the PH cDNA to a *Bam*HI site 3' of the gene (eliminated in the construction). The ATG initiation codon and TAA termination codon for translation of HPg are indicated. The *Sma*I site was lost after insertion of HPg into the vector.

Plasmid pAV4 is treated with *Eco*RI, the ends filled with Klenow fragment, and ligated to produce plasmid pAV6. This latter plasmid lacks the *Eco*RI site of pAV4.

In order to insert the cDNA encoding HPg into pAV6, a *Bam*HI/*Nae*I fragment containing the entire cDNA for HPg (and portions of the plasmid vector, pUC119) is excised from p119PN127.6[11] and is inserted between the *Bam*HI and *Sma*I sites of the viral transfer vector, pAV6. The resulting plasmid, pAV6/HPg, contains the AcMNPV PH promoter linked to the HPg signal and mature HPg coding sequences. The HPg insert consists of (from 5' to 3') 20 bases from pUC119, 38 bases of linker and 5'-untranslated sequences, 57 bases of HPg signal sequence, 2373 nucleotides coding for the entire HPg mature sequence, 322 nucleotides of 3'-untranslated sequences plus the poly(A) sequence and a linker, and 367 bases from pUC119.

The essential features of pAV6/HPg are illustrated in Fig. 1.

Isolation of Recombinant Baculovirus

The AcTR temperature inactivation-resistant strain of AcMNPV provides the host viral DNA for recombinations. The uncloned *S. frugiperda*

[11] J. W. McLean, J. E. Tomlinson, W.-J. Kuang, D. L. Eaton, E. Y. Chen, G. M. Gless, A. M. Scanu, and R. M. Lawn, *Nature (London)* **330,** 132 (1987).

IPLB-SF-21AE cell line is employed as the host for virus growth and manipulations.[5] The procedures employed for culturing insect cells have been described.[4]

The insect cells (3×10^6), in Hink's medium[12] plus 10% fetal bovine serum (FBS), are allowed to adhere to a 60-mm culture dish. After 2 hr, the medium is removed and the cells are incubated in a 0.5-ml suspension of wild-type DNA from AcTR (0.1 μg), plus pAV6HPg transfer plasmid DNA (1–10 μg) in NaCl (0.8 g/liter)/KCl (0.37 g/liter)/Na$_2$HPO$_4 \cdot$2H$_2$O (0.125 g/liter)/dextrose (1 g/liter)/HEPES–NaOH (5 g/liter), pH 7.2. After incubation overnight, the DNA suspension is removed and the cells are incubated for 5 days in Hink's medium/10% FBS.

Progeny virus from the above cotransfections are examined in plaque assays.[13] Virus from occlusion body-negative plaques (OB$^-$), identified using a dissecting microscope, are purified by two additional rounds of plaque assays. Southern blots of DNA from one of the three viral recombinants (Pg3A) indicate that the HPg cDNA is inserted in the viral DNA.

To prepare viral DNAs, cells (3×10^6) are allowed to attach to 60-mm petri dishes and are then infected with three different plaque-purified OB$^-$ viruses at a multiplicity of infection (MOI) of 10. At 72 hr postinfection (pi), cells are resuspended by gentle pipetting and are pelleted at 3500 g, for 20 min at 16°. The cell pellet is suspended and lysed in 4 M guanidinium isothiocyanate/0.5% sarkosyl/5 mM sodium citrate/0.1 mM 2-mercaptoethanol. The resulting solution is extracted four times with phenol–chloroform, and the DNA is precipitated twice with absolute ethanol.

The DNA from three baculovirus–HPg recombinants and wild-type virus DNA and the original pAV6HPg transfer vector DNA are digested with *Eco*RI and subjected to Southern analysis.[14] The digested samples are separated on an 0.85% agarose gel and are electroblotted onto a nylon membrane using a Bio-Rad (Richmond, CA) transfer unit and the Bio-Rad protocol for nondenatured gels. The original HPg-containing plasmid (p119PN127.6) and pUC13 are labeled with ^{32}P by nick translation and are used as hybridization probes, as follows. Filters are incubated at 37° for 3 hr in prehybridization fluid,[15] washed with H$_2$O, and then incubated for 18 hr at 37° with prehybridization fluid containing denatured probe. The hybridization solution is removed and nonspecific radioactivity is

[12] W. F. Hink, *Nature (London)* **226**, 466 (1970).

[13] M. D. Summers and G. E. Smith, *Tex., Agric. Exp. St. [Bull.]* **1555** (1987).

[14] E. M. Southern, *J. Mol. Biol.* **98**, 503 (1975).

[15] T. Maniatis, E. F. Fritsch, and J. Sambrook, "Molecular Cloning: A Laboratory Manual." Cold Spring Harbor Lab., Cold Spring Harbor, NY, 1982.

eliminated by gently rocking the filter for 45 min at 37° in 10 mM EDTA/ 0.2 % (w/v) sodium dodecyl sulfate (SDS), pH 8.0. The wash solution is changed four times. Hybridized bands are visualized by autoradiography.

Expression, Purification, and Deglycosylation of Recombinant Human Plasminogen

Spodoptera frugiperda SF-9 cells are maintained as monolayers in serum-free Excell 400 medium (JRH Biosciences, Lenexa, KS) and are infected with recombinant baculoviruses at multiplicities of four plaque-forming units/cell. For the purpose of the studies to be summarized herein, the culture media are collected under different temporal conditions to examine glycosylation of r-HPg produced at different baculovirus infection times. We have previously demonstrated that virtually all of the r-HPg is secreted into the media at all infection times chosen.[4] Specifically, infection of the insect cells with the recombinant baculovirus is allowed to proceed for 20 hr, after which the culture medium is collected and replaced with fresh medium. This cell-conditioned medium is collected at 60 and/ or 72 hr, and again replaced with fresh medium, which is collected again at 96 hr. This procedure allows examination of glycosylation events in r-HPg that occurred between infection time windows of 0–20 hr, 20–60 hr (and/ 20–72 hr), and 60–96 hr. After the infection times indicated above, the r-HPg is purified from the cell-conditioned media by batch purification on lysine-Sepharose affinity chromatography columns.[16] The r-HPg obtained possesses the proper molecular weight and the appropriate amino-terminal amino acid sequence (which not only indicates correct translation but also shows that the signal polypeptide has been liberated); it reacts with a battery of monoclonal antibodies against plasma HPg, it reacts with the effector molecule (ε-aminocaproic acid), and is activated to plasmin by urokinase, streptokinase, and tissue-type plasminogen activator.[4,17] All of these observations demonstrate that a proper r-HPg has been produced by the invertebrate insect cell line.

For complete N-linked deglycosylation, the desired r-HPg preparation (1 mg/ml in 10 mM sodium phosphate, pH 7.4) is treated with glycopeptidase F (GF, Boehringer Mannheim Biochemicals, Indianapolis, IN)[18] at a concentration of 2 units/ml. The reaction is allowed to proceed for 48–72 hr at 37°. These conditions are found to be suitable for removal of the Asn-289-linked carbohydrate from all of the samples investigated in this

[16] D. G. Deutsch and E. T. Mertz, *Science* **170**, 1095 (1970).
[17] D. J. Davidson, D. L. Higgins, and F. J. Castellino, *Biochemistry* **29**, 3585 (1990).
[18] A. L. Tarentino, C. M. Gomez, and T. H. Plummer, *Biochemistry* **24**, 4665 (1985).

report. The mixture is then subjected to centrifugation in molecular weight 10,000 cutoff (Centricon-10) microconcentrator tubes (Amicon, Danvers, MA) to separate the liberated oligosaccharide from the protein sample.

SDS–PAGE of the protein, followed by visualization of the bands with biotinylated wheat germ agglutinin (WGA) and avidin-peroxidase, is employed to monitor the existence of GlcNAc(β1,4)GlcNAc on the r-HPg after GF treatment. In this procedure, protein samples are separated by SDS–PAGE[19] on 9% (w/v) gels under nonreducing conditions. The resolved protein bands are transferred to Immobilon-P (Millipore, Bedford, MA) membranes according to established procedures,[20] and then incubated at 37° for 1 hr in 0.05 M Tris-HCl/0.15 M NaCl/1% (w/v) EIA grade gelatin (Bio-Rad, Richmond, CA), pH 7.4 (blocking buffer). Our exact conditions for transfer were 4° in 25 mM Tris-HCl/200 mM glycine/15% (v/v) methanol, pH 8.3, at 20 V for 12 hr. After blocking the free protein adsorption sites on the Immobilon-P sheet with the above gelatin buffer, a solution of peroxidase-labeled WGA or peroxidase-labeled CBL, containing 25 μg/ml of protein in blocking buffer, is added and allowed to incubate at 25° for 90 min. Bound lectin is visualized after addition of the peroxidase substrate, purchased as a kit from Bio-Rad (TMB peroxidase EIA substrate kit), and employed according to the protocol provided by the manufacturer.

The absence (or large diminishment) of reaction with WGA is taken as a determinant of complete deglycosylation.

Oligosaccharide Mapping

Resolution of oligosaccharide units liberated from the protein by glycopeptidase F (GF) is accomplished employing a Dionex (Sunnyvale, CA) BIO LC liquid chromatography unit, an anion-exchange column, and pulsed amperometric detection.[21]

After GF digestion of the protein, GF and remaining r-HPg are removed by centrifugation in molecular weight 10,000 cutoff (Centricon-10) microconcentrator tubes. The solution is then lyophilized and dissolved in H$_2$O. The sample (25 μl, containing oligosaccharides from approximately 150 μg of r-HPg) is injected onto a column of Carbopac PA1 (4 × 250 mm) and resolved into its component oligosaccharides employing gradient elution with solvents A (200 mM NaOH), B (1 mM NaOH), and C (1 M

[19] U. K. Laemmli, *Nature* (*London*) **227**, 680 (1970).
[20] W. H. Burnette, *Anal. Biochem.* **112**, 195 (1981).
[21] M. R. Hardy, R. R. Townsend, and Y. C. Lee, *Anal. Biochem.* **170**, 54 (1988).

sodium acetate). The column is equilibrated with a mixture (v/v) of 50% solvent A/47% solvent B/3% solvent C. This same mixture is applied for 15 min, after which a linear gradient of 50% A/47% B/3% C (start solvent) to 50% A/25% B/25% C (limit solvent) is employed up to 30 min. The limit solvent is then continued for an additional 15 min. The flow rate is 1 ml/min at room temperature. Preliminary identification of the oligosaccharides is made by comparing the elution times of the sample peaks with a library of standard N-acetyllactosaminic and oligomannosidic glycans.

An example of the resolution obtained for a variety of standard oligosaccharides (purchased from the Dionex Corporation and Oxford Glyco-Systems, Rosedale, NY) by high-pH anion-exchange chromatography (HPAEC) is provided in Fig. 2. Using this technique, dose–response

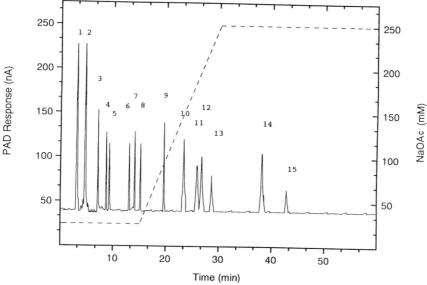

FIG. 2. Separation of standard oligosaccharides on HPAEC. The standards applied are as follows: peak 1, $Man_3GlcNAc_2$ (Fuc); peak 2, $Man_3GlcNAc_2$; peak 3, $Man_5GlcNAc_2$; peak 4, $Man_6GlcNAc_2$; peak 5, asialyl–biantennary oligosaccharide; peak 6, $Man_7GlcNAc_2$; peak 7, asialyl–triantennary oligosaccharide; peak 8, asialyl–tetraantennary oligosaccharide; peak 9, $Man_8GlcNAc_2$; peak 10, $Man_9GlcNAc_2$; peak 11, bisialyl–biantennary (Fuc) oligosaccharide; peak 12, bisialyl–biantennary oligosaccharide; peak 13, trisialyl–triantennary oligosaccharide; peak 14, trisialyl–tetraantennary oligosaccharide; peak 15, tetrasialyl–tetraantennary oligosaccharide. Man, Mannose; GlcNAc, N-acetylglucosamine; Fuc, fucose. The dashed line on the graph indicates the progress the sodium acetate concentration during the chromatographic progress. The ordinate is expressed as the arbitrary response of the pulsed amperometric detector (PAD).

curves are constructed for all standard oligosaccharides for later use in compositional analysis of GF-treated r-HPg.

Hydrolysis and Monosaccharide Compositions of Oligosaccharides

A clam (*Venus mercenaria*) liver extract containing at least 16 relevant glycosidases[22,23] is employed for total enzymatic digestions of isolated oligosaccharides collected from the HPAEC column. This extract is found to catalyze complete hydrolysis of a variety of oligosaccharides as well as glycolipids and synthetic *p*-nitrophenyl monosaccharyl substrates containing several different types of sugar linkages. The oligosaccharide to be investigated is obtained from the HPAEC column and is adjusted with sodium acetate to pH 4.0. The clam liver extract is then incubated with the sample at 37° overnight. Various concentrations of the clam glycosidase mixture are added and the results reported are based on maximal amounts of the particular monosaccharide liberated. Blank samples consist of solutions with both the oligosaccharide and the liver enzyme mixture eliminated from the incubation.

After enzymatic digestion of the oligosaccharides, the enzymes are removed by centrifugation in molecular weight 10,000 cutoff (Centricon-10) microconcentrator tubes. The solution is then lyophilized and the monosaccharides are extracted into ethanol to remove salts that interfere with subsequent liquid chromatography analysis. Because sialic acid is the only monosaccharide incompletely extracted by this procedure, we normally enzymatically (*vide infra*) desialylate (and determine the sialic acid concentration) the oligosaccharide with neuraminidase prior to digestion with the clam extract. The ethanol is then removed by evaporation, and the solids are redissolved in H_2O.

For analysis of the component monosaccharides by HPAEC, the sample (25 μl, containing monosaccharides from approximately 1 μg of glycan) is injected onto the above Carbopac PA1 (4 × 250 mm) column, in a mixture of 92% (v/v) solvent A (1 mM NaOH)/8% (v/v) solvent B (200 mM NaOH), and is washed with this same solvent mixture for 20 min. After this time a linear gradient is applied for 20 min with 92% (v/v) solvent A/8% (v/v) solvent B as the start solvent and 50% (v/v) solvent A/35% (v/v) solvent B/15% (v/v) solvent C (1 M NaOAc) as the limit solvent. The column flow rate is 1 ml/min at room temperature. Identification of the peaks is accomplished by comparison of the sample elution

[22] S. Ghosh, S. Lee, T. A. Brown, M. Basu, J. W. Hawes, D. Davidson, and S. Basu, *Anal. Biochem.* **196,** 252 (1991).
[23] D. J. Davidson and F. J. Castellino, *Biochemistry* **30,** 625 (1991).

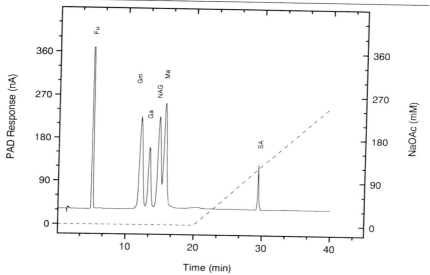

FIG. 3. Separation of monosaccharides on HPAEC. The following standards were applied to the column: Fu, fucose; Gm, glucosamine; Ga, galactose; NAG, N-acetylglucosamine; Ma, mannose; SA, sialic acid (N-acetylneuraminic acid). The dashed line on the graph indicates the progress of the sodium acetate concentration during the chromatographic progress.

times with those of standard monosaccharides. A representative standard analysis is given in Fig. 3. Dose–response curves are constructed for all monosaccharides for later use in compositional analysis of hydrolyzed oligosaccharides.

Sequential Exoglycosidase Digestions

The following exoglycosidases are added to purified oligosaccharides in different orders to samples in which the monosaccharide analyses provided tentative identifications of complex-type carbohydrate: *Arthrobacter ureafaciens* neuraminidase for determination of Sia(α2,3/6/8)Gal linkages; Newcastle disease virus (NDV) neuraminidase for analysis of Sia(α2,3)Gal linkages; *Diplococcus pneumoniae* β-galactosidase for determination of Gal(β1,4)GlcNAc linkages; *Canavalia ensiformis* (jack bean) NAc-β-D-glucosaminidase for determination of GlcNAc(β1,2/3/4/6)Man linkages; *Diplococcus pneumoniae* NAc-β-D-glucosaminidase for analysis of GlcNAc(β1,2)Man linkages; jack bean α-mannosidase for determination of Man(α1,2/3/6)Man linkages; *Aspergillus phoenicis* α-mannosidase for determination of Man(α1,2)Man linkages; *Turbo conufus* β-mannosidase for determination of Man(β1,4)GlcNAc linkages; bovine epididymis α-fucosidase for Fuc(α1,6 > 2/3/4)GlcNAc linkages. For oligosaccharides wherein total monosaccharide analysis indicates that they are present as high-mannose or truncated high-mannose structures, the following exoglycosidase additions are made: jack bean or *A. phoenicis* α-mannosidases; *T. cornufus* β-mannosidase; bovine epididymis α-fucosidase; jack bean NAc-β-D-glucosaminidase. Exoglycosidase digestions of the standard glycan with the proposed structure of each sample are conducted in parallel.

In every case, an aliquot of the total mixture is removed after each glycosidase addition in order to measure by HPLC (*vide supra*) the amount of the particular monosaccharide released from a known quantity of oligosaccharide substrate. The exact procedures used and the strategies employed in structural assignments have been described in several articles from this laboratory,[23–25] and are summarized below.

Oligosaccharide Structures Present on Recombinant Insect Cell-Expressed Human Plasminogen

Previous work from this laboratory established for the first time that a lepidopteran insect line (IPLB-SF-21AE) possessed the glycosylation machinery to assemble complex-type oligosaccharides on a protein,[26] and that this ability was related to temporal events that occurred during infection of the cells with the recombinant baculovirus.[24] The same conclusions were drawn after examination of the nature of the glycosylation of r-HPg by another lepidopteran insect cell line (IZD-MB0503), after infection with the same recombinant baculovirus/HPg construct.[25] It had been previously believed that higher order oligosaccharide processing could not occur in cells of these types, and that only complex-type precursor glycans, i.e., high-mannose oligosaccharides, were present on proteins derived from lepidopteran insect cells.[3]

In order to verify that this process was of a more general nature, we have infected a very commonly employed insect expression cell line, SF-9, which is a subclone of the IPLB-SF-21AE cell line, with the recombinant baculovirus containing the HPg cDNA, and we examined the nature of the glycan structures that were assembled on the r-HPg as a function of times of infection of these cells. An oligosaccharide fingerprint on the material released from each r-HPg, after treatment with GF, is provided in Fig. 4. Evidence for complete deglycosylation of all r-HPg preparations

[24] D. J. Davidson and F. J. Castellino, *Biochemistry* **30**, 6167 (1991).
[25] D. J. Davidson and F. J. Castellino, *Biochemistry* **30**, 6689 (1991).
[26] D. J. Davidson, M. J. Fraser, and F. J. Castellino, *Biochemistry* **29**, 5584 (1990).

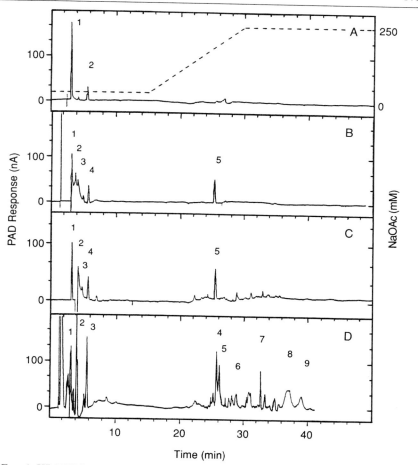

FIG. 4. HPAEC fingerprint of the oligosaccharides released from SF-9 expressed r-HPg after treatment of the protein with GF. (A) The r-HPg was purified from the conditioned culture media at 20 hr postinfection (pi) with the recombinant baculovirus. (B) As in A, except that the pi time was 20–60 hr. (C) As in A, except that the pi time was 20–72 hr. (D) As in A, except that the pi time was 60–96 hr. The sodium acetate gradient employed for the separations, shown only in A (dotted line), is identical to that employed for B–D.

was obtained from WGA blots of the deglycosylated proteins. There was no reaction with this lectin of the proteins after treatment with GF. This demonstrated that the GlcNAc(β1,4)GlcNAc structure was not present. Because this disaccharide is directly attached to Asn-289, these results demonstrate that each protein has been completely deglycosylated under the conditions of treatment with GF. Also, in all cases to be described, the total amount of oligosaccharide ultimately accounted for ranged from 0.79 to 0.87 mol of oligosaccharide per mole of r-HPg, also demonstrating the effectiveness of the enzymatic deglycosylation procedure in the case of r-HPg. Here, because r-HPg contains only one consensus sequence for N-linked glycosylation (at Asn-289), the amounts of oligosaccharides identified represent the majority of the structures present on the protein.

Figure 4A indicates that the r-HPg expressed in infected SF-9 cells at 20 hr postinfection (pi) liberates two major glycans after complete release with GF. These were tentatively identified by comparison of the elution times with standard oligosaccharides as structures **2** and **4** of Fig. 5. Monosaccharide compositions of each component were virtually identical to those of each standard glycan. The results of sequential exoglycosidase digestions of these two components are provided in Table I. Digestion of each component with (α2,3)- or (α2,6)-specific neuraminidases, β-galactosidase, NAc-β-glucosaminidase, α-fucosidase, or (α1,2)-specific mannosidase did not lead to monosaccharide release. Treatment with jack bean α-mannosidase led to release of approximately 2 and 4 mol of mannose/

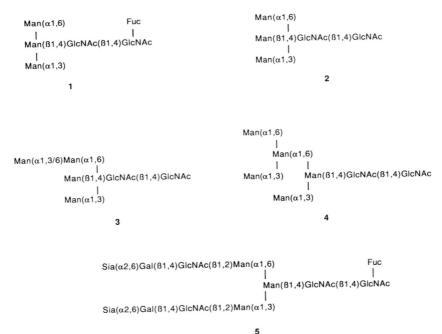

FIG. 5. Structures of the oligosaccharides (**1–10**) found in SF-9-expressed r-HPg.

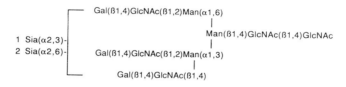

Sia(α2,6)Gal(β1,4)GlcNAc(β1,2)Man(α1,6)
|
Man(β1,4)GlcNAc(β1,4)GlcNAc
|
Sia(α2,6)Gal(β1,4)GlcNAc(β1,2)Man(α1,3)

6

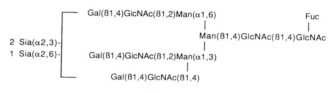

1 Sia(α2,3)-
2 Sia(α2,6)-
Gal(β1,4)GlcNAc(β1,2)Man(α1,6)
|
Man(β1,4)GlcNAc(β1,4)GlcNAc
|
Gal(β1,4)GlcNAc(β1,2)Man(α1,3)
|
Gal(β1,4)GlcNAc(β1,4)

7

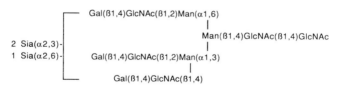

2 Sia(α2,3)-
1 Sia(α2,6)-
Gal(β1,4)GlcNAc(β1,2)Man(α1,6)   Fuc
| |
Man(β1,4)GlcNAc(β1,4)GlcNAc
|
Gal(β1,4)GlcNAc(β1,2)Man(α1,3)
|
Gal(β1,4)GlcNAc(β1,4)

8

2 Sia(α2,3)-
1 Sia(α2,6)-
Gal(β1,4)GlcNAc(β1,2)Man(α1,6)
|
Man(β1,4)GlcNAc(β1,4)GlcNAc
|
Gal(β1,4)GlcNAc(β1,2)Man(α1,3)
|
Gal(β1,4)GlcNAc(β1,4)

9

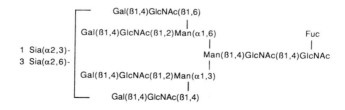

1 Sia(α2,3)-
3 Sia(α2,6)-
Gal(β1,4)GlcNAc(β1,6)
|
Gal(β1,4)GlcNAc(β1,2)Man(α1,6)   Fuc
| |
Man(β1,4)GlcNAc(β1,4)GlcNAc
|
Gal(β1,4)GlcNAc(β1,2)Man(α1,3)
|
Gal(β1,4)GlcNAc(β1,4)

1 0

FIG. 5. (*continued*)

TABLE I

Sequential Exoglycosidase Digestion of Oligosaccharides Released from
SF-9-Expressed Recombinant Human Plasminogen at 20 Hr Postinfection[a]

Fraction[b] (structure number; relative %)	Sia[c] (α2,3)/R	Gal[d] (β1,4)	GlcNAc[e] (β1,2)/R	α-Man[f] (α1,2)/R	β-Man[g]	α-Fuc[h]	GlcNAc[i]
1 (2; 72%)	—	—	—	0/1.9	1.0	—	1.9
2 (4; 28%)	—	—	—	0/3.8	1.1	—	1.9

[a] For pools containing complex carbohydrate, the sequence of enzymes added was as in footnotes c–i. For high-mannose or truncated high-mannose pools, the sequence of addition was as in footnotes e–i. Replicate analyses agreed to within 5%.

[b] The fraction numbers correspond to those in Fig. 4A. The structures are from Fig. 5.

[c] Neuraminidase (NDV) followed by neuraminidase (A. ureafaciens). This provides the number of (α2,3)-linked Sia residues followed by the number (R) of α-Sia residues in linkages other than (α2,3).

[d] (β1,4)-Galactosidase (D. pneumoniae). This provides the number of (β1,4)-linked Gal residues.

[e] NAc-β-Glucosaminidase (D. pneumoniae) followed by NAc-β-glucosaminidase (jack bean). This provides the number of (β1,2)-linked GlcNAc residues followed by the number (R) of β-GlcNAc residues in linkages other than (β1,2).

[f] α-Mannosidase (D. pneumoniae) followed by α-mannosidase (jack bean). This provides the number of (α1,2)-linked Man residues followed by the number (R) of α-Man residues in linkages other than (α1,2).

[g] β-Mannosidase (snail).

[h] α-Fucosidase (bovine epididymus).

[i] NAc-β-Glucosaminidase (jack bean).

mole of glycan for pools 1 and 2 of Fig. 4A, exactly in parallel with similar treatments of the two standard glycans. Following this, one residue of mannose is released from the remaining structures when each is next treated with β-mannosidase. On treatment with NAc-β-glucosaminidase, 2 mol of NAc-glucosamine were liberated. The results clearly allow assignment of each glycan to structures 2 and 4 of Fig. 5. Thus, at early times of infection, only high-mannose types of glycans are found on SF-9-expressed r-HPg.

From the oligosaccharide fingerprint of Fig. 4B, obtained from GF-catalyzed oligosaccharide release from r-HPg expressed in SF-9 cells at a time window of 20–60 hr pi, it is clear that other glycans are found in the mixture. At least five such components were isolated and subjected to sequential exoglycosidase analysis, with identifications made employing the same strategy as above. The data of Table II allow clear structural identifications to be made. Four of the five components are high-mannose-type oligosaccharides, and are assigned as structures 1–4 of Fig. 5. Interestingly, 1 mol of fucose is released from pool 1 in the residual glycan

TABLE II

SEQUENTIAL EXOGLYCOSIDASE DIGESTION OF OLIGOSACCHARIDES RELEASED FROM
SF-9-EXPRESSED RECOMBINANT HUMAN PLASMINOGEN AT 20–60 HR POSTINFECTION[a]

Fraction[b] (structure number; relative %)	Sia ($\alpha2,3$)/R	Gal ($\beta1,4$)	GlcNAc ($\beta1,2$)/R	α-Man ($\alpha1,2$)/R	β-Man	α-Fuc	GlcNAc
B1 (**1**; 29%)	—	—	—	0/2.0	0.9	0.9	1.9
B2 (**2**; 14%)	—	—	—	0/2.0	0.9	—	1.8
B3 (**3**; 2%)	—	—	—	0/3.0	1.0	—	1.9
B4 (**4**; 29%)	—	—	—	0/3.9	1.0	—	1.9
B5 (**6**; 26%)	0/2.0	1.8	2.0/0	0/1.8	0.9	—	1.8

[a] For footnotes, see Table I.

[b] The fraction number refers to Fig. 4B. The structures are from Fig. 5.

after β-mannosidase treatment, showing that a fucosylated high-mannose component is present. This suggests that SF-9 cells possess a fucosyltransferase that catalyzes this addition. The fifth component of this mixture is clearly identified as a fully sialylated biantennary complex-type oligosaccharide (structure **6**, Fig. 5), present in a relative yield of 26%.

At a pi time window of 20–72 hr, it is clear from Fig. 4C that additional oligosaccharide peaks appear in the chromatogram. Most of these are present at elution times that suggest that they are of complex-type structures, and are present in quantities that are insufficient for structural analysis. However, from Table III, it is clear that the bisialyl–biantennary complex-type component is further increased to a relative yield of approximately 38% of the total glycans identified.

TABLE III

SEQUENTIAL EXOGLYCOSIDASE DIGESTION OF OLIGOSACCHARIDES RELEASED FROM
SF-9-EXPRESSED RECOMBINANT HUMAN PLASMINOGEN AT 20–72 HR POSTINFECTION[a]

Fraction[b] (structure number; relative %)	Sia ($\alpha2,3$)/R	Gal ($\beta1,4$)	GlcNAc ($\beta1,2$)/R	α-Man ($\alpha1,2$)/R	β-Man	α-Fuc	GlcNAc
C1 (**1**; 22%)	—	—	—	0/1.9	1.0	1.0	2.0
C2 (**2**; 11%)	—	—	—	0/2.0	0.9	—	1.9
C3 (**3**; 6%)	—	—	—	0/2.7	1.0	—	1.9
C4 (**4**; 23%)	—	—	—	0/4.1	1.0	—	2.0
C5 (**6**; 38%)	0/1.8	1.9	2.0/0	0/2.0	0.9	—	1.9

[a] For footnotes, see Table I.

[b] The fraction number refers to Fig. 4C. The structures are from Fig. 5.

TABLE IV

SEQUENTIAL EXOGLYCOSIDASE DIGESTION OF OLIGOSACCHARIDES RELEASED FROM
SF-9-EXPRESSED RECOMBINANT HUMAN PLASMINOGEN AT 60–96 HR POSTINFECTION[a]

Fraction[b] (structure number; relative %)	Sia (α2,3)/R	Gal (β1,4)	GlcNAc (β1,2)/R	α-Man (α1,2)/R	β-Man	α-Fuc	GlcNAc
D1 (**1**; 7%)	—	—	—	0/2.0	1.0	0.9	1.9
D2 (**2**; 6%)	—	—	—	0/1.9	0.9	—	1.8
D3 (**4**; 5%)	—	—	—	0/4.0	1.0	—	2.0
D4 (**5**; 13%)	0/2.0	2.1	1.9/0	0/1.9	1.0	0.9	1.9
D5 (**6**; 12%)	0/1.9	2.1	2.0/0	0/1.8	0.9	—	1.9
D6 (**7**; 6%)	1.0/2.0	3.0	1.7/1.2	0/1.8	0.9	—	2.0
D7 (**8**; 30%)	1.8/1.0	3.1	2.0/0.9	0/1.9	1.0	0.9	1.9
D8 (**9**; 11%)	2.0/1.1	2.9	1.9/1.0	0/1.8	1.0	—	2.0
D9 (**10**; 10%)	1.0/3.0	3.8	1.1/2.9	0/1.8	1.0	1.0	2.0

[a] For footnotes, see Table I.

[b] The fraction number refers to Fig. 4D. The structures are from Fig. 5.

The final time frame investigated was 60–96 hr pi, and the data of Fig.
4D and Table IV illustrate that the preponderance of structures (82%) are
of the complex type. A variety of such structures exist with different
numbers of outer branch arms, different linkages of sialic acid, and fucosylated and nonfucosylated forms. In all cases, the strategy described
above with sequential exoglycosidase digestions revealed the structures
clearly, except for one aspect of component 10 wherein only one-half of
the theoretical amount of (β1,2)-glucosaminidase was released. This is as
expected if the structure contains the structure GlcNAc(β1,2)[Glc-
NAc(β1,6)]Man. In this case the (β1,2)-linked GlcNAc would not be
released by the *D. pneumoniae* NAc(β1,2)-glucosaminidase due to steric
hindrance of the enzyme.[27]

In conclusion, we have observed that in three lepidopteran insect
cell lines, IPLB-SF-21AE,[24,26] IZD-MB0503,[25] and SF-9, infected with a
recombinant baculovirus carrying the cDNA for HPg in its genome, the
r-HPg produced contained substantial amounts of a wide variety of complex-type oligosaccharide, with at least four outer branches. Sialylation
and fucosylation also occur on many of the glycoforms r-HPg expressed
in this manner. In all cases, the nature of the oligosaccharides assembled
on r-HPg depends on the time of infection of the cells with the virus, with
later infection times leading to greatly increased levels of complex-type

[27] K. Yamashita, T. Ohkura, H. Yoshima, and A. Kobata, *Biochem. Biophys. Res. Commun.*
100, 226 (1981).

oligosaccharides on this glycoprotein, suggesting that the infective process leads to increased functionality of the necessary glycosyltransferases and/or mannosidases[28] required for complex-type glycan processing. We have not as yet found hybrid-type and GlcNAc-bisecting glycans on r-HPg in any of our studies. It is possible that such structures are present in some of the minor uncharacterized components of oligosaccharide released from r-HPg (see Fig. 4D). If this is not the case, it is possible that the nature of the r-HPg substrate does not favor the formation of such structures, and/or our cell culture conditions are not optimal for these structures to assemble in a stable manner. In any case, manipulation of invertebrate cell cultures can lead to virtually exclusive levels of mammalian-type complex oligosaccharide processing, and these results provide an excellent system in which to investigate the factors that determine the nature of the types of oligosaccharides assembled on glycoproteins.

[28] D. J. Davidson and F. J. Castellino, *Biochemistry* **30**, 9811 (1991).

[11] Human α₂-Plasmin Inhibitor

By Nobuo Aoki, Yoshihiko Sumi, Osamu Miura, and Shinsaku Hirosawa

Introduction

α₂-Plasmin inhibitor (α₂PI)[1] is a plasma glycoprotein that rapidly inactivates plasmin proteolytic activity.[2] The α₂PI gene contains 10 exons and 9 introns distributed over ~16 kilobases (kb) of DNA,[3] and the gene is located on chromosome 17p13.[4] α₂PI is produced in the liver as a precursor form (Pro-α₂PI) with a prepeptide (signal peptide) of 27 amino acids and a propeptide of 12 amino acids.[5] The mature plasma protein is composed

[1] The name α₂-antiplasmin is also used, but the International Union of Pure and Applied Chemistry recommends the use of the name α₂-plasmin inhibitor.
[2] M. Moroi and N. Aoki, *J. Biol. Chem.* **251**, 5956 (1976).
[3] S. Hirosawa, Y. Nakamura, O. Miura, Y. Sumi, and N. Aoki, *Proc. Natl. Acad. Sci. U.S.A.* **85**, 6836 (1988).
[4] A. Kato, S. Hirosawa, S. Toyota, Y. Nakamura, H. Nishi, A. Kimura, T. Sasazuki, and N. Aoki, *Cytogenet. Cell Genet.* **62**, 190 (1993).
[5] Y. Sumi, Y. Ichikawa, Y. Nakamura, O. Miura, and N. Aoki, *J. Biochem. (Tokyo)* **106**, 703 (1989).

of 452 amino acids.[6,7] Its molecular weight, deduced from the cDNA sequence and the carbohydrate content (14%),[2] is ~58,000, whereas the molecular weight estimated by SDS–gel electrophoresis is 67,000.[2] The cause of the discrepancy is not known. Its concentration in human plasma is estimated to be 6.9 ± 0.6 mg/100 ml,[8] which is estimated to be ~1.2 μM, assuming that the molecular weight is 58,000. $\alpha_2 PI$ is a serine proteinase inhibitor (serpin) that can inhibit several different "serine" proteinases, but it mainly functions as the primary inhibitor of plasmin-mediated fibrinolysis.[9–11]

Congenital deficiency of $\alpha_2 PI$ results in a lifelong severe hemorrhagic tendency due to premature degradation of hemostatic plugs by the physiologically occurring fibrinolytic process.[12–14]

Three Functions

$\alpha_2 PI$ molecule has three functional sites: the reactive site, plasminogen-binding site, and cross-linking site. The reactive site, located at Arg-364,[7] can form a covalent bond with the active site serine of a protease, causing the enzyme to lose its proteolytic activity.[2,15] The plasminogen-binding site, located within 20 amino acid residues of the carboxyl-terminal end,[16,17] binds to the lysine-binding sites of plasminogen, to which fibrin is also bound. Hence, $\alpha_2 PI$ competitively inhibits the binding of plasminogen to fibrin,[18] thus retarding the initiation of the fibrinolytic process because plasminogen binding to fibrin plays an important role in the initiation of fibrinolysis.[19] The high affinity of the plasminogen-binding site for plasmin accelerates the complex formation of plasmin and $\alpha_2 PI$,[20] resulting in a rapid inhibition of plasmin. When blood clots, part (about 20–30%)

[6] M. Tone, R. Kikuno, A. Kume-Iwaki, and T. Hashimoto-Gotoh, *J. Biochem. (Tokyo)* **102,** 1033 (1987).
[7] W. E. Holmes, L. Nelles, H. R. Lijnen, and D. Collen, *J. Biol. Chem.* **262,** 1659 (1987).
[8] Y. Sakata and N. Aoki, *J. Clin. Invest.* **65,** 290 (1980).
[9] N. Aoki, M. Moroi, M. Mastuda, and K. Tachiya, *J. Clin. Invest.* **60,** 361 (1977).
[10] N. Aoki, *Prog. Cardiovasc. Dis.* **21,** 267 (1979).
[11] N. Aoki and P. C. Harpel, *Semin. Thromb. Hemostasis* **10,** 24 (1984).
[12] N. Aoki, H. Saito, T. Kamiya, K. Koie, Y. Sakata, and M. Kobakura, *J. Clin. Invest.* **63,** 877 (1979).
[13] N. Aoki, *Semin. Thromb. Hemostasis* **10,** 42 (1984).
[14] N. Aoki, *Blood Rev.* **3,** 11 (1989).
[15] B. Wiman and D. Collen, *J. Biol. Chem.* **254,** 9291 (1979).
[16] T. Sasaki, T. Morita, and S. Iwanaga, *J. Biochem. (Tokyo)* **99,** 1699 (1986).
[17] T. Sasaki, N. Sugiyama, M. Iwamoto, and S. Isoda, *Chem. Pharm. Bull.* **35,** 2810 (1987).
[18] M. Moroi and N. Aoki, *Thromb. Res.* **10,** 851 (1977).
[19] N. Aoki, Y. Sakata, and A. Ichinose, *Blood* **62,** 1118 (1983).
[20] B. Wiman, H. R. Lijnen, and D. Collen, *Biochim. Biophys. Acta* **579,** 142 (1979).

of the α_2PI present in plasma is rapidly bound to the fibrin α-chain by an intermolecular cross-linking reaction catalyzed by activated factor XIII.[8,21-23] The cross-linking site is located at the second amino acid residue, glutamine, from the amino terminus.[22] The cross-linking of α_2PI to fibrin plays a significant role in the inhibition of fibrinolysis because it markedly stabilizes the fibrin clot against fibrinolysis.[24]

Non-Plasminogen-Binding Form

A partially degraded form of α_2PI (65,000 form), which lacks a peptide in the carboxyl-terminal region that contains the plasminogen-binding site, reacts less readily with plasmin than does the native plasminogen-binding form of α_2PI (67,000 form).[25] This non-plasminogen-binding form is present as a minor component (about 30% of α_2PI) in normal plasma.[26]

Assay

Immunochemical Assay

Frequently employed immunochemical assays for α_2PI, such as single radial immunodiffusion, electroimmunoassay, and enzyme-linked immunosorbent assay (ELISA) using polyclonal antibodies, measure not only free α_2PI but also the plasmin–α_2PI complex. These methods cannot be used to measure the free α_2PI in plasma from patients with an activated fibrinolytic system, such as those with disseminated intravascular coagulation or those receiving thrombolytic therapy in whom the plasma levels of the plasmin–α_2PI complex are often markedly increased.

Production of proper monoclonal antibodies enabled us to develop the following ELISA method for assaying free α_2PI separately from the plasmin–α_2PI complex.[27,28] Two monoclonal antibodies, JTPI-1 and

---

[21] T. Tamaki and N. Aoki, *Biochim. Biophys. Acta* **661**, 280 (1981).

[22] T. Tamaki and N. Aoki, *J. Biol. Chem.* **257**, 14767 (1982).

[23] S. Kimura and N. Aoki, *J. Biol. Chem.* **261**, 15591 (1986).

[24] Y. Sakata and N. Aoki, *J. Clin. Invest.* **69**, 536 (1982).

[25] I Clemmensen, S. Thorsen, S. Müllertz, and L. C. Petersen, *Eur. J. Biochem.* **120**, 105 (1981).

[26] C. Kluft and N. Los, *Thromb. Res.* **21**, 65 (1981).

[27] J. Mimuro, Y. Koike, Y. Sumi, and N. Aoki, *Blood* **69**, 446 (1987).

[28] Mouse monoclonal antibodies JTPI-1, -2, and -3 are available from Diagnostics Marketing Department, Medical and Home Health care Division, Teijin Ltd., Iino Bldg., Uchisaiwai-Cho 2-1-1, Chiyoda-Ku, Tokyo 100. Kits for the simplified two-step ELISA methods, TD-80 and TD-80C, for free α_2PI and the plasmin–α_2PI complex, respectively, are also available from the same firm.

JTPI-2,[28] were used for the assay. The former recognizes the reactive site of α_2PI and the latter recognizes an epitope in the C-terminal region of α_2PI.[27] Polyvinyl chloride microtiter plates were coated with monoclonal antibody JTPI-1[28] at 10 μg/ml in 50 mM carbonate buffer, pH 9.6, for 16 hr at 4°. After washing the coated wells five times with phosphate-buffered saline (0.02 mol/liter sodium phosphate, 0.135 mol/liter NaCl, pH 7.4) containing 1% (w/v) bovine serum albumin (BSA) and 0.05% Tween 20 (washing buffer), samples were added to the coated wells and incubated at 37° for 4 hr. The wells were then washed in the same way, and the alkaline phosphatase-conjugated monoclonal antibody JTPI-2 at 10 nM in washing buffer was added to each well and incubated at 37° for 4 hr. After washing, hydrolysis of p-nitrophenyl phosphate (1 mg/ml in 97 mM diethanolamine, 0.49 mM MgCl$_2$, pH 9.8) by the microtiter plate-bound alkaline phosphatase was performed at 25°, and the absorbance at 405 nm was monitored on an ELISA analyzer. Prior to the assay, the samples were diluted with washing buffer so that the α_2PI concentrations of the diluted samples fell within the linear portion (1 to 3 nM of α_2PI) of the standard curve; plasma was usually diluted 400-fold. The assay was not influenced by the presence of plasmin–α_2PI complex up to 6 nM in the diluted samples. Even if all the α_2PI in plasma is converted to the plasmin–α_2PI complex, the concentration of the complex ($<$2.5 nM) in dilute plasma (1 : 400 dilution) would be well below the range of detection by this assay.

The plasmin–α_2PI complex can be measured by using antiplasminogen polyclonal antibody as a coating (solid-phase) antibody and the enzyme-conjugated JTPI-2 as a liquid-phase antibody in a similar way.[27] Total α_2PI antigen (free α_2PI plus plasmin–α_2PI complex) can be measured by using JTPI-3[28] as a solid-phase antibody and the enzyme-conjugated JTPI-2 as a liquid-phase antibody. The monoclonal antibodies and kits for these ELISA methods are commercially available.[28]

Functional Assay

Antifibrinolytic activity of α_2PI is assayed by the fibrin clot lysis method.[2] An aliquot of the sample is diluted in the test tube (12 × 105 mm) to 0.75 ml with cold 0.67 M phosphate buffer (pH 7.4) and is mixed with 0.05 ml of 2% Cohn fraction I (as a source of fibrinogen and plasminogen). The mixture is cooled in an ice bath. To the mixture are added 0.1 ml of urokinase solution (300 units/ml) and 0.1 ml of thrombin solution (20 units/ml) successively. The tube is quickly shaken to mix the reagents well and placed in a 37° water bath. A stopwatch is started at the time of the addition of thrombin, and the lysis of the clot formed is observed. A standard curve is constructed by plotting the clot lysis time obtained at various concentrations of urokinase against units of urokinase on double-logarithmic paper. A straight line results over the range of 6 to 30 units of urokinase. Using this standard curve, clot lysis time obtained with a sample containing the inhibitor is converted to values for urokinase activity. An amount of inhibitor that decreases the activity of urokinase to 50% of its original value (from 30 units to 15 units) is defined as 1 unit of the inhibitor. The assay is specific to α_2PI, and other plasma inhibitors, such as α_2-macroglobulin (α_2M), do not affect the assay.[29]

Inhibitory activity of α_2PI on plasmin is assayed by measuring the immediate decrease of plasmin activity after addition of a test sample to a fixed amount of plasmin. For assay of α_2PI in plasma, around 0.1 caseinolytic units/ml of plasmin and 50 times dilution of plasma are usually used. The results are expressed in terms of a percentage of normal control plasma. Plasmin activity is measured using synthetic peptide substrates such as the chromogenic substrate *H*-D-valyl-L-leucyl-L-lysine *p*-nitroanilide dihydrochloride (S2251; Kabi-Vitrum, Stockholm, Sweden) for a spectrophotometric assay[30,31] or a fluorogenic substrate *H*-D-valyl-leucyl-lysine 5-amidoisophthalic acid, dimethyl ester, ditrifluoroacetate (Dade Protopath) for a fluorescent assay.[32] An assay kit (Coatest Antiplasmin Kit) using the chromogenic substrate is available from Kabi Diagnostica (Stockholm, Sweden). For assay of residual plasmin activity, assay of the initial rate of hydrolysis of the substrate is preferable to the end-point assay, which is substantially affected by antiplasmin activity of α_2M present in plasma samples.[33] Because immediate inhibition of plasmin is measured by these methods (plasmin is incubated with a sample for less than 20 sec or for 1 min before addition of the chromogenic substrate or the fluorogenic substrate, respectively) and the non-plasminogen-binding form cannot immediately inhibit plasmin, the values obtained by these methods mainly represent the concentration of the plasminogen-binding form.

To measure the overall activity of plasminogen-binding and non-plasminogen-binding forms, the samples should be incubated with plasmin for a longer period (5–10 min) without inhibitory activity of α_2M on plasmin.[31] To abrogate antiplasmin activity of α_2M in plasma, plasma samples should be preincubated with monomethylamine hydrochloride (0.1 mol/liter) for

[29] N. Aoki, M. Moroi, and K. Tachiya, *Thromb. Haemostasis* **39,** 22 (1978).
[30] P. Fiberger, M. Knos, S. Gustavsson, L. Aurell, and G. Claeson, *Haemostasis* **7,** 138 (1978).
[31] K. Naito and N. Aoki, *Thromb. Res.* **12,** 1147 (1978).
[32] D. E. Lawson, G. A. Mitchell, and R. M. Huseby, *Thromb. Res.* **14,** 323 (1979).
[33] T. Matsuda, M. Ogawara, R. Miura, T. Seki, T. Matsumoto, Y. Teramura, and K. Nakamura, *Thromb. Res.* **33,** 379 (1984).

5 min at 37°.[33] Treatment of plasma samples with monomethylamine also abolishes the effect of α_2M on the end-point assay of plasmin and is recommended for every assay for antiplasmin activity of α_2PI.[33]

The cross-linking activity of α_2PI can be measured as follows.[27,34] α_2PI is radioiodinated using immobilized lactoperoxidase oxidase (Enzymo-bead, Bio-Rad Laboratories, Richmond, CA) and $Na^{125}I$ (New England Nuclear, Boston, MA). Radiolabeled α_2PI (24 μl; 25 μg/ml), 10 μl of fibrinogen (25 mg/ml) containing factor XIII (Kabi-Vitrum), 10 μl of thrombin (20 units/ml), and 56 μl of Tris-buffered saline containing $CaCl_2$ (25 mM), bovine serum albumin (2%), and aprotinin (20 units/ml) are mixed and incubated at 37°. After 30 min, the clot formed is squeezed with a stick and thoroughly washed with Tris-buffered saline (0.05 mol/liter Tris, 0.1 mol/liter NaCl, pH 7.4) containing bovine serum albumin (2%), EDTA (5 mM), aprotinin (10 units/ml), and iodoacetamide (1 mM). From the radioactivity remaining in the clot and the total radioactivity of the initial reaction mixture, the percentage of α_2PI cross-linked to fibrin is calculated.

Purification

Conventional Methods

α_2PI has been purified by a conventional multistep method, which includes lysine-Sepharose column chromatography, ammonium sulfate fractionation, ion-exchange chromatography, affinity chromatography on plasminogen-coupled Sepharose, and hydroxyapatite chromatography.[2] The purified α_2PI obtained by the method is stable on storage if the product is kept in a concentrated form at neutral pH (pH 6–8), and it can be frozen or lyophilized without loss of activity.[35] Although this original method allows the recovery of high-quality α_2PI, the low overall yield (10–20%) and the prolonged time required for purification, which can be more than 1 week, make this method unsuitable for routine use. Although several modifications of the original method have been made,[36] none of the modified methods have been successful in increasing the yield without sacrificing the purity or the activity of the product. α_2PI preparations obtained by these conventional methods contain only the native plasminogen-binding form because affinity chromatography on plasminogen-coupled Sepharose is used as the most efficient step in these methods. The non-plasminogen-

[34] J. Mimuro, S. Kimura, and N. Aoki, J. Clin. Invest. 77, 1006 (1986).
[35] Loss of activity on lyophilization was erroneously reported in the original paper (Ref. 2).
[36] B. Wiman, this series, Vol. 80, p. 395.

binding form was obtained by storing the purified plasminogen-binding form of α_2PI at room temperature for 6 days, during which the molecules spontaneously transformed to the non-plasminogen-binding form by cleavage at the C-terminal portion.[16] The non-plasminogen-binding form was further purified by gel filtration on Sephadex G-100, followed by affinity chromatography on plasminogen-coupled Sepharose.[16]

Immunoaffinity Purification

α_2PI can be purified directly from human plasma by immunoaffinity chromatography using a monoclonal antibody.[37] The method is simple, efficient, and easily performed. Both forms, the plasminogen-binding and nonbinding forms, can be isolated together from plasma; subsequently, they can be separated by affinity chromatography on plasminogen-coupled Sepharose.

Coupling Antibody to Sepharose. The antibody used is anti-α_2PI monoclonal antibody JTPI-1,[28] which binds to the reactive site of α_2PI and inhibits antiplasmin activity by interfering with α_2PI-plasmin complex formation.[27] The affinity of JTPI-1 for the preformed α_2PI–plasmin complex was markedly lower than that for free α_2PI in plasma[27]; hence free α_2PI can be selectively adsorbed from plasma to solid-phase JTPI-1.

The antibody is coupled to Sepharose by the following procedure[38]: JTPI-1[28] (5 mg) is dissolved in 15 ml of coupling buffer (0.5 M NaCl, 0.1 M NaHCO$_3$, pH 8.3) and added to 0.5 g of CNBr-activated Sepharose CL-4B (Pharmacia, Piscataway, NJ). The coupling is continued overnight at 4° with gentle stirring. After blocking the remaining reactive sites with 0.2 M glycine solution, pH 8.3, for 5 hr, the gel is washed sequentially with the coupling buffer and 0.1 M citric acid buffer, pH 4.0, containing 0.5 M NaCl. The gel is then equilibrated with phosphate-buffered saline (PBS: 0.01 M potassium phosphate, 0.15 M NaCl, pH 7.4) and stored at 4°.

Immunoaffinity Chromatography. Plastic tubes and columns are used throughout the purification procedures, and all the steps are carried out at 4°. Venous blood is drawn from normal individuals into 0.1 volume of 3.8% sodium citrate. Plasma is separated by centrifugation at 2000 g for 30 min, and aprotinin powder is added to plasma to a concentration of 10 units/ml. Plasma is filtered through filter paper (Toyo Filter Paper No. 6, Toyo Roshi Co., Tokyo) before application to chromatography column. The size of the column should be as small as indicated in the following

---

[37] Y. Sumi, Y. Koike, Y. Ichikawa, and N. Aoki, *J. Biochem. (Tokyo)* **106**, 192 (1989).
[38] R. Axén, J. Poráth, and S. Ernbäck, *Nature (London)* **214**, 1302 (1967).

example, and the use of a larger column should be avoided because a prolonged elution time for α_2PI from a larger column increases the risk of activity loss by prolonged exposure of the protein to acidic pH or denaturing properties of the eluant used. When a larger amount of α_2PI is needed, multiple columns can be used at the same time.

Plasma (50 ml) is applied at a flow rate of 10 ml/hr onto the JTPI-1-coupled Sepharose column (1.5 × 7.5 cm) equilibrated with PBS, and then the column is washed with PBS containing 0.05% Tween 20. The flow rate is kept at approximately 20 ml/hr. Washing is continued until the absorbance at 280 nm is nearly zero. Washing is further continued with PBS without Tween 20 until the absorbance value at 280 nm has returned to zero. Elution of α_2PI can then be performed by two different buffers: One is 0.2 M glycine-HCl, pH 2.5, and the other is 50% (v/v) ethylene glycol, 0.05% Tween 80 in PBS, pH 7.4. When the former is used, the eluate should be collected into 2-ml fractions, each in a tube containing 0.25 ml of 1 M Tris-HCl, pH 8.3, so that pH of the eluate is neutralized immediately after it comes out of the column, because α_2PI is labile at acidic pH. The peak fractions are combined (about 14 ml), dialyzed against PBS, and then concentrated to approximately 1 ml by ultrafiltration using a Centricon-30 (Amicon, Danvers, MA). When 50% (v/v) ethylene glycol, 0.05% Tween 80 in PBS, pH 7.4, is used for elution, the eluate is immediately dialyzed against PBS to remove ethylene glycol and Tween, and then it is concentrated by ultrafiltration. After concentration, further dialysis may be necessary to remove the remaining ethylene glycol. After elution, the column is washed with 6 M guanidine hydrochloride, pH 3.1, and then reequilibrated with PBS for further use. The recovery of α_2PI by this method is around 50%, and the specific activity of the immunopurified α_2PI (3.0 ± 0.1 units/μg, N = 5, by the antifibrinolytic assay) is the same as that of the best preparation of conventionally purified α_2PI. Sodium dodecyl sulfate polyacrylamide gel electrophoresis (SDS–PAGE) of the immunopurified α_2PI under the reduced condition gives one major band associated with a minor band, corresponding to the 67,000 plasminogen-binding form and the 65,000 non-plasminogen-binding form, respectively. The minor band is less than 30% of the total α_2PI. The two forms can be separated by subsequent chromatography on plasminogen-Sepharose, which is used in the conventional method. The concentrated immunopurified α_2PI (1 ml) is applied on a plasminogen-coupled Sepharose 4B column (Pharmacia, Sweden) (1 × 20 cm) equilibrated with PBS. The gel column is washed with PBS at a flow rate at 4 ml/hr. The breakthrough fraction contains the non-plasminogen-binding form, and the plasminogen-binding form can be eluted from the column with PBS containing 20 mM 6-aminohexanoic acid.

Recombinant Expression of Human α_2PI

Holmes et al.[39] constructed an expression vector for human α_2PI using a cDNA fragment coding for the signal peptide of tPA and cDNA encoding mature α_2PI. The recombinant α_2PI expressed in Chinese hamster ovary cells using this expression vector had three additional amino acids at the N-terminal end, corresponding to the carboxyl-terminal three amino acids of the tPA sequence employed to secrete α_2PI, and it had no cross-linking activity to fibrin, which is an important function of mature plasma α_2PI.

We have constructed an expression vector for human α_2PI with a cDNA fragment coding for the carboxyl-terminal half of mature α_2PI and a fragment of the α_2PI gene that codes for the signal peptide and the amino-terminal half of mature α_2PI.[40] The recombinant α_2PI expressed by the vector in a baby hamster kidney (BHK) cell line retained the propeptide of 12 amino acids at the amino terminal, and it possessed only one-third of the cross-linking activity of mature plasma α_2PI.[5] The proteolytic processing enzyme in BHK cells may have failed to recognize the Pro-Asn peptide bond between the propeptide and the mature α_2PI. In order to express mature α_2PI in BHK cells, we changed the amino acid sequence of the propeptide processing site to a serine protease type of processing sequence. Because the processing site (-4 to -1 position) of the propeptide sequence of serine proteases is represented by basic amino acids, Arg or Lys, and because especially the -4 and -1 positions are highly occupied with Arg,[41] we replaced four amino acids of the propeptide processing site (-4 to -1) of α_2PI with arginine residues. Based on the cassette mutation method, we constructed an expression vector for α_2PI with the mutated propeptide.

Construction of Expression Vector

The expression vector for propeptide-mutated α_2PI, pPI906, is composed of a pSV2 plasmid containing SV40 promoter sequences, chemically synthesized DNA including the modified propeptide sequence, and cDNA sequence of α_2PI. The vector was constructed as shown in Fig. 1 via pSV2007, pSV2057, and pSV057.

Chemical Synthesis of DNA. All deoxyribonucletides shown in Fig. 2 (YS1–YS6), which code for the prepropeptide and the amino-terminal portion of α_2PI, were synthesized by an automatic DNA synthesizer (Ap-

[39] W. E. Holmes, H. R. Lijnen, and D. Collen, *Biochemistry* **26**, 5133 (1987).

[40] O. Miura, S. Hirosawa, A. Kato, and N. Aoki, *J. Clin. Invest.* **83**, 1598 (1989).

[41] A. K. Bentley, D. J. G. Rees, C. Rizza, and G. G. Brownlee, *Cell (Cambridge, Mass.)* **45**, 343 (1986).

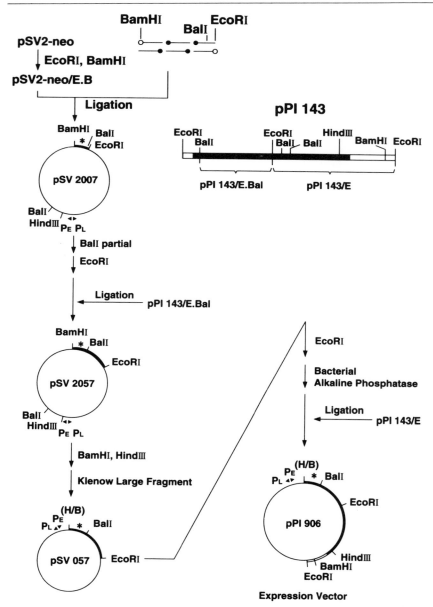

FIG. 1. Construction of the expression vector pPI906 encoding the α_2PI with mutated propeptide; there are four consecutive arginine residues at positions -1 to -4. The BamHI–EcoRI fragment containing the mutation was synthesized (see Fig. 2) and ligated to pSV2neo/E.B. The BalI–EcoRI fragment (pPI143/E.Bal) and EcoRI–EcoRI fragment (pPI143/E) were isolated from plasmid pPI143, the α_2PI cDNA clone, and ligated to pSV2007 and pSV057, respectively. The site of mutation is indicated by an asterisk. PE, SV40 early promoter; PL, SV40 late promoter.

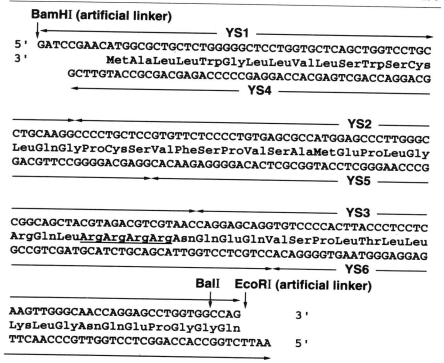

BamHI (artificial linker)

┌──────────────────────── YS1 ─────────────────────►

5' GATCCGAACATGGCGCTGCTCTGGGGGCTCCTGGTGCTCAGCTGGTCCTGC
3'   MetAlaLeuLeuTrpGlyLeuLeuValLeuSerTrpSerCys
         GCTTGTACCGCGACGAGACCCCCGAGGACCACGAGTCGACCAGGACG
◄──────────────────────── YS4 ────────────────────

┌────────────────────────────────── YS2 ─────────

CTGCAAGGCCCCTGCTCCGTGTTCTCCCCTGTGAGCGCCATGGAGCCCTTGGGC
LeuGlnGlyProCysSerValPheSerProValSerAlaMetGluProLeuGly
GACGTTCCGGGGACGAGGCACAAGAGGGGACACTCGCGGTACCTCGGGAACCCG
───────────────────────────────►◄──── YS5 ───────

┌──────────────────────────────── YS3 ─────────

CGGCAGCTACGTAGACGTCGTAACCAGGAGCAGGTGTCCCCACTTACCCTCCTC
ArgGlnLeu**ArgArgArgArg**AsnGlnGluGlnValSerProLeuThrLeuLeu
GCCGTCGATGCATCTGCAGCATTGGTCCTCGTCCACAGGGGTGAATGGGAGGAG
────────────────────────────────────►◄──── YS6 ────

BalI   EcoRI (artificial linker)

┌────────────────────────────
AAGTTGGGCAACCAGGAGCCTGGTGGCCAG   3'
LysLeuGlyAsnGlnGluProGlyGlyGln
TTCAACCCGTTGGTCCTCGGACCACCGGTCTTAA   5'
───────────────────────────────►

FIG. 2. Synthetic oligonucleotide cassette for propeptide-mutated α_2PI. The *Bam*HI site and *Eco*RI site are artificial linkers for cloning into the plasmid. The corresponding amino acid sequence is presented between the nucleotide sequences, and mutated amino acids (-1 to -4) are underlined.

plied Biosystems, model 380A). The oligonucleotides were purified by electrophoresis in 10% polyacrylamide gel containing 7 M urea.

Construction of pSV2007. The synthesized oligonucleotides were cloned in pSV2*neo* as pSV2007. An amount of 1 μg of each oligonucleotide, YS2, YS3, YS4, and YS5, was dissolved in 48 μl of reaction buffer (50 mM Tris-HCl, pH 7.6, 100 mM MgCl$_2$, 5 mM dithiothreitol, 0.1 mM spermidine, 0.1 mM EDTA, 10 mM ATP). Then 2 μl of T4 polynucleotide kinase (20 units) was added and incubated at 37° for 2 hr. After the kination reaction, protein was extracted and removed with 50 μl of phenol : chloroform (1 : 1). Then 5 μl of each kinated oligonucleotide solution (YS2, YS3, YS4, and YS5) (0.1 μg), 5 μl of YS1 (0.1 μg), and 5 μl of YS6 (0.1 μg) were heated at 90° for 2 hr and annealed as shown in Fig. 1. Of this solution, 3 μl was ligated with 5 μl of *Eco*RI–*Bam*HI-digested pSV2*neo* by using a ligation kit (Takara Ligation Kit; Takara Shuzo, Kyoto, Japan) at 16° for 30 min. The recombinant obtained was named pSV2007.

Construction of pSV2057. After partially digesting 20 μg of pSV2007 with *Bal*I, this plasmid was completely digested with *Eco*RI. The largest DNA fragment of 5.1 kb (5 μg) was purified on a 0.9% agarose gel. A 0.48-kb *Bal*I–*Eco*RI fragment of the 5′ region in the cDNA of α_2PI (pPI143/E.Bal) was ligated with this *Bal*I–*Eco*Ri-digested pSV2007 fragment by using the ligation kit, yielding plasmid pSV2057.

Construction of pSV057. To shorten the distance between the SV40 early promoter and the translational start signal (ATG) in the inserted oligonucleotide for the cDNA sequence, plasmid pSV2057 was digested with *Bam*HI and *Hin*dIII. After purification of the 3.5-kb fragment by 0.9% agarose gel electrophoresis, both ends were filled-in with *E. coli* DNA polymerase I (Boehringer Mannheim, Mannheim, Germany), and then both ends were ligated by using the ligation kit to produce plasmid pSV057.

Construction of pPI906. To add the remaining coding sequence for α_2PI to pSV057, a 1.7-kb fragment of *Eco*RI pPI143/E [containing the 3′ half of the coding region, 3′ noncoding region, and poly(A) additional signal] was isolated from pPI143. Plasmid pSV057 was digested with *Eco*RI and treated with bacterial alkaline phosphatase. Then pPI143/E was inserted into the *Eco*RI site of pSV057, yielding plasmid pPI906.

Transfection of BHK Cells

Cotransfection of BHK cells with the expression vector pPI906 and selection marker plasmid pSV2*dhfr* was performed by the calcium phosphate method using a Cell Phect transfection kit (Pharmacia). The transformed cells were cultured for selection in 10% fetal calf serum–Dulbecco's modified Eagle's medium (FCS–DMEM) containing 250 μM methotrexate (MTX) for 2 weeks. Of 40 stable MTX-resistant clones isolated, 18 clones were positive for α_2PI secretion as analyzed by ELISA. The highest producing clone, BHK/mu-4, which expresses 1.5 μg/ml of α_2PI in the cultured medium, was used for further analyses.

Purification and Analysis of Expressed α_2PI

A volume of 500 ml of the culture media of the α_2PI producing clone was concentrated to 80 ml in an Amicon concentrator (Amicon Corporation, Danvers, MA) and dialyzed against PBS. The dialyzed concentrate was subjected to immunoaffinity chromatography as described in this chapter. The purified recombinant α_2PI migrated as a single band on SDS–PAGE with an estimated molecular weight of 67,000, that is the same as native mature plasma α_2PI. The NH_2-terminal amino acid sequence of the recombinant α_2PI was determined by a gas-phase sequencer (Applied Biosystems). On each cycle of Edman degradation, two amino acid peaks were detected, and the signal ratio was nearly $1:1$. These sequences were determined to be MEPLGRQL and NQEQVSPL, which correspond to the amino-terminal sequences of the pro type (-12 to -5) and the mature type ($+1$ to $+8$) of α_2PI, respectively. This result shows that half of the pro-α_2PI produced in BHK cells was correctly recognized and cleaved by the proteolytic processing enzyme and secreted as mature α_2PI, while the rest of the pro-α_2PI escaped cleavage by the enzyme and was secreted as pro-α_2PI. The inhibitory activity on plasmin of the recombinant α_2PI was nearly the same as that of purified native (plasminogen-binding form) plasma α_2PI,[5] whereas the cross-linking activity was approximately two-thirds that of plasma α_2PI.

[12] Molecular Interactions between Tissue-Type Plasminogen Activator and Plasminogen

By H. Roger Lijnen and Désiré Collen

Introduction

Tissue-type plasminogen activator (t-PA) converts the proenzyme plasminogen into the active enzyme plasmin. Human t-PA is a serine proteinase with M_r about 70,000, composed of one polypeptide chain containing 527 amino acids. The t-PA molecule contains four domains: (1) a 47-residue-long (residues 4–50) amino-terminal region (F domain) that is homologous with the finger domains mediating the fibrin affinity of fibronectin; (2) residues 50–87 (E domain) that are homologous with human epidermal growth factor; (3) two regions comprising residues 87–176 and 176–262 (K_1 and K_2 domains) that share a high degree of homology with the five kringles of plasminogen; and (4) a serine proteinase domain (residues 276–527) with the active site residues His-322, Asp-371, and Ser-478. Limited plasmic hydrolysis of the Arg^{275}-Ile^{276} peptide bond converts the molecule to a two-chain activator held together by one interchain disulfide bond. Two-chain t-PA has a 5- to 10-fold higher reactivity with low-M_r substrates and inhibitors than does single-chain t-PA. The specific activity of t-PA is approximately 500,000 IU/mg, by comparison with the Second International Reference Preparation for t-PA (code 86/670; National Institute for Biological Standards and Control, UK).

Human plasminogen is a single-chain glycoprotein with M_r 92,000, consisting of 791 amino acids and containing five homologous triple-loop

METHODS IN ENZYMOLOGY, VOL. 223

structures, or kringles. Native plasminogen has NH_2-terminal glutamic acid (Glu-plasminogen), but is easily converted by limited plasmic digestion to modified forms with NH_2-terminal lysine, valine, or methionine, commonly designated Lys-plasminogen. This conversion occurs by hydrolysis of the Arg^{67}-Met^{68}, Lys^{76}-Lys^{77}, or Lys^{77}-Val^{78} peptide bonds. Lys-plasminogen is 5- to 10-fold more susceptible to activation than Glu-plasminogen by all known plasminogen activators. Plasminogen is converted to plasmin by cleavage of the Arg^{561}-Val^{562} peptide bond. The plasmin molecule is a two-chain trypsinlike serine proteinase with an active site composed of His-603, Asp-646, and Ser-741.[1,2]

t-PA is a poor enzyme in the absence of fibrin, but the presence of fibrin strikingly enhances the activation rate of plasminogen. The kinetic data of Hoylaerts et al.[3] support a mechanism in which fibrin provides a surface to which t-PA and plasminogen adsorb in a sequential and ordered way, yielding a cyclic ternary complex. Fibrin essentially increases the local plasminogen concentration by creating an additional interaction between t-PA and its substrate. Early fibrin digestion by plasmin may accelerate fibrinolysis by increasing the binding of both t-PA and plasminogen.

Kinetic Analysis of Plasminogen Activation by t-PA

Plasminogen activation by t-PA may be modulated by several kinetically relevant plasmin-mediated side reactions. Proteolytic degradation of fibrin by plasmin is associated with an increased binding of both plasminogen and t-PA, resulting in a positive-feedback effect on plasminogen activation. Furthermore, the fibrin affinity of single-chain and two-chain t-PA was found by some authors to be similar, but by others to be significantly different. Therefore, kinetic investigation of the activation of plasminogen by t-PA may be hampered by side reactions caused by generated plasmin. These include the conversion of single-chain t-PA to its two-chain derivative, the conversion of Glu-plasminogen to Lys-plasminogen, and the proteolytic degradation of fibrin. Moreover, plasminogen may exist in different conformational states (normal or metastable), with different sensitivities to activation by plasminogen activators. Furthermore, differently glycosylated t-PA moieties differ in the kinetics of fibrin-dependent plasminogen activation. These considerations may to some extent explain the greatly different kinetic constants obtained in different studies (Table I).

[1] D. Collen and H. R. Lijnen, Biochem. Pharmacol. **40**, 177 (1990).
[2] H. R. Lijnen and D. Collen, Thromb. Haemostasis **66**, 88 (1991).
[3] M. Hoylaerts, D. C. Rijken, H. R. Lijnen, and D. Collen, J. Biol. Chem. **257**, 2912 (1982).

TABLE I

KINETIC PARAMETERS FOR ACTIVATION OF PLASMINOGEN BY t-PA IN THE ABSENCE OR PRESENCE OF FIBRIN STIMULATOR

Proteinase	Without fibrin		With fibrin			Ref.
	K_m (μM)	k_{cat} (sec^{-1})	Stimulator	K_m (μM)	k_{cat} (sec^{-1})	
Single-chain t-PA						
Glu-plasminogen	>100	ND	Solid phase	2.42	0.22	Rijken et al.[4]
	4.9	0.0013	Fibrin monomer	0.46	0.084	Rånby[5]
	83	0.07	CNBr-digested fibrinogen	0.18	0.28	Zamarron et al.[6]
Lys-plasminogen	4.0	0.024	Solid phase	0.049	0.194	Takada et al.[7]
	0.17	0.003	Fibrin monomer	0.036	0.185	Rånby[5]
	0.40	0.062		—	—	Takada et al.[8]
Two-chain t-PA						
Glu-plasminogen	65	0.06	Solid phase	0.16	0.10	Hoylaerts et al.[3]
	>100	ND	Solid phase	1.07	0.10	Rijken et al.[4]
	7.6	0.0078	Fibrin monomer	0.18	0.12	Rånby[5]
	0.053	0.002	CNBr-fragment 2 of fibrinogen	0.003	0.058	Nieuwenhuizen et al.[9]
Lys-plasminogen	1.20	0.015	Solid phase	0.063	0.359	Takada et al.[7]
	19	0.20	Solid phase	0.02	0.20	Hoylaerts et al.[3]
	0.30	0.025	Fibrin monomer	0.035	0.22	Rånby[5]
	0.030	0.006	CNBr-fragment 2 of fibrinogen	0.010	0.22	Nieuwenhuizen et al.[9]

Plasminogen Activation by t-PA in Absence of Stimulation

Activation of Glu-plasminogen by single-chain or two-chain t-PA obeys Michaelis–Menten kinetics.[3-9] Activation of Lys-plasminogen by t-PA also obeys Michaelis–Menten kinetics according to most studies. Rånby[5] has, however, reported that activation of Lys-plasminogen by both single-chain and two-chain t-PA obeys Michaelis–Menten kinetics only at substrate concentrations below 1 μM. At higher substrate concentrations, the experimental data (activation rate versus substrate concentration) could more accurately be fitted to the Wong–Hanes equation. Activation of low-M_r plasminogen by t-PA also obeys Michaelis–Menten kinetics, with k_{cat} equal to that of Glu-plasminogen or Lys-plasminogen, but with K_m, respectively, fivefold lower or twofold higher.[8]

Experimental Conditions. Different amounts of Glu-plasminogen (final concentration up to 100 μM) are incubated at 37° with t-PA (fixed enzyme concentration of 10–20 nM) in 0.1 M phosphate buffer, pH 7.4, containing 0.01% Tween 80. At different time intervals (0–10 min), samples (10 μl) are removed from the incubation mixtures, diluted at least 50-fold, and generated plasmin is measured with a chromogenic substrate (i.e., D-Val-Leu-Lys-pNA, S-2251, at a final concentration of 0.3 to 1.0 mM). Plasminogen concentrations are increased until the rate of plasmin generation is saturated, i.e., does not further increase proportionally with increase of the plasminogen concentration.

Analysis. A calibration curve is constructed with purified plasmin (of which the concentration is determined by active site titration, i.e., with p-NPGB); therefore the hydrolysis of the chromogenic substrate (same final concentration and incubation milieu as in the kinetic measurements) is measured with different concentrations of plasmin, and the change in absorbance at 405 nm ($\Delta A_{405\ nm}$/min) is plotted versus the plasmin concentration. The rate of hydrolysis of the chromogenic substrate is proportional to the plasmin concentration. The change in $A_{405\ nm}$ measured experimentally at different time points is then converted to plasmin concentration (in nanomoles/liter) using this calibration curve. Under initial rate conditions, plots of generated plasmin versus time are linear, and allow determination of the initial activation rate v (nM/sec) for each concentration of plasmino-

[4] D. C. Rijken, M. Hoylaerts, and D. Collen, *J. Biol. Chem.* **257,** 2920 (1982).

[5] M. Rånby, *Biochim. Biophys. Acta* **704,** 461 (1982).

[6] C. Zamarron, H. R. Lijnen, and D. Collen, *J. Biol. Chem.* **259,** 2080 (1984).

[7] A. Takada, Y. Sugawara, and Y. Takada, *Haemostasis* **18,** 117 (1988).

[8] A. Takada, Y. Takada, and Y. Sugawara, *Thromb. Res.* **49,** 253 (1988).

[9] W. Nieuwenhuizen, M. Voskuilen, A. Vermond, B. Hoegee-de Nobel, and D. W. Traas, *Eur. J. Biochem.* **174,** 163 (1988).

gen ([P]) used. Over a range of substrate concentrations three to five times above and below the Michaelis constant K_m, double-reciprocal plots of v versus [P] are linear according to the Michaelis–Menten equation. From this Lineweaver–Burk plot, K_m and the catalytic rate constant k_{cat} or k_2 can be determined, and the second-order rate constant (k_{cat}/K_m), a measure of the catalytic efficiency of the enzyme, can be calculated.

Comments. Kinetics of plasminogen activation should be measured in buffers devoid of chloride ions, as these were reported to affect activatability of plasminogen in purified systems. If low t-PA concentrations are used, traces of detergent (i.e., 0.01% Tween 80) should be included in the buffers to avoid loss of t-PA by adsorption to surfaces. To stabilize generated plasmin in the incubation mixture, glycerol may be included.

Catalytic amounts of t-PA should be used, in order to keep the enzyme : substrate ratio as low as possible (i.e., below 10%). At high product : substrate (plasmin : plasminogen) ratios, plots of generated plasmin versus time will deviate from linearity (downward). At low plasmin concentration, plasmin-mediated side reactions, including conversion of Glu-plasminogen to Lys-plasminogen and conversion of single-chain t-PA to two-chain t-PA, will be minimal. Significant conversion of Glu-plasminogen to Lys-plasminogen will be apparent by deviation from linearity (upward) of plots of generated plasmin versus time, due to the higher susceptibility of Lys-plasminogen to activation.

Samples from the incubation mixtures should be diluted before addition of chromogenic substrate in order to interrupt the activation process, to determine the generated plasmin at exact time points, and to eliminate effects of the chromogenic substrate on the activation rate.

The quality of the plasminogen used is important. Availability of homogeneous preparations of Glu- or Lys-plasminogen should be confirmed (e.g., by NH$_2$-terminal amino acid sequence analysis or SDS–PAGE). Furthermore, plasminogen preparations should be devoid of ε-aminocaproic acid or other lysine analogs, which are used in the isolation of plasminogen by chromatography on lysine-Sepharose. These lysine analogs indeed significantly affect plasminogen activation by all known plasminogen activators. Activatability of plasminogen preparations can be highly variable and should be evaluated, for instance, as follows. Plasminogen (final concentration 10–20 mg/ml), in phosphate-buffered saline containing 25% glycerol, is activated at 0° with streptokinase ($\approx$1500 U/mg plasminogen) and generated plasmin is measured after 2 to 4 hr by active site titration with *p*-NPGB. Plasminogen preparations adequate for kinetic analysis should be activated for ≥80% under these conditions and should be fully activated after overnight incubation at 0°.

Plasminogen Activation by t-PA in Presence of Fibrinlike Stimulators

Several studies have shown that the activation of plasminogen by t-PA is significantly enhanced in the presence of fibrin.[3-9] Kinetic analysis of plasminogen activation by t-PA in the presence of fibrin is complicated by the fact that fibrinlike material both stimulates the activation step and acts as a substrate of the formed enzyme plasmin. In the study by Hoylaerts *et al.*,[3] these two phenomena could be dissociated in time and measured separately by interruption of the activation process at different time points by addition of a mixture of tranexamic acid and zinc chloride. Under these conditions true initial activation rates could be measured, and the activation of both Glu-plasminogen and Lys-plasminogen by two-chain t-PA was found to obey Michaelis–Menten kinetics with kinetic constants as summarized in Table I. Thus, Hoylaerts *et al.*[3] found a Michaelis constant of 0.16 μM for the activation of Glu-plasminogen in the presence of fibrin as compared to 65 μM in the absence of fibrin, whereas the catalytic rate constant was comparable in the presence or the absence of fibrin (0.10 or 0.06 sec^{-1}, respectively). The catalytic efficiency (k_{cat}/K_m) for the activation of plasminogen thus increases about 1500-fold in the presence of fibrin. Nieuwenhuizen *et al.*,[9] however, reported that fibrin affects both the K_m (20-fold decrease) and the k_{cat} (30-fold increase) of plasminogen activation by t-PA.

Rijken *et al.*,[4] using a system in which lysis of fibrin was prevented by an excess of aprotinin, which is a potent plasmin inhibitor, found a K_m of 1.1 μM. Rånby[5] found that during the degradation of fibrin, the K_m decreased, compatible with the formation of a more stable ternary complex at later phases of fibrinolysis. This finding was further elaborated by Norrman *et al.*,[10] who showed that there were two phases in the activation of Glu-plasminogen by t-PA in the presence of fibrin: the first-phase K_m was 1.05 μM and k_{cat} was 0.15 sec^{-1}, whereas the second-phase K_m was 0.07 μM and k_{cat} was 0.14 sec^{-1}. Transition from the first to second phase occurred after limited plasmin-mediated fibrin digestion and was associated with a 15-fold decrease in K_m.

More recently, it has been reported that *in vitro* the activation of plasminogen by t-PA is also stimulated by soluble cofactors such as polylysine, fibrin monomer, and CNBr-digested fibrinogen. The use of soluble cofactors permits a more thorough kinetic analysis, because initial activation rates can be measured under true equilibrium conditions. From studies with active site mutagenized plasminogen and plasmin-resistant t-PA, it was suggested that the mechanism of stimulation of plasminogen activation

[10] B. Norrman, P. Wallén, and M. Rånby, *Eur. J. Biochem.* **149,** 193 (1985).

with single-chain t-PA by CNBr-digested fibrinogen is different from that by soluble fibrin monomer (desAAfibrin). This may be due to a higher sensitivity of the latter to further plasmic degradation during the activation process. These studies also indicated that, if generation of plasmin is not adequately prevented, conversion of single-chain to two-chain t-PA and further degradation of the fibrin stimulator may indeed significantly influence the initial activation rate.[11] Because of the better solubility of CNBr-digested fibrinogen than of fibrin monomer, several authors have preferred the use of this fibrinlike stimulator to evaluate the effect of fibrin on plasminogen activation by different t-PA moieties. We will therefore describe such a system in more detail, using experimental conditions designed to minimize the above-mentioned plasmin-mediated side reactions.

Experimental Conditions. Different amounts of plasminogen (i.e., final concentration 0.05–1.0 μM for Glu-plasminogen) are incubated at 37° in 0.1 M phosphate buffer, pH 7.4, containing 0.01% Tween 80, with different amounts of CNBr-digested fibrinogen (i.e., final concentration 0.05–2.0 μM) prior to addition of t-PA (i.e., final concentration 2–5 nM). At different time intervals (0–10 min) samples are removed from the incubation mixtures, diluted 20- to 50-fold, and generated plasmin is measured with a chromogenic substrate (i.e., S2251 at a final concentration of 0.3 to 1.0 mM).

Analysis. Initial activation rates (v) are obtained from linear plots of generated plasmin concentration versus activation time, as described above. Under the experimental conditions described above, double-reciprocal plots of the initial activation rate ($1/v$) versus the plasminogen concentration ($1/[P]$) are linear for each concentration of CNBr-digested fibrinogen, indicating that the activation obeys Michaelis–Menten kinetics (illustrated in Fig. 1A). Plots of $1/v$ versus the inverse of the concentration of CNBr-digested fibrinogen ($1/[f]$) are also linear for each concentration of plasminogen (Fig. 1B). The ordinate intercepts in Fig. 1B represent, for each plasminogen concentration, the initial rates of plasmin generation at infinite fibrin concentration. These intercepts are replotted in Fig. 1A versus the plasminogen concentration (∞f), yielding a straight line of which the abscissa intercept determines the Michaelis constant ($1/K_m$) of plasminogen activation by t-PA at infinite fibrin concentration (at which all plasminogen activator is assumed to be bound to fibrin). The ordinate intercept of this replot (which coincides with the intercepts of the other lines in Fig. 1A) gives the maximal activation rate and thus allows determination of the catalytic rate constant (k_{cat}) for plasminogen activation by

---

[11] H. R. Lijnen, B. Van Hoef, F. De Cock, and D. Collen, *Thromb. Haemostasis* **64**, 61 (1990).

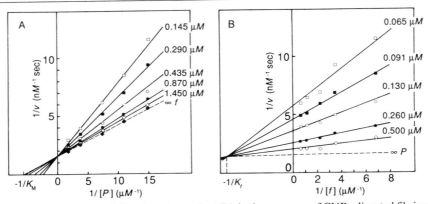

FIG. 1. Activation of Glu-plasminogen by t-PA in the presence of CNBr-digested fibrinogen. (A) Lineweaver–Burk plots of $1/v$ versus $1/[P]$ for the activation of plasminogen
($0.065-0.500 \mu M$) by t-PA (60 nM melanoma t-PA) in the presence of increasing concentrations of CNBr-digested fibrinogen ($0.145-1.45 \mu M$). The dotted line represents $1/v$ at infinite
concentration of CNBr-digested fibrinogen (data obtained from ordinate intercepts in B).
(B) Lineweaver–Burk plots of $1/v$ versus $1/[f]$, obtained by replotting the data of A against
plasminogen concentration. The dotted line represents $1/v$ at infinite plasminogen concentration (ordinate intercept of A). Reproduced with permission from C. Zamarron, H. R. Lijnen,
and D. Collen, *J. Biol. Chem.* **259**, 2080 (1984).

fibrin-bound t-PA. From the intersection point ($-1/K_f$) in Fig. 1B, the
dissociation constant of the fibrin–t-PA complex can be determined (yielding values ranging between 0.14 and 0.65 μM for different t-PA moieties[3,6]).

    Comments. The comments made above for plasminogen activation
in the absence of stimulator remain applicable. Furthermore, arginine,
frequently added to t-PA solutions to improve solubility, must be removed
because it may interfere with the interaction of t-PA with fibrin.

    Dilution of samples removed from the incubation mixtures before mea-
surement of generated plasmin is preferred over continuous monitoring
in the presence of a chromogenic substrate. Competition between the
fibrinlike stimulator and the chromogenic substrate for plasmin in the
incubation mixture may result in underestimation of the plasmin concentration. On ≥20-fold dilution of samples, the concentration of fibrinlike
material is reduced to ≤0.1 μM, at which concentration competition with
the chromogenic substrate for plasmin can be neglected (i.e., K_I for fibrinogen fragments on hydrolysis of S2251 by plasmin is about 1 μM).

Quantitation of Stimulatory Effect of Fibrinlike Material on
Plasminogen Activation by t-PA

    If a kinetic analysis of the effect of fibrin on the activation rate of
plasminogen by t-PA has been performed as described above (i.e., range

of concentrations of both plasminogen and fibrin stimulator, with fixed enzyme concentration), the stimulation factor by fibrin can be obtained as the ratio of the catalytic efficiency (k_{cat}/K_m) in the presence and the absence for fibrin. Thus, in the study of Zamarron et al.,[6] the catalytic efficiency of single-chain t-PA for activation of Glu-plasminogen in the presence of fibrin (1.56 μM^{-1} sec^{-1}) is 1940-fold higher than in the absence of fibrin (0.008 μM^{-1} sec^{-1}). Although different kinetic parameters were obtained in the studies summarized in Table I, activation of Glu-plasminogen by single-chain or two-chain t-PA was always found to be stimulated between 460- and 3770-fold by fibrinlike material, as compared to 75- to 1000-fold for Lys-plasminogen. Similar kinetic analyses of the effect of fibrinogen on the activation of plasminogen by t-PA have revealed a stimulation factor of 12- to 16-fold for activation of Glu-plasminogen or Lys-plasminogen by two-chain t-PA in the presence of saturating concentrations of fibrinogen.[3] Using only one fibrinogen concentration (1.5 μM), Takada et al.[7] have also found that activation of Glu-plasminogen by single-chain or two-chain t-PA is stimulated 13- or 16-fold, respectively.

The procedure outlined above to quantitate the stimulatory effect of fibrin on plasminogen activation by t-PA requires measurements at different concentrations of both substrate and stimulator, which is relatively laborious and time-consuming. To compare the stimulatory effect of fibrin on plasminogen activation by different t-PA moieties (i.e., a set of recombinant domain deletion/substitution mutants of t-PA), this procedure may be simplified as follows.[12]

Experimental Conditions. The initial activation rate (v in nM sec^{-1}) of plasminogen by t-PA in the absence of fibrin is determined as described above. The effect of fibrin on the activation rate of plasminogen by t-PA is evaluated as described above, but using only one plasminogen concentration (i.e., final concentration 1.0 μM).

Analysis. The initial activation rate of plasminogen by t-PA is stimulated in a concentration-dependent manner by addition of CNBr-digested fibrinogen (Fig. 2A). The activation rate at infinite concentration of fibrin is determined from the inverse of the ordinate intercept in a double-reciprocal plot of the activation rate versus the concentration of CNBr-digested fibrinogen (Fig. 2B). The stimulation factor of CNBr-digested fibrinogen on plasminogen activation is obtained as the ratio of the initial activation rate in the presence of infinite fibrin concentration and the initial activation rate in the absence of fibrin (expressed for the same concentration of substrate and enzyme). In the example shown in Fig. 2, the activation rate at infinite fibrin concentration equals 5.5 nM sec^{-1} as

[12] H. R. Lijnen, L. Nelles, B. Van Hoef, F. De Cock, and D. Collen, *J. Biol. Chem.* **265,** 5677 (1990).

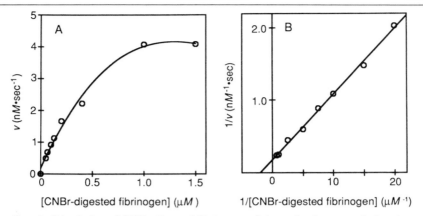

FIG. 2. Stimulation of CNBr-digested fibrinogen of the activation rate of plasminogen by t-PA. (A) The initial activation rate of plasminogen (final concentration 1 μM) by recombinant t-PA (final concentration 10 nM) is plotted against the concentration of CNBr-digested fibrinogen (0–1.5 μM). (B) Double-reciprocal plot of the activation rate versus the concentration of CNBr-digested fibrinogen.

compared to 0.05 nM sec^{-1} in the absence of fibrin, resulting in a stimulation factor of 110.

Comments. In order to allow direct comparison, initial activation rates in the absence and the presence of fibrin have to be determined at the same concentration of plasminogen and t-PA. To convert initial activation rates (v) to catalytic efficiencies of t-PA (k_{cat}/K_m), the values of v would have to be multiplied by a factor that relates to the K_m according to the formula $v/[E][S] = k_{cat}/(K_m + [S])$. In the absence of fibrin, with a high K_m value (65 to 100 μM) and [S] of 1.0 μM, this correction factor is negligible; in the presence of fibrin, however, assuming a K_m value of 0.2 μM, this correction factor would be 6 for a substrate concentration of 1.0 μM. The stimulatory effect of fibrin on the catalytic efficiency of t-PA moieties may thus be greater than reflected by its effect on the initial activation rates. Furthermore, this approach is valid only when the reaction studied obeys Michaelis–Menten kinetics.

[13] Cellular Receptor for Urokinase-Type Plasminogen Activator: Protein Structure

By Niels Behrendt, Michael Ploug, Ebbe Rønne, Gunilla Høyer-Hansen, and Keld Danø

Introduction

The urokinase pathway of plasminogen activation is assumed to play an important role in extracellular proteolytic events in a number of tissue degradation processes occurring under normal as well as pathological conditions, including cancer invasion.[1-3] Therefore, the existence of a specific cell surface binding site for the urokinase-type plasminogen activator (uPa) has aroused great interest. This receptor, uPAR, was first found on human monocyte-derived cells[4,5] and later on a variety of other cell types of malignant and nonmalignant origin.[6,7] It binds active uPA as well as the zymogen pro-uPA with a high affinity ($K_D \approx 10^{-9}-10^{-11} M$, depending on the cell type[8]).

A number of *in vitro* studies have provided clues as to possible roles of the uPAR–uPA interaction. On certain cell types, receptor-bound uPA is discretely localized at cell–cell and focal cell–substratum contact sites.[9-11] Concomitant binding of the zymogen pro-uPA to uPAR and of plasminogen to unidentified binding sites at cell surfaces leads to a strong enhancement in plasmin generation compared to the liquid-phase situa-

[1] K. Danø, P. A. Andreasen, J. Grøndahl-Hansen, P. Kristensen, L. S. Nielsen, and L. Skriver, *Adv. Cancer Res.* **44,** 139 (1988).

[2] O. Saksela and D. B. Rifkin, *Annu. Rev. Cell Biol.* **4,** 93 (1988).

[3] J. Pöllänen, R. W. Stephens, and A. Vaheri, *Adv. Cancer Res.* **57,** 273 (1991).

[4] J.-D. Vassalli, D. Baccino, and D. Belin, *J. Cell Biol.* **100,** 86 (1985).

[5] M. P. Stoppelli, A. Corti, A. Soffientini, G. Cassani, F. Blasi, and R. K. Associan, *Proc. Natl. Acad. Sci. U.S.A.* **82,** 4939 (1985).

[6] A. Bajpai and J. B. Baker, *Biochem. Biophys. Res. Commun.* **133,** 994 (1985).

[7] L. S. Nielsen, G. M. Kellerman, N. Behrendt, R. Picone, K. Danø, and F. Blasi, *J. Biol. Chem.* **263,** 2358 (1988).

[8] F. Blasi, N. Behrendt, M. V. Cubellis, V. Ellis, L. R. Lund, M. T. Masucci, L. B. Møller, D. P. Olson, N. Pedersen, M. Ploug, E. Rønne, and K. Danø, *Cell Differ. Dev.* **32,** 247 (1990).

[9] C. A. Hébert and J. B. Baker, *J. Cell Biol.* **106,** 1241 (1988).

[10] J. Pöllänen, O. Saksela, E.-M. Salonen, P. Andreasen, L. Nielsen, K. Danø, and A. Vaheri, *J. Cell. Biol.* **104,** 1085 (1987).

[11] J. Pöllänen, K. Hedman, L. S. Nielsen, K. Danø, and A. Vaheri, *J. Cell Biol.* **106,** 87 (1988).

tion.[12] These and other observations have led to the speculation that uPAR provides specific areas on surfaces with the potential for preferential plasminogen activation and extracellular proteolysis.[13] uPAR also takes part in internalization processes, involving complexes of uPA and the plasminogen activator inhibitors PAI-1[14] or PAI-2,[15] by an as yet unknown mechanism.

Various agents have been found to regulate the cellular expression of uPAR[5,16-20] and thus influence the above processes. Thus, the stimulation of U937 cells with phorbol 12-myristate 13-acetate (PMA) leads to a profound increase in the uPAR number per cell,[5] which has been traced back to an increased gene transcription.[21] This increase is accompanied by a decrease in the affinity of uPA binding[19]; however, the mechanism is unknown.

Work during the last few years has led to the purification of the human uPAR,[7,22] the isolation and sequencing of complete cDNAs encoding the human[23] and mouse[24] proteins, and the assignment of the human uPAR gene to the long arm of chromosome 19.[25] The human uPAR, isolated from PMA-treated U937 cells, is a single-polypeptide chain, highly glycosylated membrane protein with an M_r of $\approx$55,000–60,000.[7,22] Further molecular characterization has shown that the protein is COOH-terminally processed

[12] V. Ellis, N. Behrendt, and Keld Danø, this volume [14].

[13] K. Danø, N. Behrendt, L. R. Lund, E. Rønne, J. Pöllänen, E.-M. Salonen, R. W. Stephens, H. Tapiovaara, and A. Vaheri, in "Cancer Metastasis" (V. Schirrmacher and R. Schwartz-Albiez, eds.), p. 98. Springer-Verlag, Berlin, 1989.

[14] M. V. Cubellis, T. C. Wun, and F. Blasi, EMBO J. 9, 1079 (1990).

[15] A. Estreicher, J. Mülhauser, J.-L. Carpentier, L. Orci, and J. D. Vassalli, J. Cell Biol. 111, 783 (1990).

[16] D. Boyd, G. Florent, G. Murano, and M. Brattain, Biochim. Biophys. Acta 947, 96 (1988).

[17] J. C. Kirchheimer, Y.-H. Nong, and H. G. Remold, J. Immunol. 141, 4229 (1988).

[18] A. Estreicher, A. Wohlwend, D. Belin, and J.-D. Vassalli, J. Biol. Chem. 264, 1180 (1989).

[19] R. Picone, E. L. Kajtaniak, L. S. Nielsen, N. Behrendt, M. R. Mastronicola, M. V. Cubellis, M. P. Stoppelli, S. Pedersen, K. Danø, and F. Blasi, J. Cell Biol. 108, 693 (1989).

[20] L. R. Lund, J. Rømer, E. Rønne, V. Ellis, F. Blasi, and K. Danø, EMBO J. 10, 3399 (1991).

[21] L. R. Lund, E. Rønne, A. L. Roldan, N. Behrendt, J. Rømer, F. Blasi, and K. Danø, J. Biol. Chem. 266, 5177 (1991).

[22] N. Behrendt, E. Rønne, M. Ploug, T. Petri, D. Løber, L. S. Nielsen, W.-D. Schleuning, F. Blasi, E. Appella, and K. Danø, J. Biol. Chem. 265, 6453 (1990).

[23] A. L. Roldan, M. V. Cubellis, M. T. Masucci, N. Behrendt, L. R. Lund, K. Danø, E. Appella, and F. Blasi, EMBO J. 9, 467 (1990).

[24] P. Kristensen, J. Eriksen, F. Blasi, and K. Danø, J. Cell Biol. 115, 1763 (1991).

[25] A. D. Børglum, A. Byskov, A. L. Roldan, P. Ragno, P. Triputti, G. Cassani, K. Danø, F. Blasi, L. Bolund, and T. A. Kruse, Am. J. Hum. Genet. 50, 492 (1992).

and anchored to the membrane by a glycosylphosphatidylinositol moiety.[26] The amino acid sequence consists of three homologous repeats likely to represent individual domains, the binding determinant for uPA being located in the NH$_2$-terminal domain.[27,28] Each of the domains shows homology to the murine Ly-6 antigens, the human membrane inhibitor of reactive lysis CD59, and a squid glycoprotein Spg-2.[27] Several reviews[8,13,29] cover some of these findings. Here, we describe a number of methods that we have found convenient for the detection, purification, and molecular characterization of this protein.

Detection and Assay

The detection methods listed here rely on the high-affinity binding of uPAR to the amino-terminal fragment (ATF) of uPA (i.e., residues 1–135, which contain the receptor-binding determinant[5,30]) or to uPA inactivated with the active site titrant diisopropyl fluorophosphate (DFP). The advantage of these uPAR-binding reagents is their inability to bind plasminogen activator inhibitors, unlike active uPA. Two of the methods (i.e., chemical cross-linking and ligand-blotting) lead to electrophoretic identification of the active component whereas the radioligand binding assay with whole cells relies on binding alone. We are not aware of any biochemically characterized component other than uPAR that has similar binding characteristics. However, for a secure biochemical identification, the present methods may be combined with other techniques based on independent criteria (see sections on glycosylation and antibodies, below).

Reagents for Detection

For DFP treatment of uPA, the two-chain M_r 55,000 form of the human enzyme (Serono) is dissolved at a concentration of 1 mg/ml in 0.1 M Tris-

[26] M. Ploug, E. Rønne, N. Behrendt, A. L. Jensen, F. Blasi, and K. Danø, *J. Biol. Chem.* **266,** 1926 (1991).

[27] N. Behrendt, M. Ploug, L. Patthy, G. Houen, F. Blasi, and K. Danø, *J. Biol. Chem.* **266,** 7842 (1991).

[28] E. Rønne, N. Behrendt, V. Ellis, M. Ploug, K. Danø, and G. Høyer-Hansen, *FEBS Lett.* **288,** 233 (1991).

[29] M. Ploug, N. Behrendt, D. Løber, and K. Danø, *Semin. Thromb. Hemostasis* **17,** 183 (1991).

[30] E. Appella, E. A. Robinson, S. J. Ulrich, M. P. Stoppelli, A. Corti, G. Cassani, and F. Blasi, *J. Biol. Chem.* **262,** 4437 (1978).

HCl, 0.1% (w/v) Tween 80, pH 8.1. A fresh, 500 mM stock solution of DFP (Sigma, St. Louis, MO) in 2-propanol is added to yield a DFP concentration of 5 mM. (Precautions must be taken when handling the acutely toxic DFP, in order to avoid any contact, including the inhalation of vapors.) The solution is incubated at 37° for 4 hr, with a further addition of the same amount of DFP after the first 2 hr. The reaction is terminated by extensive dialysis at 0° against the same buffer as used during DFP treatment. When treated in this manner, uPA retains less than 0.02% of its enzymatic activity but is still capable of binding uPAR; however, side reactions due to the high DFP concentration used cannot be excluded and any minor effects on the binding properties have not been studied. For radioiodination of ATF[31] or DFP-treated uPA, 20 μg of each protein is labeled with 1 mCi of ^{125}I using the Iodogen procedure as published previously.[22] For labeling of DFP-treated uPA with biotin, the protein is treated with biotin amidocaproate N-hydroxysuccinimide ester (Sigma) according to a published method.[32]

Method 1: Radioligand Binding Assay

Before the assay, the cells to be analyzed are washed and acid treated as described in the section on purification, below. Cells (10^6 per tube), radiolabeled ATF, and, in certain samples, competing proteins (see below) are mixed in a total volume of 300 μl. PBS (10 mM sodium phosphate, 140 mM NaCl, pH 7.4) with 0.1% bovine serum albumin (PBS/BSA) is used for suspension of cells, for dilution of reagents, and for all subsequent washing steps. For the construction of a binding curve, the concentration of labeled ligand is varied from 0.05 to 10 nM. For each ATF concentration, parallel samples are incubated in the absence and presence of a large excess (700 nM) of unlabeled uPA. For the higher ATF concentrations, a significant part of the radioactivity may bind to the plastic tube in a semisaturable manner, which makes separate, cell-free control samples necessary in some cases. To ensure specific binding, related proteins with no receptor cross-reactivity (e.g., tPA or plasminogen[4,5,7]) may be included in some samples. The samples are incubated for 1 hr at 4° with vigorous shaking and then diluted to a volume of 1.3 ml, after which the cells are recovered by mild centrifugation. The cell pellet is washed twice in 1 ml of buffer. After the final centrifugation, the bound radioactivity is mea-

---

[31] ATF has been made available to us by kind gifts from Drs. G. Cassani (Lepetit, Italy) and A. Mazar (Abbott); however, methods for the production and purification of this fragment have been published.[30]

[32] J.-L. Guesdon, T. Ternynck, and S. Avrameas, *J. Histochem. Cytochem.* **27**, 1131 (1979).

sured in a gamma counter and compared to the nonbound radioactivity. The specific binding is taken as the amount displaceable by simultaneous incubation with excess uPA. With U937 cells (not subjected to PMA treatment), the specific binding accounts for more than 95% of the total binding obtained in the range of 0–0.3 nM labeled ligand. Scatchard plotting of the binding data obtained with this cell type reveals a single class of receptors with $(1–5) \times 10^4$ molecules per cell and a K_d in the range of 0.2–0.5 nM.[7] The convenience of this assay depends strongly on the cell type, adherent cells being difficult to handle. ^{125}I-labeled, DFP-treated uPA can be used for binding experiments in the same manner but tends to yield higher nonspecific binding levels than is found with ATF. As an alternative to this method, the labeled ligand (ATF or DFP-treated uPA) can be used at a low, fixed concentration, the binding curve being constructed by titration with increasing amounts of nonlabeled competitor.[5,7]

Method 2: Chemical Cross-Linking Assay

For the preparation of samples, cell lysates or isolated detergent phases thereof are produced as detailed in the section on purification, below. However, the applicability of Triton X-114 phase separation is found to vary strongly among cell types, necessitating the use of a high Triton X-114 concentration (3%) in some cases. The final samples should include 0.25% CHAPS, which is a zwitterionic detergent, 3-[(3-cholamidopropyl) dimethylammonio]-1-propane sulfonate, for prevention of renewed phase separation. Alternatively, total cell lysates not to be used for phase separation can be prepared by replacing Triton X-114 with CHAPS (0.5–1%) directly in the lysis buffer. The samples are diluted 5- to 100-fold in PBS with 0.1% Tween 80 (PBS/Tween) and incubated with 1 nM ^{125}I-labeled ATF. Parallel samples may include competing or noncompeting proteins as above, and a control is made with buffer instead of cell material. After 1 hr at 4°, the samples are transferred to room temperature. The homobifunctional cross-linker, N,N'-disuccinimidyl suberate (DSS; Pierce, Rockford, IL; 2 mM final concentration), is immediately added in the form of a 40 mM stock solution in dimethyl sulfoxide (DMSO). After 15 min at room temperature, ammonium acetate (10 mM final concentration) is added in order to block any remaining unreacted cross-linker. The samples are left for at least 10 min at room temperature and finally analyzed by SDS–PAGE[33] and autoradiography of the gel. uPAR activity is revealed

---

[33] For SDS–PAGE, 6–16% gradient slab gels are routinely used, employing the system of U. K. Laemmli [*Nature (London)* **227**, 680 (1970)].

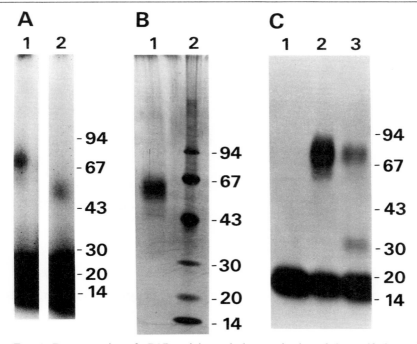

Fig. 1. Demonstration of uPAR activity and characterization of the purified protein. (A) Demonstration of uPAR activity within a crude mixture by chemical cross-linking to ^{125}I-labeled ATF. Lane 1, A total lysate of PMA-treated U937 cells, produced using 0.5% CHAPS, was subjected to cross-linking analysis followed by SDS–PAGE under reducing conditions. uPAR activity is revealed by the formation of an M_r 70,000–80,000 radiolabeled protein conjugate. Lane 2, After cross-linking, a sample of the material shown in lane 1 was subjected to enzymatic deglycosylation with peptide N-glycosidase F and electrophoresed on the same gel. The labeled conjugate is turned into a product of $M_r \approx 50,000$. (B) SDS–PAGE of purified uPAR. Lane 1, Approximately 1 μg of purified protein was analyzed under reducing conditions; lane 2, M_r marker proteins. The gel was silver stained. (C) Activity of purified uPAR and its ligand-binding domain. The following samples were subjected to chemical cross-linking with ^{125}I-labeled ATF and analyzed by SDS–PAGE under reducing conditions. Lane 1, Buffer control; lane 2, purified uPAR; lane 3, a digest of purified uPAR, generated by treatment with chymotrypsin (40 ng/ml) for 7 hr at 37°. Liberation of the active, ligand-binding domain is revealed by the formation of an M_r 32,000 conjugate. The M_r 70,000–80,000 conjugate represents residual, intact protein. Numbers on the right-hand side of each gel represent the molecular weight ($M_r \times 10^{-3}$) of marker proteins. The data are from Behrendt et al.[22,27]

by the formation of a labeled conjugate with an apparent M_r in the range of 55,000–80,000 (Fig. 1A, lane 1), this variability being due to differences in N-linked carbohydrate.[22] Identical results are found using reduced and nonreduced samples for SDS–PAGE. The cross-linking assay can also be carried out with labeled DFP-treated uPA as the ligand, leading to a conjugate of M_r 90,000–110,000. The signal (i.e., the fraction of label found in the form of protein conjugate) is dose dependent and a dynamic range can be found that allows semiquantitative activity comparison of samples within the same experiment.[7] However, the role of factors such as the total protein concentration and differences in affinity known to exist among uPAR variants[19] has not been studied in detail. Cross-linking can also be performed on whole cells, preincubated with ^{125}I-labeled ATF and washed as above, in the absence of detergent.[7] In this case, cells are washed again after cross-linking, followed by detergent lysis and electrophoretic analysis.

Method 3: Ligand Blotting

Detergent phases of cell lysates (see purification, below) are resolved by SDS–PAGE under nonreducing conditions. The gel is electroblotted[34] onto a polyvinylidene difluoride (PVDF) membrane (Millipore, Bedford, MA). After blotting, the membrane is treated with blocking buffer (1% skimmed milk powder in 10 mM Tris-HCl, 150 mM NaCl, 0.1% Tween 20, pH 8.0) for 1 hr, followed by incubation for 2 hr with biotin-labeled, DFP-treated uPA (a 4 nM solution in blocking buffer). The membrane is washed with TBS/Tween (10 mM Tris-HCl, 150 mM NaCl, 0.1% Tween 20, pH 8.0), and the bound ligand is localized by incubation of the membrane for 1 hr with streptavidin-coupled alkaline phosphatase (Dakopatts, Copenhagen, Denmark; 1000-fold diluted in blocking buffer), followed by washing and visualization of phosphatase activity by a standard procedure employing nitroblue tetrazolium and 5-bromo-4-chloro-3-indolyl phosphate.[35] The electrophoretic mobility of uPAR (apparent M_r 40,000–60,000) is revealed directly by this method, which allows direct comparison to, for example, Western blotting studies. Ligand blotting works only with nonreduced samples and is most conveniently performed with detergent phases obtained by Triton X-114 phase separation (see below); in unfractionated lysates the high total protein concentration leads to dominant unspecific staining of the more abundant components. The sensitivity is 10- to 50-fold less than that obtained with chemical cross-linking to radiolabeled ligand.

---

[34] A "semi-dry" electroblotting apparatus (Kem-En-Tec, Denmark) has been used routinely, following a blotting procedure developed for this equipment [J. Kyhse-Andersen, *J. Biochem. Biophys. Methods* **10**, 203 (1984)].

[35] E. Harlow and D. Lane, *in* "Antibodies: A Laboratory Manual," p. 407. Cold Spring Harbor Lab., 1988. Cold Spring Harbor, New York, 1988.

Purification

Phorbol Ester Treatment of U937 Cells

Treatment of U937 cells (a human histiocytic lymphoma cell line[36]) with PMA leads to a 20- to 100-fold increase in the number of uPAR molecules per cell.[5,7,19] PMA stock solution (100 μl, 1 mg/ml in DMSO) is added to 1 liter of cell suspension [(0.5–1) $\times$ 10^6 cells/ml] and the cell culture is incubated for 4 days. (Strict precautions must be taken to avoid any contact with PMA during addition and with the PMA-containing material during all subsequent steps). The PMA treatment results in differentiation of the U937 cells into adherent, macrophage-like cells.[37] The degree of cell adherence, however, varies among different sublines of U937 cells.

Cell Harvest and Lysis

After PMA treatment, cells are washed with PBS and harvested using a rubber scraper. Following centrifugation for 10 min at room temperature, 300–500 g, the cells are washed again, twice in PBS, and the cell pellet is gently resuspended in a low-pH buffer in order to release endogenous pro-uPA, bound to uPAR[38]; 5 ml of 50 mM glycine hydrochloride, 100 mM NaCl, pH 3.0, is used per 10^8 cells. This treatment is performed at room temperature for exactly 3 min. For termination of the treatment, the suspension is neutralized by addition of 500 mM HEPES, 100 mM NaCl, pH 7.5 (1 ml per 10^8 cells), followed by immediate isolation of the cells by centrifugation as above, in order to prevent renewed pro-uPA binding. Lysis buffer [0.1 M Tris-HCl, 1% Triton X-114 (Sigma), 10 mM EDTA, 10 μg/ml aprotinin, pH 8.1] is prepared by dissolving the detergent at 0° and is kept frozen or is freshly made. The cell pellet is transferred to ice and treated with lysis buffer (2 ml per 10^8 cells) and phenylmethylsulfonyl fluoride (PMSF; 20 μl of a 100 mM stock solution in DMSO per 10^8 cells) at 0° for 10 min. During this lysis period, the material is mixed a few times through a pipette tip. The lysate is clarified by centrifugation in a high-speed centrifuge at 0–4°, strict temperature control being critical.

[36] C. Sundström and K. Nilsson, *Int. J. Cancer* **17**, 565 (1976).
[37] K. Nilsson, K. Forsbeck, M. Gidlund, C. Sundström, T. Toetterman, J. Sallström, and P. Venge, *Haematol. Blood Transfus.* **26**, 215 (1981).
[38] M. P. Stoppelli, C. Tacchetti, M. V. Cubellis, A. Corti, V. J. Hearing, G. Cassani, E. Appella, and F. Blasi, *Cell (Cambridge, Mass.)* **45**, 657 (1986).

Detergent Phase Separation

Isolation of the pool of hydrophobic proteins is performed by temperature-dependent detergent phase separation.[39] With PMA-treated U937 cells as the raw material, this step represents an about 50-fold purification factor for uPAR. For preparative purpose, the separation is performed as follows. The clarified Triton X-114-containing lysate is warmed to 37°, the detergent thus becoming immiscible with water, and is incubated at this temperature for 10 min. The detergent phase is collected by centrifugation at 1800 g for 10 min at room temperature. The aqueous upper phase is discarded, while the lower detergent phase is washed by addition of ice-cold 0.1 M Tris-HCl, pH 8.1, to the original volume, followed by renewed warming and centrifugation as above. The final detergent phase thus obtained is made up to the original volume with the same Tris buffer, followed by addition of CHAPS (0.25% final concentration), which serves to prevent renewed phase separation during subsequent steps. Finally, any precipitated material is removed by high-speed centrifugation at 4°.

Affinity Chromatography

For preparation of an affinity matrix, DFP-treated uPA (see above) is coupled to CNBr-activated Sepharose (Pharmacia) using a standard procedure as described.[22] For each milliliter of swollen gel 2 mg protein is used. The gel material is washed with elution buffer (0.1 M acetic acid, 0.5 M NaCl, 0.1% CHAPS, pH 2.5) before use. For affinity chromatography, we have routinely used the Triton X-114 detergent phase obtained from $(5-10) \times 10^9$ cells. This material is mixed with an equal volume of PBS with 0.1% (w/v) CHAPS (PBS/CHAPS) and pumped at a rate of 4 ml/hr through a column containing 8 ml of the affinity matrix. The adsorbent is washed with several column volumes of PBS/CHAPS followed by a buffer with the same composition except for a higher (1 M) content of NaCl. The column is eluted by reverse flow, 4 ml/hr, with the elution buffer specified above. Fractions of 1 ml are collected in tubes containing an amount of neutralization buffer (0.1 M sodium phosphate, 1 M sodium carbonate, pH 9.0) sufficient to titrate the fractions to pH 7.4. Fractions containing uPAR are identified by chemical cross-linking to ^{125}I-labeled ATF. For protein analysis, it is convenient to concentrate peak fractions by extensive dialysis against 0.1% acetic acid followed by lyophilization.

The protein thus isolated has binding activity and has a high purity

---

[39] C. Bordier, *J. Biol. Chem.* **256**, 1604 (1981).

(see below). The purification yield, determined by amino acid analysis, is 10–15 μg uPAR from 6 × 10^9 PMA-treated U937 cells.

Protein Chemical Characterization

Purity and Biological Activity

After SDS–PAGE and silver staining, the purified protein shows a single, broad band, migrating with an apparent M_r of 55,000–60,000 (Fig. 1B). Identical results are found with reduced and nonreduced samples. The M_r heterogeneity is due to variation in N-linked carbohydrate (see below). The purified material shows cross-linking activity with ^{125}I-labeled ATF (Fig. 1C, lane 2), leading to a conjugate of apparent M_r 70,000–80,000. This conjugate M_r fits with simple M_r additivity; the apparent M_r of ATF is approximately 17,000. The cross-linking reaction has the expected specificity and saturability properties; i.e., it is competed for by excess uPA or DFP-treated uPA but not by tPA, plasminogen, or epidermal growth factor.[22] A cross-linking study with nonlabeled components[22] has confirmed that the binding activity is indeed possessed by the silver-stainable component.

Amino Acid Composition and Sequence

For amino acid composition or sequence analyses, the purified, concentrated material is subjected to Tricine–SDS–PAGE after a reducing sample treatment, followed by electroblotting onto a PVDF membrane, both techniques being conducted as described[40,41] in order to limit modifications such as NH$_2$-terminal blocking. The protein is localized on the membrane by Coomassie staining. In order to allow the identification of cysteine during sequencing, this amino acid is alkylated by treatment of the Coomassie-stained protein directly on the membrane with iodoacetamide.[40] Determinations of the NH$_2$-terminal sequence[22] have allowed the isolation and sequencing of uPAR cDNA.[23] The cDNA encodes a polypeptide of 313 amino acid residues, excluding the signal sequence of 22 residues. There are five potential N-glycosylation sites and a high number (28) of cysteine residues. The experimentally determined amino acid composition[22] shows significant deviations from the composition calculated from cDNA. These discrepancies are due to COOH-terminal processing occurring during glycosylphosphatidylinositol attachment[26] (see below).

[40] M. Ploug, B. Stoffer, and A. L. Jensen, *Electrophoresis* **13**, 148 (1992).
[41] M. Ploug, A. L. Jensen, and V. Barkholt, *Anal. Biochem.* **181**, 33 (1989).

N-Linked Carbohydrate

uPAR contains large amounts of N-linked carbohydrate. Indeed, the dramatic shift in apparent M_r occurring on enzymatic deglycosylation can be used as a supplementary tool for identification of the protein. This is especially convenient because the analysis does not require isolation of the receptor; instead, uPAR within a protein mixture is selectively labeled by means of chemical cross-linking to ^{125}I-labeled ATF. For deglycosylation, cross-linked samples are denatured under mildly reducing conditions (in order to allow access of the enzyme) by boiling for 3 min after addition of 0.5% SDS and either dithiothreitol (1.6 mM) or 2-mercaptoethanol (100 mM). After cooling, the denatured samples (10 μl) are adjusted to include 200 mM sodium phosphate, pH 8.6, 1.5% Triton X-100, and 0.5–1 unit of peptide N-glycosidase F[42] (N-glycanase; Genzyme, Cambridge, MA) in a total volume of 30 μl. A chelating agent such as 1,10-phenanthroline (10 mM; added from a methanol stock solution) or EDTA (10 mM) may be included for protection against contaminating metalloproteases. Deglycosylation takes place overnight at 37°, and a similarly treated sample without added enzyme is incubated in parallel. Conjugates of uPAR and ^{125}I-labeled ATF, deglycosylated in this manner, run in SDS–PAGE with an apparent M_r of only ≈50,000 (see Fig. 1A). By simple subtraction of the ligand M_r, this is in agreement with similar studies on purified uPAR. Thus, the purified, heterogeneous M_r 55,000–60,000 protein can be deglycosylated with N-glycanase to yield a product of M_r 35,000, which migrates in SDS–PAGE as a sharp band. Glycosylation is therefore responsible for the heterogeneity of the intact protein.[22] By the above method, several glycosylation variants of uPAR have been identified in crude samples.[22]

Identification of a Glycosylphosphatidylinositol Membrane Anchor in Human uPAR

For anchoring in the membrane, uPAR utilizes a glycosylphosphatidylinositol (GPI) moiety,[26] a modification also found in several other eukaryotic membrane glycoproteins.[43] Some of the diagnostic tools we have used to identify and characterize this structural feature[26] are described here.

Method 1: Treatment of Whole Cells with Phosphatidylinositol-Specific Phospholipase

The specific release of a certain membrane protein from the cell surface by treatment with the bacterial phospholipase, phosphatidylinositol-spe-

[42] A. L. Tarantino, C. M. Gomez, and T. H. Plummer, *Biochemistry* **24**, 4665 (1985).
[43] M. A. Ferguson and A. F. Williams, *Annu. Rev. Biochem.* **57**, 285 (1988).

cific phospholipase C (PI-PLC), is probably the most widely used criterion for the identification of a putative GPI anchoring. PI-PLC cleaves the glycolipid in such a way that the hydrophobic part (i.e., the diacylglycerol moiety) is removed, which consequently leads to the release of the otherwise intact protein into the extracellular medium.[43] For this type of analysis, adherent, PMA-treated U937 cells (2×10^7 cells per Petri dish, with a diameter of 10 cm) are initially subjected to acid treatment as above, followed by two washes in serum-free RPMI 1640 medium including 25 mM HEPES, pH 7.5 (Buffer A). The cells are then incubated at 37° on a shaking table with 5 ml of Buffer A, including 0.1 μg/ml PI-PLC (Boehringer Mannheim; specific activity 600 U/mg). The medium is withdrawn after 30 min and centrifuged immediately for the removal of any insoluble material (20,000 g for 5 min). After addition of CHAPS (0.1% final concentration) to this supernatant, the presence of any released uPA-binding components is analyzed by chemical cross-linking to ^{125}I-labeled ATF as described above. This method can also be applied to other adherent cells, or to cells in suspension, such as peripheral blood leukocytes.[44] A useful modification of the method is to bind the radiolabeled ligand to the acid-treated cells before they are subjected to PI-PLC. For this analysis, 1 nM of ^{125}I-labeled DFP-treated uPA is allowed to bind to the cells at 4° for 2 hr in 5 ml of Buffer A. The cells are then washed three times in the same buffer without added ligand. One dish is extracted with 5% SDS defining 100% of cell-associated radioactivity. The remaining two dishes receive 5 ml each of Buffer A and are incubated in the absence (buffer control) or presence of PI-PLC (0.6 μg/ml). Aliquots of 100 μl are withdrawn from both dishes during incubation on a shaking table at 37° and the released radioactivity is determined in the supernatants after clarification by centrifugation (20,000 g for 5 min). The percentage of cell-associated radioactivity specifically released by PI-PLC (i.e., the fraction of uPAR molecules sensitive to the enzyme) is approximately 50% for U937 cells[26] (see below).

Method 2: Change in Amphiphilicity of Purified uPAR after Treatment with Base and PI-PLC

Inability of PI-PLC to release a certain membrane protein from intact cells does not necessarily exclude the presence of a GPI anchor. There is a still growing list of GPI-anchored proteins that have been identified but that cannot be released by this enzyme because the glycolipid contains an additional acylation that blocks the catalytic mechanism.[43] This inhibi-

[44] M. Ploug, T. Plesner, E. Rønne, V. Ellis, G. Høyer-Hansen, N. E. Hansen, and K. Danø, *Blood* **79**, 1447 (1992).

tory modification can, however, be removed by partial deacylation of the protein during mild base treatment. The effect of base treatment followed by PI-PLC, or of either treatment alone, can then be studied by analytical Triton X-114 phase separation and chemical cross-linking to labeled ligand.[26] For this kind of study, 0.2 pmol of purified uPAR is suspended in 15 μl of methanol plus 15 μl of 32% ammonia (or, as a control, in 15 μl of methanol plus 15 μl of water). Treatment takes place at room temperature for 1 hr. The samples are then dried by vacuum evaporation and redissolved in 20 μl of triethylamine hydrochloride (TEA) buffer (50 mM triethylamine hydrochloride, 5 mM EDTA, 0.1% (w/v) Triton X-100, pH 7.5). The redissolved samples are incubated with either 5 μl of PI-PLC (final concentration 20 μg/ml) or 5 μl TEA buffer (buffer control) at 37° for 30 min. Each sample is thoroughly mixed with 75 μl of ice-cold 1.5% (w/v) Triton X-114 in 0.1 M Tris-HCl, 10 mM EDTA, pH 8.1, after which detergent phase separation is induced by incubation at 37° for 5 min. The oily detergent phase is collected beneath the aqueous phase by centrifugation at room temperature (10,000 g for 2 min). The two phases are washed by addition of 20 μl of 5% (w/v) Triton X-114 to 80 μl of the aqueous phase, and of 80 μl of 0.1 M Tris-HCl, pH 8.1, to the detergent phase, respectively. The detergent phase is reconstituted in 80 μl of 0.1 M Tris-HCl, pH 8.1, and both phases are clarified by addition of CHAPS to a final concentration of 0.3% (w/v). The presence of uPAR activity in the resulting samples is analyzed by chemical cross-linking to ^{125}I-labeled ATF. The combined treatment of purified uPAR with mild base followed by PI-PLC removes the glycolipid completely as judged by the altered detergent phase partitioning (i.e., shift from the detergent to the aqueous phase).

Additional, independent means of demonstrating the GPI moiety of uPAR include the identification of ethanolamine in an acid hydrolyzate of the purified protein, as well as successful metabolic labeling of uPAR with [3H]ethanolamine and *myo*-[3H]inositol.[26]

Preparation of Ligand-Binding Domain

The ligand-binding domain of uPAR, i.e., residues 1–87, can be liberated from the purified protein by mild treatment with chymotrypsin.[27] Before treatment, the affinity-purified protein is concentrated by dialysis and lyophilization as described above and redissolved at a concentration of 30 μg/ml in 0.05 M Tris-HCl, 0.05% (w/v) CHAPS, pH 8.1. Chymotrypsin (Worthington Biochemical Corporation, Freehold, NJ; 67 units/mg) is added to a final concentration of 40 ng/ml, and the sample is incubated at 37° for 7 hr after which the reaction is stopped by addition of PMSF (final concentration 1 mM; added in the form of a 20 mM stock solution in DMSO). This treatment leads to cleavage of the Tyr[87]-Ser[88] bond in uPAR with no other cleavages being detected by Tricine–SDS–PAGE and silver staining or by NH$_2$-terminal sequencing of the two fragments thus observed. The smaller fragment (i.e., residues 1–87) has an apparent M_r of 16,000 (analyzed by Tricine–SDS–PAGE) and contains N-bound carbohydrate at Asn-52; its binding activity can be demonstrated by chemical cross-linking to ^{125}I-labeled ATF (Fig. 1C, lane 3), and this activity is retained in the absence of the rest of the protein.[27] This ligand-binding fragment can be separated from the larger fragment, as well as from the residual intact protein, by analytical Triton X-114 phase separation as described above.

The experimental demonstration of an autonomous domain comprising residues 1–87 fits with a computer analysis[45] of internal homology and proposed domain structure in uPAR.[27]

Immunological Methods

Monoclonal[28] and polyclonal[22,46] antibodies against uPAR have proved useful for the detection of the protein, for the study of its structure–function relationship, and as a tool for the analysis of the ligand-binding process. A separate field of investigation, not dealt with in the present article, includes FACS analyses.[44]

Method 1: Immunoprecipitation

For radioimmunoprecipitation of uPAR, [35S]methionine-labeled, PMA-treated U937 cells[7] are lysed and subjected to Triton X-114 phase separation as described above, except that 10-fold less cell material is used per volume of lysis buffer. The detergent phase sample is incubated overnight at 4° with the appropriate antibody (purified or in the form of antiserum or hybridoma culture supernatant). The control includes an irrelevant antibody (not reactive with components in the biological material) or nonimmune serum or IgG. Protein A-Sepharose CL-4B (Pharmacia) is equilibrated with Buffer A (0.1 M Tris-HCl, 0.1% Tween 80, 0.1% BSA, pH 8.1), and 200 μl of a 50% suspension of the wet adsorbent is added to each 500-μl sample. The mixture is incubated at 4° for 2 hr on an "end-over-end" rotor. The adsorbent is then isolated and washed with Buffer A, then with Buffer B (Buffer A with 1 M NaCl), and twice

[45] This computer analysis was performed by Dr. Laszlo Patthy, Institute of Enzymology, Hungarian Academy of Sciences, Budapest.

[46] V. Ellis, T. C. Wun, N. Behrendt, E. Rønne, and K. Danø, J. Biol. Chem. 265, 9904 (1990).

with Buffer C (Buffer A without BSA). Finally, the adsorbent is boiled in the presence of sample buffer for SDS–PAGE, including reducing agent, and the resulting sample is analyzed by SDS–PAGE. Immunoprecipitation has also proved useful for mapping of the reactivities of individual monoclonal antibodies within the structure of uPAR. Using a similar procedure for analysis of chymotryptic uPAR fragments, we thus identified one antibody that precipitated exclusively the ligand-binding domain, and three other antibodies that precipitated only the larger fragment. The former antibody proved to be the only one capable of blocking the binding of DFP-treated uPA to cells.[28] These results show that the NH_2-terminal domain must be unblocked in the intact receptor for ligand binding to occur.

Method 2: Western Blotting

Three out of four tested monoclonal antibodies have been found suitable for Western blotting analysis,[28] all reacting, however, only with nonreduced uPAR. These antibodies are capable of demonstrating uPAR in samples from, e.g., human blood leukocytes.[44] The best result is obtained when using Triton X-114 detergent phases as the samples, rather than unfractionated cell lysates, in order to avoid dominant nonspecific staining on the blots. The antibodies differ in their reactivity preference toward different glycosylation variants of uPAR.[28]

Method 3: Ligand-Binding ELISA

This assay uses a catching antibody that is directed against uPAR; after binding of the sample, the ligand (i.e., DFP-treated uPA) is added exogenously and is ultimately detected by means of an anti-uPA antibody. The method is useful for two kinds of investigation. Using purified uPAR, it provides a simple, isolated system for study of the ligand-binding interaction, e.g., in the search for putative agents that interfere with binding. When using instead an unknown biological material, it serves as a detection method for uPAR activity. The wells of an ELISA plate (Nunc, Copenhagen, Denmark; 96 wells, flat-bottomed, high binding-capacity type) are coated with 100 μl of a purified monoclonal anti-uPAR antibody[47] diluted in 0.1 M Na_2CO_3, pH 9.8, by incubation overnight at 4°. The wells are then washed and the remaining protein-binding capacity is blocked by incubation with blocking buffer (1% skimmed milk powder in PBS; 200 μl in each well). In this and all subsequent steps, washing is performed using PBS with 0.1% Tween 20 and is repeated five times; incubation is

---

[47] The antibody used must be one that allows ligand binding to the bound uPAR. We have routinely used the antibody R4[28] at a concentration of 20 μg/ml.

done for 1 hr at room temperature with gentle shaking, using blocking buffer with 0.1% Tween 20 for dilution of all reagents. The wells are washed again, and 100 μl of affinity-purified uPAR ($\approx$20 ng/ml) is added and incubated as described above. Alternatively, a Triton X-114 detergent phase from a cell lysate to be analyzed is introduced at this step. After washing, incubation is performed with 200 μl of DFP-treated uPA (diluted to 5 ng/ml), followed by renewed washing. A biotin-labeled,[32] purified monoclonal anti-uPA antibody (directed against the non-receptor-binding part of the enzyme[48]) is diluted appropriately,[49] and 100 μl thereof is added to each well. After incubation and washing, 100 μl of horseradish peroxidase-conjugated avidin (Dakopatts; 5000-fold diluted) is added. Following incubation and washing as usual, the peroxidase activity is measured using a standard procedure with 1,2-phenylenediamine and H_2O_2 and reading the absorbance at 490 nm.[48] Controls include wells with single omission of reagents at each step except for the substrate solution.

This assay detects purified uPAR as well as uPAR within, e.g., the detergent phase from PMA-treated U937 cells in a dose-dependent manner, with all controls being negative. The sensitivity, estimated using purified uPAR, is better than 1 ng. The assay has been used to demonstrate the inhibitory effect of the polysulfonated dinaphthylurea derivate, suramin, against the uPA–uPAR interaction.[50] A variant procedure includes the use of ^{125}I-labeled pro-uPA as the ligand, instead of unlabeled DFP-treated uPA.[51] This modified procedure allows direct quantification of the bound ligand, in contrast to the indirect measurement using biotinylated antibody. It has been used for affinity studies with uPAR glycosylation variants, allowing direct Scatchard plotting of the experimental data.[51] Other ELISA-type assays are currently being developed which rely on a catching/recognition system with two antibodies.

Acknowledgments

Methods for ligand blotting were developed by Helene Solberg and Dorte Løber at the Finsen Laboratory.

[48] L. S. Nielsen, J. Grøndahl-Hansen, P. A. Andreasen, L. Skriver, J. Zeuthen, and K. Danø, *J. Immunoassay* **7,** 209 (1986).

[49] A concentration of 2 μg/ml is convenient with the anti-uPA clone 5 antibody.[48]

[50] N. Behrendt, E. Rønne, and K. Danø, *J. Biol. Chem.* **268,** 5985 (1993).

[51] L. B. Møller, J. Pöllänen, E. Rønne, N. Pedersen, and F. Blasi, *J. Biol. Chem.* **268,** in press (1993).

[14] Cellular Receptor for Urokinase-Type Plasminogen Activator: Function in Cell-Surface Proteolysis

By VINCENT ELLIS, NIELS BEHRENDT, and KELD DANØ

Numerous studies have delineated a system whereby the generation of cell surface proteolytic activity can be mediated by uPAR,[1–7] the cellular receptor for the urokinase-type plasminogen activator (uPA). uPAR, the structure of which is described in detail elsewhere in this volume,[8] binds both uPA and its essentially inactive zymogen precursor, pro-uPA. Pro-uPA can be activated when it is receptor bound, and receptor-bound uPA is catalytically active.[1,2] The concomitant binding of pro-uPA to uPAR and plasminogen to as yet unidentified cellular binding sites results in the generation of plasmin on the cell surface, thereby forming a completely cell-associated reciprocal zymogen activation system.[4,5] A feedback amplification of these reactions has the overall effect that uPA-catalyzed plasminogen activation mediated by uPAR on the cell surface proceeds with a much greater efficiency than uPA-catalyzed plasminogen activation in the solution phase.[4,7] For these reasons it has been speculated that *in vivo* uPA-catalyzed plasminogen activation is a completely cell-associated phenomenon, dependent on the presence of uPAR,[9] in a manner somewhat analogous to the dependence of the tPA-catalyzed system on the presence of fibrin.

Initial observations on the activity of the receptor-bound uPA system relied on caseinolytic plaque methods,[1] whereby agar gels containing casein and plasminogen are overlaid on cells, and the formation of lysis zones, or plaques, is detected. Such methods, although remarkably sensitive (lysis zones around single cells can be observed), are nonquantitative

[1] J.-D. Vassalli, D. Baccino, and D. Belin, *J. Cell Biol.* **100**, 86 (1985).
[2] M. V. Cubellis, M. L. Nolli, G. Cassani, and F. Blasi, *J. Biol. Chem.* **261**, 15819 (1986).
[3] E. F. Plow, D. E. Freaney, J. Plescia, and L. A. Miles, *J. Cell Biol.* **103**, 2411 (1986).
[4] V. Ellis, M. F. Scully, and V. V. Kakkar, *J. Biol. Chem.* **264**, 2185 (1989).
[5] R. W. Stephens, J. Pöllänen, H. Tapiovaara, K.-C. Leung, P.-S. Sim, E.-M. Salonen, E. Rønne, N. Behrendt, K. Danø, and A. Vaheri, *J. Cell Biol.* **108**, 1987 (1989).
[6] E. S. Barnathan, A. Kuo, L. Rosenfeld, K. Kariko, M. Leski, F. Robbiati, M. L. Nolli, J. Henkin, and D. B. Cines, *J. Biol. Chem.* **265**, 2865 (1990).
[7] V. Ellis, N. Behrendt, and K. Danø, *J. Biol. Chem.* **266**, 12752 (1991).
[8] N. Behrendt, M. Ploug, E. Rønne, G. Høyer-Hansen, and K. Danø, this volume [13].
[9] V. Ellis and K. Danø, *Semin. Thromb. Hemostasis* **17**, 194 (1991).

and therefore unsuitable for studies on kinetic parameters of the enzymatic function of this system.

The requirement for quantitative assays to investigate the functional effects of cell surface uPAR on the activity of the uPA-catalyzed plasminogen activation system presents a number of difficulties, particularly with respect to sensitivity and specific quantification. The low receptor number on most cell types dictates the use of sensitive techniques, the need for which is compounded by the necessity to observe the reactions over relatively short time periods, both to ensure minimal dissociation of the receptor-bound ligand and to negate possible deleterious effects of the generated plasmin, e.g., degradation of uPAR. Spectrophotometric methods using synthetic peptide substrates to detect generated plasmin have in our hands proved to be the most suitable for this purpose.

The methods presented here, although developed specifically for the study of uPA and plasmin activity in relation to uPAR, may also have a more general applicability in the study of other cell surface-associated protease systems, particularly those that may have an intimate involvement with the plasminogen activation system, e.g., the matrix metalloproteases.

Principle

Cells possessing receptor-bound uPA or pro-uPA are incubated simultaneously with plasminogen and the plasmin-specific peptide substrate H-D-Val-Leu-Lys-7-amido-4-methylcoumarin, which is continuously hydrolyzed by the generated plasmin to liberate the highly fluorescent 7-amido-4-methylcoumarin (AMC) group; the AMC group is quantified spectrofluorometrically either by continuous recording or by measurement at timed intervals. The average plasmin concentration between each time point is then calculated from the first derivative of this data, i.e., dF/dt, thereby allowing plots of active plasmin concentration versus time to be constructed. The adoption of continuous substrate hydrolysis methods allows accurate reaction progress curves to be made over shorter time periods than is possible using discontinuous methods, although at the expense of sensitivity. This is, however more than compensated for by the use of substrates liberating the fluorophore AMC, which give sensitivities approximately 100-fold higher than with equivalent chromogenic substrates, such as those liberating p-nitroaniline.

Two different procedures have been developed for quantitative activity determinations in the uPA-catalyzed, cell-associated plasminogen activation system; one is suitable for cells in suspension (and has been successfully applied to U937 cells[7,10,11] and blood leukocytes,[12] and the other is for adherent cells in monolayer culture (and has been applied to a variety of cells, including A549 lung carcinoma cells,[13] HT-1080 fibrosarcoma cells,[14] and MCF-7 mammary carcinoma cells). The former has more general applicability in the study of interactions between components of the uPAR-mediated system, for example, using the cell line U937 as a model, while the latter is of use for studying the specific properties of individual adherent cell lines.

Determination of uPAR-Mediated Plasminogen Activation by Cells in Suspension

U937 cells are washed three times in RPMI-1640 buffered in 25 mM HEPES, pH 7.4, and resuspended in RPMI-1640 buffered in 25 mM HEPES, pH 7.4, containing 2 mg/ml fatty acid-free bovine serum albumin (BSA). Prior to this final resuspension the cells may be acid washed to dissociate ligand endogenously bound to uPAR (as described elsewhere in this volume[8]), but it should be noted that this procedure has been shown to alter the plasminogen-binding characteristics of the cells and consequently the kinetics of plasminogen activation by receptor-bound uPA[7] (see Table I). The cells (approximately 1×10^7/ml) are then saturated with either uPA[15] or pro-uPA by incubation with 1–3 nM ligand for 20 min at 37° (these concentrations of ligand may not achieve true saturation under these conditions; however, the use of higher ligand concentrations is often associated with an unacceptably high degree of nonspecific ligand

[10] V. Ellis, T.-C. Wun, N. Behrendt, E. Rønne, and K. Danø, *J. Biol. Chem.* **265,** 9904 (1990).

[11] E. Rønne, N. Behrendt, V. Ellis, M. Ploug, K. Danø, and G. Høyer-Hansen, *FEBS Lett.* **228,** 233 (1991).

[12] M. Ploug, T. Plesner, E. Rønne, V. Ellis, G. Høyer-Hansen, N. E. Hansen, and K. Danø, *Blood* **79,** 1447 (1992).

[13] L. R. Lund, J. Rømer, E. Rønne, V. Ellis, F. Blasi, and K. Danø, *EMBO J.* **10,** 3399 (1991).

[14] V. Ellis and K. Danø, *Fibrinolysis* **6**(4), 27 (1992).

[15] uPA is obtained commercially from Serono, Denens, Switzerland, and is greater than 90% high-molecular-weight active uPA. Pro-uPA from a number of sources has been successfully employed in these experiments. Natural pro-uPA was from either human kidney cell cultures (Sandoz Forschungsinstitut, Vienna, Austria; Abbott Laboratories, North Chicago, IL) or human tumor cell lines (HT-1080 fibrosarcoma cells), and recombinant pro-uPA was expressed in *Escherichia coli* (Grünenthal GmbH, Stolberg, Germany).

TABLE I
EFFECT OF CELL SURFACE RECEPTOR BINDING ON KINETIC CONSTANTS FOR
uPA-CATALYZED PLASMINOGEN ACTIVATION[a]

Type	K_m (μM)	k_{cat} (sec^{-1})[b]	k_{cat}/K_m (μM^{-1} sec^{-1})
U937 cells			
$-$Trx[c]	0.67 ± 0.30	0.12 ± 0.06	0.165 ± 0.048
$+$Trx	$>15^d$	$>0.85^d$	0.056 ± 0.011
U937 cells (acid-washed)			
$-$Trx	0.11 ± 0.06	0.22 ± 0.15	2.23 ± 0.88
$+$Trx	$>15^d$	$>0.65^d$	0.042 ± 0.015
In solution			
$-$Trx	25	0.73	0.029
$+$Trx	25	1.85	0.074

[a] Data shown, from Ellis et al.,[7] are for the activation of the native form of plasminogen with the Glu[1] NH$_2$ terminus; however, essentially similar effects have been observed with Lys[77]-plasminogen.[7]

[b] Concentration of cell-bound enzyme was determined in parallel incubations with ^{125}I-labeled uPA.

[c] Data are shown in the presence and absence of 1 mM tranexamic acid (Trx), an antagonist of plasminogen binding.

[d] Individual kinetic constants could not be determined accurately in the presence of tranexamic acid because the double-reciprocal plots had intercepts close to zero; therefore, lower limits for these constants are shown.

binding) and washed three times in PBS containing 2 mg/ml BSA.[16] Alternatively, in the case of pro-uPA, rather than being preincubated with the cells, the ligand may be included in the reaction mixture at a concentration of 1 nM; under these conditions a proportion of the pro-uPA rapidly binds to uPAR, plasminogen activation is initiated on the cell surface, and the reactions subsequently spread to the solution phase, providing a simple, sensitive procedure for determining the occurrence of the initial cell surface-dependent reactions, e.g., in the determination of the effects of potential binding antagonists. The cells are then incubated at a final density of 1×10^6/ml in 0.05 M Tris-HCl, pH 7.4, 0.1 M NaCl in the presence of 0.2 mM H-D-Val-Leu-Lys-AMC[17] and 20 μg/ml human Glu-plasminogen (purified according to Danø and Reich,[18] a modification of the original

[16] The presence of standard cell culture media during the measurement of plasminogen activation must be avoided due to the presence of millimolar concentrations of lysine in these media, which will interfere with both the activation of plasminogen and its cellular binding.

[17] Available from Bachem Feinchemikalien AG, CH-4416 Bubendorf, Switzerland.

[18] K. Danø and E. Reich, Biochim. Biophys. Acta 566, 138 (1979).

method[19]). This is most conveniently done in disposable plastic fluorimeter cuvettes, which can be gently stirred in order to maintain the cells in suspension by means of a micromagnetic stirrer built into the cuvette holder of the fluorimeter. Routinely a four-position holder is used, allowing four samples to be measured concurrently at 1-min intervals. The excitation and emission wavelengths generally employed are 380 and 480 nm, respectively, with 5-nm slit widths for optimum signal/noise ratio. The emission wavelength may be reduced to 460 nm to give an approximately twofold increase in sensitivity, or the sensitivity may be lowered in order to extend the measuring range by increasing the emission wavelength to 500 nm. The collected data are subsequently converted to their first derivative (dF/dt), and the plasmin concentration at each time point is determined by comparison with calibrations made using active-site-titrated plasmin[20] (see Fig. 1). Providing that the cell density in the cuvette does not exceed approximately 1×10^6 cells/ml, there is no interference due to turbidity. Above this density the effect is dependent on the cell type and can be corrected for with calibration curves constructed using free AMC in cell suspensions of the required density.

Applications of Method

This method has been used to demonstrate a number of basic properties and characteristics of this system: (1) The potentiation of the initiation of the system by binding of pro-uPA to uPAR. With pro-uPA as the bound ligand the rate of plasmin generation increases rapidly with time, after an initial lag period, due to the feedback activation of pro-uPA by the generated plasmin, the overall rate of which is much higher than in the solution phase (Fig. 2).[4] (2) The effect of antagonists of pro-uPA and plasminogen binding (ATF[21] and lysine analogs, such as 6-aminohexanoic acid and tranexamic acid, respectively), both of which abolish the potentiation of plasmin generation.[4] (3) The kinetics of plasminogen activation by receptor-bound uPA, which are markedly different from those of the solution-phase reaction (Table I). The primary effect is a large reduction in the K_m for plasminogen activation, which occurs as a consequence of the cellular binding of plasminogen.[7] This binding can be reversed by lysine analogs, and under these conditions the activity of receptor-bound uPA and solution-phase uPA cannot be distinguished kinetically.[7] (4) The relationship between uPAR structure and functional activity using anti-uPAR mono-

---

[19] D. G. Deutsch and E. T. Mertz, *Science* **170**, 1095 (1970).
[20] T. Chase, Jr. and E. Shaw, this series, Vol. 19, p. 20.
[21] ATF, the uPAR-binding NH$_2$-terminal fragment of uPA, normally comprising residues 1–135 or 6–135.

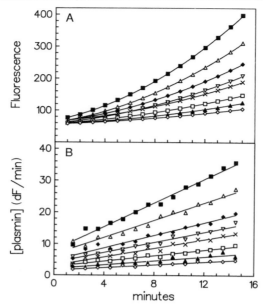

FIG. 1. Determination of plasmin generation by uPAR-bound uPA on U937 cells. After saturation with uPA and washing, cells were incubated with varying concentrations of Glu-plasminogen ($\Diamond$, 0.09; $\blacktriangle$, 0.15; $\square$, 0.22; $\times$, 0.35; $\triangledown$, 0.52; $\blacklozenge$, 0.78; $\triangle$, 1.16; $\blacksquare$, 1.73 μM) and H-D-Val-Leu-Lys-AMC. (A) Continuous substrate hydrolysis measured by recording fluorescence intensity at 1-min intervals using excitation and emission wavelengths of 380 and 480 nm, respectively. (B) Rate of change of fluorescence between each measurement, equivalent to plasmin concentration, plotted against time. The slopes of these lines are converted to rates of plasmin generation by reference to calibrations made using active-site-titrated plasmin, which under these conditions gives a dF/min of 46.5 nM^{-1}. The lines shown in B intercept the ordinate at values greater than zero due to minor residual plasmin activity in the plasminogen preparation, which in this case was $7.7 \times 10^{-3}\%$. The data shown are from Ellis et al.[7]

clonal antibodies[8] (Fig. 2).[11] (5) The effect of the specific plasminogen activator inhibitors, PAI-1 and PAI-2, on receptor-bound uPA.[10] Varying concentrations of the inhibitors were incubated with the uPA-saturated cells and the inhibition was monitored from the decrease in the rate of plasmin generation with time. The second-order association rate constants for the interactions were calculated using standard procedures for the analysis of the kinetics of substrate reactions during enzyme modification by irreversible inhibitors.[22] As shown in Table II, receptor binding has only a very limited effect on the inhibition of uPA by the plasminogen

[22] C. L. Tsou, Adv. Enzymol. **61**, 381 (1988).

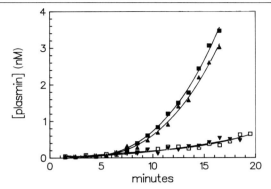

FIG. 2. Potentiation of plasmin generation on binding of pro-uPA to uPAR on U937 cells. In these experiments pro-uPA was added to the cells at time zero, i.e., with no preincubation. Data are shown in the presence (■) and absence (□) of cells, and in the presence of cells preincubated with 2 μg/ml of the anti-uPAR antibodies R2 (▲), which has no effect on uPA binding, and R3 (▼), which has been shown to inhibit uPA binding to uPAR. The latter antibody completely abolishes the cell-dependent potentiation, thereby demonstrating its mediation by uPAR. The data shown are from Rønne et al.[11]

TABLE II

EFFECT OF CELL SURFACE RECEPTOR BINDING
ON SECOND-ORDER ASSOCIATION RATE
CONSTANTS FOR INHIBITION OF uPA BY PAI-1
AND PAI-2[a]

Location of uPA	Association rate constant $(M^{-1} \text{ sec}^{-1})$	
	PAI-1[b]	PAI-2
In solution	7.9×10^6	5.3×10^5
Receptor bound[c]	4.5×10^6	3.3×10^5

[a] Data shown are from Ellis et al.[10]

[b] PAI-1 used in these experiments was complexed with NH_2-terminal fragments of vitronectin, but similar data have also been obtained with PAI-1 complexed with intact vitronectin (V. Ellis and T.-C. Wun, unpublished observations, 1990) and SDS–reactivated PAI-1.[10]

[c] uPA was bound to uPAR on U937 cell, the specificity of the binding being demonstrated by competition with a monospecific anti-uPAR polyclonal antibody.

activator inhibitors. (6) To demonstrate uPA-catalyzed plasminogen activation on peripheral blood leukocytes, and to demonstrate a correlation between a reduced expression or absence of uPAR on these cells and a reduced capacity for plasminogen activation. This latter study[12] was performed using leukocytes from patients affected by paroxysmal nocturnal hemoglobinuria (PNH); these cells are unable to express glycosylphosphatidylinositol-anchored membrane proteins on their surface.

Determination of uPAR-Mediated Plasminogen Activation by Adherent Cells

This method is generally applicable to adherent cells. It is of use in studying the overall plasmin-generating characteristics of individual cell lines, primarily reflecting their uPAR content and degree of endogenous saturation with pro-uPA, and the possible effects of various regulators of the expression of these proteins, e.g., cytokines, growth factors, and hormones. This method therefore relies on endogenously secreted pro-uPA as the source of receptor-bound ligand. The procedure described is for HT-1080 fibrosarcoma cells (American Type Culture Collection, Rockville, MD, CCL 121) but is of equal application to other cell types.

After harvesting by trypsinization the cells are seeded into 24-well culture trays (Falcon multiwell plates, well area 2 cm^2) at a density of 1×10^5/well, and cultured to confluence for 24 hr in 10% fetal calf serum, 10 μg/ml aprotinin in Dulbecco's modified Eagle's medium (DMEM). The cells are subsequently cultured for a further 24 hr in serum-free DMEM/aprotinin. These precautions are taken to ensure that the receptor-bound endogenous ligand is essentially completely in the nascent pro-uPA form, both circumventing possible interactions of the active enzyme with plasminogen activator inhibitors secreted by the cells and allowing the study of pro-uPA-mediated reactions.

In this procedure each well represents an individual time point in the determination. Therefore a number of wells (usually 12) are washed three times in DMEM buffered in 25 mM HEPES, pH 7.4, and simultaneously incubated with 0.2 mM H-D-Val-Leu-Lys-AMC and 20 μg/ml human Glu-plasminogen in 0.05 M Tris-HCl, pH 7.4, 0.1 M NaCl (200 μl/well, which is the minimum volume necessary to cover the cell layer). At timed intervals, usually 3 min, a 150-μl aliquot is taken from a well and added to an equal volume of buffer in a microfluorimeter cuvette. A fluorescence measurement is taken immediately (Fig. 3A), followed by another at a 1-min interval (Fig. 3A, inset). This allows two parameters to be derived directly from the measurements. First, from the initial fluorescence measurements taken at each 3-min interval, a curve is constructed that allows the calculation of the total generation of plasmin (as in the preceding

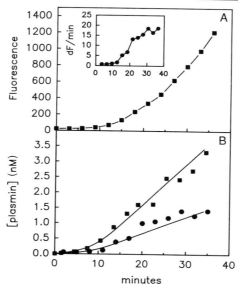

FIG. 3. Plasminogen activation on HT-1080 fibrosarcoma cells. Cells, maintained at confluence in serum-free medium containing aprotinin in 24-well culture plates, were incubated with Glu-plasminogen and H-D-Val-Leu-Lys-AMC, as described in the text. (A) Fluorescence intensity at timed intervals by sampling from individual wells, representing substrate hydrolysis by the total generated plasmin. The increase in this fluorescence intensity over the following 1 min, due to solution phase plasmin activity transferred to the measuring cuvette, is shown in the inset. (B) Data converted to plasmin concentrations for the generation of total (■) and free (●) plasmin. Under these conditions total plasmin activity gives $dF/$min values of 23.2 nM^{-1} (compared to 46.5 in Fig. 1, due to 50% dilution of plasmin in the measuring cuvette) and free plasmin 13.3 nM^{-1} (due to plasmin and substrate dilution).

procedure) (Fig. 3B). Second, from the rate of change in fluorescence during the 1-min measuring period, the generation of free plasmin activity, i.e., the proportion of plasmin generated by receptor-bound uPA and subsequently dissociated into the solution phase (Fig. 3B) can be determined by reference to plasmin calibration curves made under these conditions. The concentration of cell-bound plasmin can therefore be derived indirectly by subtraction.

Using this method care must be taken to avoid mechanical detachment of the cells from the culture wells during the washing procedures. With some cell lines detachment can also be a problem during the assay, due to plasmin proteolysis of the extracellular matrix components necessary for cell adherence. This can either take the form of the detachment of whole "sheets" of cell, or of individual cells "rounding up" and detaching from the substratum. The latter morphological change is usually observed only when high concentrations of plasmin have been generated. Such effects can often be minimized by employing lower plasminogen concentrations and/or shorter measuring times, but, providing that no cells are transferred to the fluorimeter cuvettes, the effects on the measurements are negligible.

Determination of Effect of Isolated uPAR on Plasminogen Activation

The effect of isolated uPAR (either from PMA-stimulated U937 cells[8] or recombinant COOH-terminally truncated uPAR[23]) on uPA-catalyzed plasminogen activation has been observed (Fig. 4) using the following procedure.

uPA (0.01 nM) is preincubated with varying concentrations of uPAR (0.5–25 nM) in a total volume of 50 μl for 30 min at 37°. A solution containing Glu-plasminogen and H-D-Val-Leu-Lys-AMC is added (15 μl) to give final concentrations of 0.3 μM and 0.6 mM, respectively. All measurements of the effect of purified uPAR are performed in 0.05 M Tris-HCl, pH 7.4, 0.1 M NaCl containing 0.1% Triton X-100. After a further 30-min incubation plasmin activity is inhibited by the addition of aprotinin (80 μg/ml), and the AMC accumulated by the generated plasmin is determined by measurement of the fluorescence intensity. Residual uPA activity, which has a linear relationship to the total fluorescence, is calculated by reference to calibration curves made with varying dilutions of uPA.

A saturable, partial (maximally approximately 40%) inhibition of uPA-catalyzed plasminogen activation by isolated uPAR was demonstrated using this method, as shown in Fig. 4. No effect of isolated uPAR was observed, however, on the reaction of uPA with low-molecular-weight peptide substrates, suggesting that the partial inhibition caused by isolated uPAR is primarily due to steric effects.[7] It is not known whether this effect has any functional significance; however, it does serve to emphasize the fundamental point that uPAR does not act as a classical nonenzymatic "cofactor" for uPA activity. The further resolution of the enzyme kinetic mechanisms underlying this effect awaits the availability of larger quantities of isolated uPAR, presumably from recombinant sources.

Due to the saturability of the inhibitory effect of isolated uPAR it is possible, under the experimental conditions described, to derive an approximation to the dissociation constant for the uPA–uPAR interaction in solution (Fig. 4, inset). With uPAR isolated from PMA-stimulated U937 cells this K_d was determined as 1.6 nM. A similar K_d was determined with the COOH-terminally truncated, water-soluble mutant of uPAR using this

[23] M. T. Masucci, N. Pedersen, and F. Blasi, *J. Biol. Chem.* **266**, 8655 (1991).

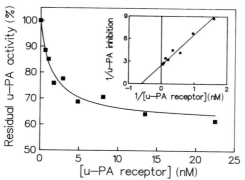

FIG. 4. Effect of isolated uPAR on uPA-catalyzed plasminogen activation. uPAR purified from PMA-stimulated U937 cells was incubated with uPA prior to determination of plasminogen activating activity. uPA activity expressed as percentage activity compared to uPA in the absence of uPAR. (Inset) Data replotted in a double-reciprocal manner, with residual uPA activity expressed fractionally. The intercepts of this plot give a maximum inhibitory effect at infinite uPAR concentration of 41% and half-maximal inhibition at 1.6 nM uPAR. The experimental conditions used, with uPAR in a large molar excess over uPA, allow this parameter to be used as an approximation to the K_d for the uPA–uPAR interaction. The data shown are from Ellis et al.[7]

method. This method may therefore have a more general applicability in the determination of the functional integrity of uPAR and uPAR mutants produced by recombinant DNA techniques.

Acknowledgments

These studies have been supported by grants from the Danish Biotechnology Program and the Danish Cancer Society.

[15] Vampire Bat Salivary Plasminogen Activator

By STEPHEN J. GARDELL and PAUL A. FRIEDMAN

Vampire bats (*Desmodus rotundus*) are absolutely dependent on a diet of fresh blood, a requirement necessitating that a vampire bat overcome unique and formidable obstacles to the acquisition of a hearty meal.[1] For

[1] C. M. Hawkey, *in* "Natural History of Vampire Bats" (A. M. Greenhall and U. Schmidt, eds.), p. 133. CRC Press, Boca Raton, FL, 1988.

instance, blood flow from the wounds inflicted on their victims must be sustained and the ingested blood must be maintained in a fluid state.

A plasminogen activator (PA) was previously identified in vampire bat saliva[2] and was partially characterized.[3] It was hypothesized that this protein plays a pivotal role in facilitating feeding by catalyzing the formation of plasmin, a proteolytic enzyme that readily lyses blood clots.

We investigated the PA present in vampire bat saliva (BatPA) to characterize its structure and evaluate its ability to activate plasminogen.[4] The amino acid sequence of BatPA is homologous to that of tissue-type plasminogen activator (tPA).[4] However, the activity of BatPA, in contrast to tPA, displayed a strict requirement for the presence of a fibrin cofactor.[4,5] This remarkable selectivity of BatPA for fibrin-bound plasminogen may be desirable for the treatment of thrombotic complications.[6]

Assay Methods

BatPA activity can be measured using a variety of assays that have been developed for plasminogen activators. The catalytic activity of BatPA is directly determined by a colorimetric assay using a low-molecular-weight chromogenic substrate (Method A). The ability of the enzyme to activate plasminogen can be monitored indirectly using a chromogenic substrate readily cleaved by plasmin (Method B),[7] or by monitoring for the release of radiolabeled fibrin degradation products (Method C).[8] BatPA activity can also be detected by zymographic analysis after denaturing polyacrylamide gel electrophoresis (method D)[9]; this latter procedure yields the molecular weight values of the active species.

Method A: Direct Amidolytic Assay

Reagents

Spectrozyme tPA: Chromogenic substrate (methylsulfonyl-D-cyclohexyltyrosylglycylarginine *p*-nitroaniline acetate) is purchased from American Diagnostica Inc. (Greenwich, CT)

[2] C. M. Hawkey, *Br. J. Haematol.* **13**, 1014 (1967).

[3] T. Cartwright, *Blood* **43**, 317 (1974).

[4] S. J. Gardell, L. T. Duong, R. E. Diehl, J. D. York, T. R. Hare, R. B. Register, J. W. Jacobs, R. A. F. Dixon, and P. A. Friedman, *J. Biol. Chem.* **264**, 17947 (1989).

[5] S. J. Gardell, T. R. Hare, P. W. Bergum, G. C. Cuca, L. O'Neill-Palladino, and S. M. Zavodny, *Blood* **76**, 2560 (1990).

[6] S. J. Gardell, D. R. Ramjit, I. I. Stabilito, T. Fujita, J. J. Lynch, G. C. Cuca, D. Jain, S. Wang, J.-S. Tung, G. E. Mark, and R. J. Shebuski, *Circulation* **84**, 244 (1991).

[7] M. Ranby and P. Wallen, *Prog. Fibrinolysis* **5**, 233 (1981).

[8] W. D. Sawyer, A. P. Fletcher, N. Alkjaersig, and S. Sherry, *J. Clin. Invest.* **39**, 426 (1960).

[9] A. Granelli-Piperno and E. Reich, *J. Exp. Med.* **148**, 223 (1978).

HBS buffer: 10 mM HEPES (Calbiochem, La Jolla, CA), 150 mM NaCl (Sigma Chemical Co., St. Louis, MO), 0.02% (v/v) Triton X-100 (Sigma Chemical Co.), pH 7.4

Procedure. Aliquots of samples containing BatPA are added to HBS buffer containing 0.5 mM Spectrozyme tPA. The reactions are incubated at 37° and the hydrolysis of the substrate is detected by monitoring for the increase in absorbance at 405 nm, reflecting the release of *p*-nitroaniline from the substrate.

Method B: Coupled Plasminogen Activation Assay

Reagents

HBS buffer: Same as above

Human fibrinogen: Grade L fibrinogen is purchased from American Diagnostica Inc. as lyophilized powder and resolubilized in H$_2$O. Fibrinogen concentration is determined using an extinction coefficient (1%, 1 cm, 280 nm) of 15.5[10]

Human plasminogen: Glu-type plasminogen is purchased from American Diagnostica Inc. as a lyophilized powder and resolubilized in H$_2$O. Plasminogen concentration is determined using an extinction coefficient (1%, 1 cm, 280 nm) of 17.0[11]

Human thrombin: Obtained from Sigma as a frozen solution. Activity concentration is approximately 4000 NIH units per mg protein.

Hirudin: Obtained as a lyophilized powder from American Diagnostica Inc. and resolubilized in H$_2$O

Spectrozyme Pl: Amidolytic chromogenic substrate obtained from American Diagnostica Inc. (H-D-norleucylhexahydrotyrosyllysine *p*-nitroanilide diacetate) is resuspended with H$_2$O to yield a 10 mM solution and is stored at −60°

Single-Chain tPA activity standard: 10 μg of single-chain tPA purified from Bowes melanoma cell line and lyophilized with bovine serum albumin is obtained from American Diagnostica Inc. The fibrinolytic activity, expressed in international units (IU), is determined by the supplier using the human melanoma tPA reference standard, lot 83/517, of the National Institute of Biological Standards and Controls, London, England. The sample is resuspended with 100 μl of 1 M KHCO$_3$ under gentle agitation for 20 min at 25°. The activity concentration is 50,800 IU/ml. Aliquots of the single-chain tPA activity standard are stored at −60°

[10] H. A. Scheraga and M. Laskowski, Jr., *Adv. Protein Chem.* **12,** 1 (1957).
[11] B. N. Violand and F. J. Castellino, *J. Biol. Chem.* **251,** 3906 (1976).

Procedure. HBS buffer (180 μl) containing fibrinogen (0.39 μM) and Glu-plasminogen (0.66 μM) is added to the wells of a microtitration plate (Linbro/Titertek; Flow Laboratories, McLean, VA). Thrombin (10 μl, 0.26 U/ml) is added to each well and the plate is incubated at 37° for 30 min. Hirudin (10 μl, 200 U/ml) is added to each well and the plate is incubated in a THERMOmax microplate reader (Molecular Devices, Menlo Park, CA) for 30 min at 37° with intermittent shaking (using the Automix function). Then 30 μl of 3.3 mM Spectrozyme Pl and 20 μl of a sample containing BatPA are added and the plasmin-mediated release of *p*-nitroaniline from the chromogenic substrate is measured spectrophotometrically (A_{405}–A_{490}) using the THERMOmax microplate reader set at 37°. The reactions are typically monitored for 1 hr and the absorbance data are collected at 18-sec intervals. The absorbance data are analyzed as a function of t^2 by reiterative linear least-squares regression using the TFFIT computer program[12] to yield the velocity of plasminogen activation. The plasminogen activator assay can be calibrated using serial dilutions of the single-chain tPA standard (final concentration in the 20-μl sample aliquot ranges from 0 to 10 IU/ml).

Method C: Lysis of Radiolabeled Plasma Clots

Reagents

Human plasma: Citrated fresh-frozen plasma is purchased from Biological Specialities, Lansdale, PA

^{125}I-Labeled human fibrinogen: Fibrinogen is labeled with Iodogen Iodination Reagent (Pierce, Rockford, IL) and ^{125}I (Amersham, Arlington Heights, IL). Unincorporated ^{125}I is removed by Sephadex G-25 (Pharmacia LKB, Piscataway, NJ) gel filtration chromatography run in the presence of 0.05 M Tris-HCl, 0.1 M NaCl, 0.025 M sodium citrate, pH 7.4. The final specific activity of the labeled fibrinogen is approximately 0.4 μCi/μg and the clottability is greater than 85%

Procedure. Aliquots (95 μl) of human plasma spiked with radiolabeled fibrinogen (5 × 10^3 cpm/μl) are dispensed into 96-well microtiter dishes. To each well is added 5 μl of a solution containing thrombin (0.2 NIH units/μl) and CaCl$_2$ (0.3 M). The samples are mixed and incubated for 30 min at 37°. Solutions of BatPA are diluted in 0.01 M HEPES, 0.1 M NaCl, pH 7.3, to yield final concentrations between 0.5 and 15 nM. A control sample consisting of buffer alone is also examined. Reactions, typically run in triplicate, are initiated by pipetting 100 μl of buffer or BatPA dilution into the wells containing the radiolabeled plasma clots. The reactions are

---

[12] S. D. Carson, *Comput. Programs Biomed.* **19,** 151 (1985).

incubated at 37° and, at various times, 5-μl aliquots are withdrawn and counted with a Gamma 5500B counting system (Beckman Instruments Inc., Fullerton, CA). Percent lysis is determined by calculating the ratio of cpm released over total cpm after correction for release of cpm in the "buffer-only"-containing wells.

Method D: Zymographic Analysis

Reagents and Procedure. Aliquots of BatPA are electrophoresed on an acid urea 8% polyacrylamide gel.[13] The gel is bathed in 0.01 M NaH$_2$PO$_4$, 0.14 M NaCl, pH 7.4, for 30 min with constant agitation, and overlaid onto a plasminogen-containing fibrin indicator gel. The components of the indicator gel are agarose (1% w/v) (SeaKem LE, FMC Bioproducts, Rockland, ME), human Glu-plasminogen (6 μg/ml) (American Diagnostica), bovine fibrinogen (2 mg/ml) (Calbiochem), bovine thrombin (0.06 NIH units/ml) (Sigma), and 0.01 M sodium phosphate, 0.14 M NaCl, pH 7.4. Alternatively, sodium dodecyl sulfate polyacrylamide gel electrophoresis (SDS–PAGE) followed by bathing in 2.5% (v/v) Triton X-100[14] can be used in place of acid urea electrophoresis and the nondetergent wash. Contact between the turbid fibrin indicator gel and BatPA in the polyacrylamide gel results in clear zones indicative of PA activity and fibrinolysis. The positions of the clear zones can be correlated with the migration of protein standards during the denaturing gel electrophoresis to yield the apparent molecular weight values of the active species.

Purification Procedure

Vampire bat saliva has previously been used as a source of the plasminogen activator.[2,3] However, we found the collection of vampire bat saliva to be painstaking and grossly inadequate. We subsequently discovered that vampire bat salivary glands served as a relatively accessible and plentiful source of BatPA and used them for our purification attempts.

Materials. The principal and accessory submaxillary glands from vampire bats (*D. rotundus*) captured in Venezuela are supplied by Dr. C. Rupprecht (Wistar Institute, Philadelphia, PA). These salivary glands are immediately frozen on dry ice following dissection and subsequently stored at −60°. Cellulose phosphate P11 is obtained from Whatman (Maidstone, Kent, England) and prepared according to the manufacturer's recommended protocol. tPA inhibitor from *Erythrina* seeds (ETI) is pur-

[13] S. Panyim and R. Chalkley, *Arch. Biochem. Biophys.* **130**, 337 (1969).
[14] E. G. Levin and D. J. Loskutoff, *J. Cell Biol.* **94**, 631 (1982).

chased from American Diagnostica Inc. and coupled to CNBr-activated Sepharose 4B (Pharmacia LKB) according to the manufacturer's instructions. Centriprep-30 concentrator devices are obtained from Amicon Division, W.R. Grace & Co. (Danvers, MA). Tris base is obtained from Boehringer Mannheim Biochemicals (Indianapolis, IN). Sodium acetate is purchased from Fisher Scientific (Fair Lawn, NJ). All other reagents and chemicals are of analytical grade.

Procedure. The vampire bat salivary glands are homogenized at 0° in 10 volumes (w/v) of 0.05 M Tris-HCl, 0.5 M NaCl, pH 7.5, using a Dounce manual tissue grinder (Wheaton Scientific, Millville, NJ). The glandular extract is clarified by centrifugation at 16,000 g for 20 min at 4°. The supernatant fraction is saved and stored on ice. The pellet is reextracted twice, each time with 10 volumes of the Tris-NaCl extraction buffer. The three supernatant fractions are combined, diluted 10-fold with 0.02 M Tris-HCl, 0.01% Tween 80, pH 7.2, and applied to a cellulose phosphate cation-exchange column preequilibrated with 0.02 M Tris-HCl, 0.05 M NaCl, 0.01% Tween 80, pH 7.2. BatPA activity binds quantitatively to the column under these conditions. The column is exhaustively washed with equilibration buffer and BatPA is then eluted with a linear 0.05–1 M NaCl gradient. The BatPA-containing fractions are identified with the coupled plasminogen activator assay, pooled, and applied to an affinity column containing immobilized ETI; this affinity chromatography step has been previously used to purify tPA.[15] The ETI column is repeatedly washed with 0.05 M Tris-HCl, 0.5 M NaCl, 0.01% Tween 80, pH 7.5, and the BatPA is eluted with 0.05 M sodium acetate, 0.5 M NaCl, 0.1% Tween 80, pH 4.0. Fractions (0.5 ml) are collected and the pH is immediately adjusted upward by the addition of 5 μl of 2 M Tris base. The BatPA-containing fractions are identified by the coupled plasminogen activator assay, pooled, concentrated with a Centriprep-30 device, and diafiltered into 0.05 M sodium acetate, 0.01% Tween 80, pH 5.0. BatPA is stored at −60° as a frozen solution or as a lyophilized powder.

Typical results for the purification of BatPA from vampire bat salivary glands are shown in Table I. The wet weight of the glands used for this purification attempt is 840 mg. The final product represents a 67% recovery of BatPA activity found in the gland extract. Homogeneous BatPA is isolated after an approximate 80-fold purification relative to the gland extract. In this representative purification, approximately 0.62 mg of BatPA is purified from 72.7 mg of total protein extracted from the vampire bat salivary glands.

[15] C. Heussen, F. Joubert, and E. B. Dowdle, *J. Biol. Chem.* **259**, 11635 (1984).

TABLE I
PURIFICATION OF BatPA FROM VAMPIRE BAT SALIVARY GLANDS

Purification step	Volume (ml)	Protein[a] (mg)	Activity[b] (IU)	Recovery (%)	Specific activity (IU/mg)	Purification (fold)
Glandular extract	43	72.7	39,300	100	541	1
Cellulose phosphate	112	12.3	33,600	85	2732	5.1
ETI-Sepharose	1.1	0.62	26,400	67	42,580	78.7

[a] Protein concentration was determined with the Micro BCA protein assay reagent (Pierce, Rockford, IL) using bovine serum albumin as the standard.
[b] Activity was determined with the coupled plasminogen activator assay and calibrated against the single-chain tPA activity standard.

Properties

Structural Properties

Analysis of the purified BatPA preparation by SDS–PAGE in the presence of dithiothreitol revealed at least three protein species that exhibited apparent relative molecular weight values of 53,600, 45,700, and 42,400 (Fig. 1, lane 1). Fractionation of the BatPA preparation by SDS–PAGE in the absence of reducing agents yielded a protein pattern similar to that observed under reducing conditions. The relative amounts of the three different protein bands varied from one glandular preparation to another (data not shown). Treatment of the BatPA preparation with endoglycosidase F (Boehringer Mannheim Biochemicals) increased the mobilities of each protein band.[4] These results demonstrate that the glandular BatPA preparation is composed of a heterogeneous group of glycoproteins. Furthermore, the observed gross heterogeneity is not abolished by treatment with glycosidases, thus suggesting that the differences arise from dissimilarities in protein structure. This latter inference is firmly established by the experiments detailed below.

Treatment of the glandular BatPA preparation with [³H]diisopropylfluorophosphate (Dupont NEN, Boston, MA) completely inactivated plasminogen activator activity. Furthermore, all three protein bands fractionated by SDS–PAGE were radiolabeled (Fig. 1, lane 3). These data establish that each of the three protein bands represents a serine protease. Convincing evidence that each protein band represents a plasminogen activator species is provided by zymographic analysis in which the three protein

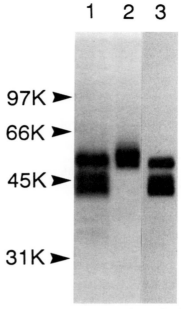

FIG. 1. SDS–polyacrylamide gel electrophoresis of vampire bat salivary plasminogen activator. Samples, pretreated with dithiothreitol, were electrophoresed on an SDS-denaturing 12% polyacrylamide gel (NOVEX, San Diego, CA) and were analyzed by protein staining with Coomassie brilliant blue R-250 (lanes 1 and 2) or by autoradiography (lane 3). Lane 1, BatPA purified from the vampire bat salivary glands (50 pmol); lane 2, recombinant DNA-derived BatPA-2 (20 pmol); lane 3, glandular BatPA treated with [³H]diisopropyl fluorophosphate (21,000 cpm). The polyacrylamide gel containing ³H-labeled BatPA was treated with EN³HANCE autoradiography enhancer (NEN Research Products, Boston MA) and placed in contact with film (X-OMAT AR, Eastman Kodak Company, Rochester, NY) for 64 hr. Molecular weight markers are indicated.

bands each gave rise to lytic zones on the fibrin indicator gel (Fig. 2, lane 2). The lytic zones were not observed if the fibrin indicator gel did not contain plasminogen (data not shown).

The glandular BatPA preparation was fractionated by SDS–PAGE, electrophoretically transferred to Protein Blott Paper (Applied Biosystems, Menlo Park, CA), and the NH₂ termini of the protein species in the three protein bands were determined with an Applied Biosystems Model 470 gas-phase sequencer, essentially as described elsewhere.[16] The NH₂-terminal amino acid sequences of the individual BatPA species are presented in Table II. Note the two similar but nonidentical sequences con-

[16] P. Matsudaira, *J. Biol. Chem.* **262,** 10035 (1987).

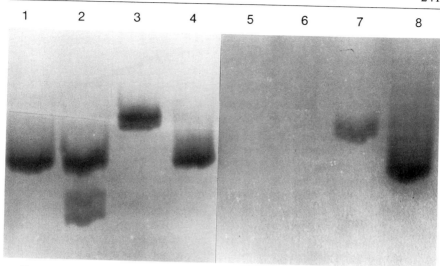

FIG. 2. Zymographic analysis of plasminogen activator activity. Samples were not pre-treated with a reducing agent prior to electrophoresis on an acid urea polyacrylamide gel. Lanes 1–4, Fibrin indicator gel; lanes 5–8, delipidated milk indicator gel. Lanes 1 and 5, rBatPA-2 (4.5 pmol); lanes 2 and 6, glandular BatPA (9 pmol); lanes 3 and 7, rtPA (7.4 pmol); lanes 4 and 8, urokinase (3.6 pmol). The dark zones represent localized areas of plasmin formation and subsequent clearing due to degradation of fibrin or delipidated milk.

tained within the upper SDS–PAGE band. In this preparation, there was an approximate fivefold excess of species I over species II. All of the sequences began with the same tripeptide sequence, Ala-Tyr-Gly, and subsequently diverged. Interestingly, the deduced NH_2-terminal se-

TABLE II

NH_2-TERMINAL SEQUENCES DERIVED FROM BATPA PREPARATION ISOLATED FROM VAMPIRE BAT SALIVARY GLANDS[a]

SDS–PAGE band	Sequence														
	1				5					10					15
Upper															
I	A	Y	G	V	A	–	K	D	E	I	T	Q	M	I	Y
II	A	Y	G	V	A	–	R	D	E	K	T	Q	M	I	Y
Middle	A	Y	G	S	–	S	E	L	R	–	F	N	G	G	T
Lower	A	Y	G	D	P	–	A	T	–	Y	K	D	Q	C	V

[a] Amino acid residues are indicated by single-letter code. The positions denoted by a dash are amino acid residues that were not identified.

quences contained within the upper, middle, and lower SDS–PAGE bands are in register with different regions of tPA (see below).

A vampire bat salivary gland cDNA library was screened with oligonucleotide probes designed from the NH_2-terminal amino acid sequences of the glandular BatPA preparation.[4] Presented in Fig. 3 is a schematic representation of the predicted amino acid sequence of a protein, BatPA-2, corresponding to species II contained in the upper SDS–PAGE band (see lane 1 of Fig. 1). The overall identity shared between BatPA-2 and tPA is 82% and encompasses several structural domains previously identified in tPA.[17] These include fibronectin finger, epidermal growth factor, kringle, and proteinase domains. There are two conspicuous structural differences between BatPA-2 and tPA: (1) BatPA-2 contains a single kringle domain whereas tPA contains two; the kringle domain in BatPA-2 is more similar to kringle 1 than kringle 2 of tPA and (2) BatPA-2 contains amino acid substitutions at the plasmin-sensitive processing site in tPA; hence, BatPA is a stable single-chain plasminogen activator. The amino acid sequence of BatPA-2 contributes two potential N-linked glycosylation sites: one each in the kringle and proteinase domains.

cDNAs encoding the BatPA species present in the middle and lower SDS–PAGE bands (see lane 1 of Fig. 1) were also isolated and sequenced. The species in the middle SDS–PAGE band (F^-BatPA) is a BatPA-2 derivative lacking the fibronectin finger domain. The species in the lower SDS–PAGE band (FG^-BatPA) is essentially a BatPA-2 derivative lacking both the fibronectin finger and epidermal growth factor domains (there are scattered amino acid differences between FG^-BatPA and BatPA-2, most notably in the kringle domain).

A cDNA corresponding to species I (see Table II) in the upper SDS–PAGE band, BatPA-1, was not identified in our original cDNA screen,[4] but was later described by Schleuning and co-workers.[18] We subsequently isolated the BatPA-1 cDNA from a library constructed with RNA extracted from vampire bat salivary glands that were excised soon after the vampire bats were captured.[19] The amino acid residues in BatPA-2 that are dissimilar in BatPA-1 are indicated by the filled circles in Fig. 3. Our original salivary gland cDNA library was prepared with RNA from vampire bats held in captivity for several days during shipment to our laboratory. A plausible explanation for the apparent absence of the

[17] D. Pennica, W. E. Holmes, W. J. Kohr, R. N. Harkins, G. A. Vehar, C. A. Ward, W. F. Bennett, E. Yelverton, P. H. Seeburg, H. L. Heyneker, D. V. Goeddel, and D. Collen, *Nature* (*London*) **301**, 214 (1983).
[18] J. Kratzschmar, J. Haendler, B. Langer, W. Boidol, P. Bringmann, A. Alagon, P. Donner, and W. D. Schleuning, *Gene* **105**, 229 (1991).
[19] S. J. Gardell and L. O'Neill-Palladino, unpublished results (1991).

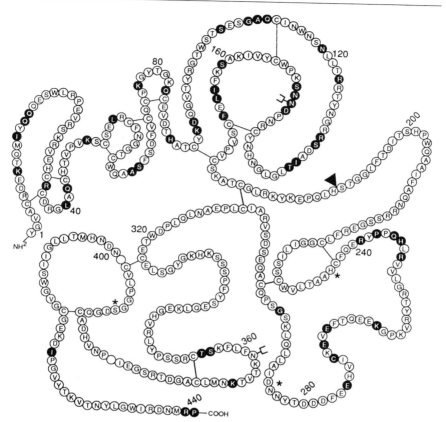

FIG. 3. Schematic representation of the deduced protein structure of BatPA-2. The amino acids deduced from the BatPA-2 cDNA are indicated by single-letter code. The amino acid residues in the pre and pro regions have been omitted. The proposed disulfide bridges depicted by straight lines between cysteine residues correspond to those postulated for human tPA (11) and delineate the finger domain (residues 1–43), epidermal growth factor domain (residues 44–84), kringle domain (residues 92–173), and proteinase domain (residues 189–441). The closed circles depict those residues in BatPA-2 that differ in BatPA-1. The position in BatPA-2 corresponding to the plasmin-sensitive processing site in tPA is indicated (▼). Active site residues are marked by asterisks. The postulated N-linked glycosylation sites in BatPA-2 are indicated by attached brackets.

BatPA-1 species in our initial vampire bat salivary gland cDNA library is that the level of RNA encoding BatPA-1 may be dramatically altered by the hunger and/or stress imposed on the animals during their captivity and transport.

Enzymatic Properties

Zymographic analysis demonstrated that BatPA-1 and/or BatPA-2, F⁻BatPA, and FG⁻BatPA display PA activity (Fig. 2, lane 2). When the zymographic indicator gel contained delipidated milk (18 mg/ml; Carnation Company, Los Angeles, CA), instead of fibrin, none of the three bands gave rise to zones of clearing (lane 6, Fig. 2). In contrast, tPA gave rise to prominent signals on both the fibrin and delipidated milk indicator gels (lanes 3 and 7, Fig. 2, respectively), as did UK (lanes 4 and 8, Fig. 2, respectively). This analysis revealed that each of the unique BatPA species present in the glandular preparation lacked PA activity unless fibrin was present.

The ability to bind tightly to fibrin is a property displayed by BatPA-1 and BatPA-2 but not F⁻BatPA and FG⁻BatPA (Fig. 4). Lane 1 (Fig. 4) shows the characteristic triplet band profile of glandular BatPA. Lane 2 (Fig. 4) shows the BatPA species that remained in the supernatant fraction after forming a fibrin clot in the presence of glandular BatPA and subsequently pelleting the fibrin clot. Note the selective disappearance of the upper band (Fig. 4), consistent with the binding of BatPA-1 and BatPA-2 to fibrin.

The heterogeneous nature of the BatPA preparation isolated from the vampire bat salivary glands precluded rigorous evaluation of its activity. Our efforts to carefully examine enzymatic activity utilized recombinant BatPA-2 (rBatPA-2) that was produced by expression of its cDNA in CV-1 African Green monkey cells.[20] rBatPA-2 was purified from conditioned media using the ETI-Sepharose affinity chromatography step.

SDS–PAGE of rBatPA-2 (Fig. 1, lane 2) revealed a doublet that migrated similarly with the upper band displayed by glandular BatPA (Fig. 1, lane 1). A single band was observed following the treatment of rBatPA-2 with endoglycosidase, thus indicating that the observed heterogeneity was due to the presence of carbohydrate (data not shown). rBatPA-2 displayed two other properties shared with the glandular BatPA species represented by the upper SDS–PAGE band: (1) PA activity was evident with the fibrin indicator gel (Fig. 2, lane 1) but not the delipidated milk indicator gel (Fig. 2, lane 5) and (2) relatively tight binding to fibrin (Fig. 4; compare lanes 3 and 4).

Active Site Titration

Quantitation of the rBatPA-2 preparation was based primarily on amino acid composition analysis. Nevertheless, the deduced estimate for protein

---

[20] J.-S. Tung, L. O'Neill-Palladino, S. J. Gardell, G. C. Cuca, M. Silberklang, R. W. Ellis, and G. E. Mark, unpublished results (1990).

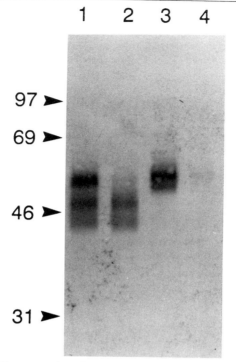

FIG. 4. Fibrin binding properties of BatPA. The glandular BatPA preparation (13 pmol) (lanes 1 and 2) or rBatPA-2 (6 pmol) (lanes 3 and 4) was added to 200 μl of 0.01 M HEPES, 0.14 M NaCl, 0.01% Tween 80, pH 7.4 (lanes 1 and 3) or 200 μl of the same buffer also containing human fibrinogen (2 mg/ml) (lanes 2 and 4). Following the addition of human thrombin (5 U/ml, final concentration), the samples were incubated for 30 min at 37° and centrifuged at 100,000 g for 20 min at 4°. Aliquots (3.5 μl) of the supernatant fractions were electrophoresed on a SDS-denaturing 12% polyacrylamide gel, transferred to nitrocellulose (Scheicher & Schuell, Keene, NH; 0.4 μm pore size), and analyzed by Western immunoblotting using, sequentially, rabbit anti-BatPA antisera, biotinylated protein G (Pierce), and streptavidin–alkaline phosphatase (GIBCO-BRL, Gaithersburg, MD). The reagents used for color development were 5-bromo-4-chloro-3-indolyl phosphate and nitro blue tetrazolium (both from GIBCO-BRL). The molecular weight markers ($\times$ 10^{-3}) are indicated.

concentration was corroborated by an active site titration protocol used for the measurement of tPA.[21]

Reagents

HN buffer: 0.05 M HEPES, 0.1 M NaCl, pH 7.4
4-Methylumbelliferyl-p-guanidinobenzoate (MUGBE): Obtained from

[21] T. Urano, S. Urano, and F. J. Castellino, *Biochem. Biophys. Res. Commun.* **150**, 45 (1988).

Sigma and resuspended in dimethylformamide to yield a 5 mM solution

4-Methylumbelliferol (4-MU): Obtained from Sigma and resuspended in 1 mM NaOH. The concentration of 4-MU is determined using an absorptivity of 17,000 M^{-1} cm^{-1} at 360 nm[22]

Procedure. HN buffer (3 ml) containing MUGBE (20 μM) is chilled to 4° in the thermostatted cuvette holder of a Perkin-Elmer (Norwalk, CT) LS-3 fluorescence spectrophotometer attached to a Brinkmann RM6 refrigerated water bath. rBatPA-2 is added (100–500 pmol) and the hydrolysis of MUGBE is assayed using excitation and emission wavelengths of 365 and 445 nm, respectively. The reaction is continuously monitored until a constant rate of product formation is achieved. The linear portion of the curve is extrapolated to zero time, and the deduced burst displacement at zero time is measured. The concentration of rBatPA-2 is calculated by comparison of the burst value with a standard curve generated with 4-MU.

Activation of Glu-Plasminogen by rBatPA-2 and rtPA

rBatPA-2 displays virtually no activity toward Glu-plasminogen in the absence of a fibrin cofactor as judged by its exceedingly low k_{cat}/K_m value (Table III). In contrast, rtPA exhibits a k_{cat}/K_m value that is more than 100-fold greater than rBatPA-2. CNBr-degraded fibrinogen has been shown to be a potent cofactor for tPA.[23] Interestingly, CNBr-degraded fibrinogen failed to appreciably stimulate the plasminogen activator activity of rBatPA-2. On the other hand, fibrin II is a potent cofactor for both rBatPA-2 and rtPA; indeed, in the presence of this cofactor, the corresponding k_{cat}, K_m and k_{cat}/K_m values displayed by these two PAs are similar. The overall fold increases in the PA activities of rBatPA-2 and rtPA in the presence of fibrin II relative to "no cofactor" are 44,000 and 340, respectively. Hence, rBatPA-2 appears to be approximately 130-fold more selective than rtPA toward fibrin-bound plasminogen.

Lysis of Plasma Clots Mediated by rBatPA-2

The addition of rBatPA-2 to human plasma clots containing ^{125}I-labeled human fibrinogen promotes the release of radiolabeled fibrin degradation products (Fig. 5). The rates of clot lysis mediated by rBatPA-2 steadily increase as the dose is progressively augmented from 0.5 to 15 nM. Similar lytic rates are observed when equimolar doses of rtPA are assayed.[5]

[22] W. R. Sherman and E. Robins, *Anal. Chem.* **40**, 803 (1968).

[23] J. H. Verheijen, E. Mullaart, G. T. G. Chang, C. Kluft, and G. Wijngaards, *Thromb. Haemostasis* **48**, 266 (1982).

TABLE III
ACTIVATION OF GLU-PLASMINOGEN BY rBatPA-2 AND rtPA[a]

Cofactor	Enzyme	$10^4 \times k_{cat}$ (sec^{-1})	K_m (μM)	k_{cat}/K_m (M^{-1} sec^{-1})
None	rBatPA-2	0.02 ± 0.0005	0.58 ± 0.39	4 ± 1
	rtPA	31 ± 2	6.7 ± 0.7	468 ± 37
CNBr-degraded	rBatPA-2	0.06 ± 0.0003	0.84 ± 0.11	8 ± 1
fibrinogen	rtPA	100 ± 5	0.07 ± 0.02	$136{,}000 \pm 6800$
Fibrin II	rBatPA-2	260 ± 10	0.15 ± 0.07	$174{,}000 \pm 66{,}000$
	rtPA	210 ± 60	0.13 ± 0.03	$158{,}000 \pm 38{,}000$

[a] The kinetics of Glu-plasminogen activation were determined as described under Assay Method B except that the plasminogen concentration was varied from 0.015 to 2.5 μM and the PA concentration was 400 pM. The derived velocities were converted to [p-nitroaniline]/min^2 (using an extinction coefficient of 9700 M^1 cm^{-1}) and plotted against [plasminogen]. The plots were then analyzed by nonlinear regression analysis using Enzfitter Software (Biosoft, Cambridge, England) to yield the Michaelis–Menten constants. Fibrinogen was omitted in the "no cofactor" determinations and CNBr-degraded fibrinogen (American Diagnostica) was substituted for fibrinogen where indicated.

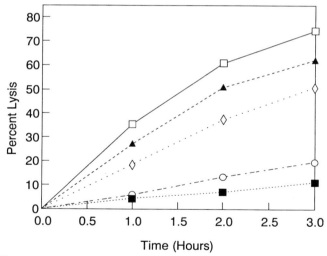

FIG. 5. Plasma clot lysis mediated by rBatPA-2. Various amounts of rBatPA-2 were added to ^{125}I-labeled fibrin containing human plasma clots. The samples were incubated at 37° and aliquots of the "bathing solution" were removed at the indicated times and monitored for the presence of radiolabeled fibrin degradation products. The rBatPA-2 concentrations are as follows: ■, 0.5 nM; ○, 1 nM; ◆, 5 nM; ▲, 10 nM; □, 15 nM. The data were adjusted for the background release of counts per minute in a control sample (no rBatPA-2); the counts per minute released from the control plasma clot were (in %) 3.8 ± 2.5, 5.3 ± 0.6, and 5.9 ± 0.3 at 1, 2, and 3 hr, respectively. Each time point represents the mean of four duplicate samples; the maximal standard deviation was less than 1.2% in all instances and averaged 0.5%. Reproduced with permission from Gardell et al.[4]

Susceptibility to Inactivation by Plasminogen Activator Inhibitor Type I

Plasminogen activator inhibitor type I (PAI-1), a serpin class inhibitor found in plasma in low amounts, rapidly inactivates tPA and urokinase.[24] We showed that PAI-1 is also a relatively potent inhibitor of rBatPA-2; however, in contrast to tPA, the interaction between rBatPA-2 and PAI-1 is not enhanced in the presence of fibrinogen.[5]

Reagents and Procedure. PAI-1 purified from HT-1080 fibrosarcoma cells (American Diagnostica, Inc.) is activated[25] by treatment with 4 M guanidine hydrochloride (Pierce) followed by renaturation using a NAP-10 gel-filtration column (Pharmacia LKB). The concentration of activated PAI-1 is determined by quantitative neutralization of two-chain tPA in a direct amidolytic assay using S2288 (Kabi-Vitrum, Stockholm, Sweden). rBatPA-2 or rtPA (1 nM) is added to cuvettes containing activated PAI-1 (2 to 10 nM), a fluorogenic substrate [N-Methoxysuccinyl-L-isoleucylglycl-L-Arginine-4-methylcovmaryl-7-amide (MeoSuc-Ile-Gly-Arg-AMC, Enzyme Systems Products, Livermore, CA) and L-Pyroglutamylglycyl-L-Arginine-4-methylcoumaryl-7-amide (Pyr-Gly-Arg-MCA, Peninsula Laboratories, Belmont, CA) for rtPA and rBatPA-2, respectively], and, where indicated, human plasminogen-free fibrinogen (American Diagnostica, Inc.) (0.3 μM) in HBS buffer containing dimethyl sulfoxide (1%, v/v). The concentration of the fluorogenic substrate is at least twice the experimentally determined K_m value. The signal is monitored with a Perkin-Elmer LS-3 fluorescence spectrophotometer (Perkin-Elmer, Norwalk, CT) at excitation and emission wavelengths of 380 and 460 nm, respectively. The time-dependent inhibition of rBatPA-2 or rtPA at varying concentrations of PAI-1 is recorded as a family of progress curves indicative of slow, tight binding inhibition.[26] The data from each curve (fluorescence, F, as a function of time, t) are fit to the integrated first-order rate equation, $F = v_s t + (v_o - v_s)(1 - e^{-kt})/k$, by nonlinear regression using GraphPAD In-Plot software (GraphPAD Software, San Diego, CA) to yield the apparent constant k, where $k = k_2 + [k_1/(1 + s/K_m)]I$. A plot of K as a function of PAI-1 concentration (I) yields a straight line with a slope equal to $k_1/(1 + s/K_m)$, from which the second-order inhibition rate constant, k_1, is derived.

Result. The apparent second-order rate constant for the interaction between rBatPA-2 and PAI-1 was determined to be $4.8 \times 10^6 \ M^{-1} \ \text{sec}^{-1}$ (Table IV). The corresponding k_1 value for rtPA was $3.0 \times 10^6 \ M^{-1} \ \text{sec}^{-1}$, which agreed with published values.[27] We examined the effect of fibrinogen

---

[24] E. Sprengers and C. Kluft, *Blood* **69**, 381 (1987).

[25] C. M. Hekman and D. J. Loskutoff, *J. Biol. Chem.* **260**, 11581 (1985).

[26] J. W. Williams and J. F. Morrison, this series, Vol. 63, p. 437.

[27] S. Thorsen, M. Philips, J. Selmer, I. Lecander, and B. Astedt, *Eur. J. Biochem.* **175**, 33 (1988).

TABLE IV

INHIBITION RATE CONSTANTS FOR INACTIVATION OF
rBatPA-2 AND rtPA BY PAI-1[a,b]

Activator	k_1 (M^{-1} sec^{-1})	
	$-$ FBG	$+$ FBG
rBatPA-2	$(4.84 \pm 0.21) \times 10^6$	$(4.37 \pm 0.25) \times 10^6$
rtPA	$(3.00 \pm 0.01) \times 10^6$	$(9.49 \pm 0.72) \times 10^6$

[a] Reproduced with permission from Gardell et al.[5]
[b] Second-order inhibition rate constants (k_1) were determined as described in the text. Where indicated, the reactions contained 100 μg/ml human fibrinogen (FBG). Each value is the mean $\pm$ standard deviation.

on the interaction between these plasminogen activators and PAI-1 because (1) fibrinogen affects the enzymatic activity of tPA and (2) plasma, the relevant pharmacologic milieu, contains ample quantities of fibrinogen. The k_1 value for the interaction between rtPA and PAI-1 was increased 3.2-fold in the presence of fibrinogen whereas the corresponding rate constant for rBatPA-2 and PAI-1 was unchanged. The prediction that rtPA would be inactivated by PAI-1 in plasma more rapidly than rBatPA-2 has been supported by the results of in vitro studies.[5]

[16] Probing Structure–Function Relationships of Tissue-Type Plasminogen Activator by Oligonucleotide-Mediated Site-Specific Mutagenesis

By Edwin L. Madison and Joseph F. Sambrook

I. Introduction

Tissue-type plasminogen activator (t-PA), a 66-kDa member of the chymotrypsin family of serine proteases, catalyzes the rate-limiting step in the endogenous fibrinolytic cascade: conversion of the zymogen plasminogen into the active enzyme plasmin.[1–3] Plasmin, a serine protease of

[1] H. R. Lijnen and D. Collen, Thromb. Haemostasis 66, 88 (1991).
[2] R. D. Gerard, K. R. Chien, and R. S. Meidell, Mol. Biol. Med. 3, 449 (1986).
[3] K. Danø, P. A. Andreason, J. Grondahl-Hansen, P. Kristensen, L. S. Nielsen, and L. Skriver, Adv. Cancer Res. 44, 139 (1985).

relatively broad specificity, efficiently degrades the fibrin that forms the meshwork of a thrombus or blood clot.[1,2] The fibrinolytic cascade, therefore, provides an important counterbalance *in vivo* to the blood coagulation cascade.

Consistent with its important regulatory role, the activity of t-PA is tightly regulated *in vivo*. Unlike other plasminogen activators, such as urokinase or streptokinase, t-PA both binds to and is dramatically stimulated by fibrin. This property of t-PA has been extensively studied and shown to result from a large increase in the affinity of the enzyme for its substrate. In the absence of fibrin, the K_m of t-PA for Glu-plasminogen is 65 μM; in the presence of fibrin; however, this K_m decreases to a value of 0.16 μM.[4-6] *In vivo*, where plasminogen is present at a concentration of approximately 2 μM, this property of t-PA is important because it effectively restricts the production of plasmin by t-PA to sites of fibrin deposition, where it is needed, and consequently prevents systemic activation of plasminogen, which would lead to a rapid depletion of circulating fibrinogen.

The activity of circulating t-PA is also held in check by specific protease inhibitors present in human plasma. The most important endogenous inhibitor of t-PA is endothelial cell plasminogen activator inhibitor type 1 (PAI-1), a 50-kDa member of the serpin, or serine protease inhibitor, gene superfamily.[7-9] PAI-1 is believed to function as a suicide substrate of t-PA. t-PA and PAI-1 very rapidly form a specific 1 : 1 inactive complex that is stable to boiling in SDS. Because the concentration of PAI-1 in human plasma is significantly higher than that of t-PA, endogenous t-PA circulates almost exclusively as a complex with PAI-1.[10] Another mechanism adopted to regulate negatively the activity of t-PA *in vivo* is the physical removal or clearance of this enzyme from the circulation. Clearance of t-PA from the circulation is mediated by the liver[11] and appears to involve at least three distinct receptors. One of these receptors is present on hepatic endothelial cells and recognizes the mannose-rich oligo-

---

[4] M. Hoylaerts, D. C. Rijken, H. R. Lijnen, and D. Collen, *J. Biol. Chem.* **257,** 2912 (1982).

[5] M. Ranby, *Biochim. Biophys. Acta* **704,** 461 (1982).

[6] D. C. Rijken, M. Hoylaerts, and D. Collen, *J. Biol. Chem.* **257,** 2920 (1982).

[7] R. Carrell and G. Boswell, in "Proteinase Inhibitors" (A. J. Barett and G. S. Salvesan, eds.), p. 403. Elsevier, Amsterdam, 1986.

[8] H. Pannekoek *et al., EMBO J.* **5,** 2539 (1986).

[9] D. Ginsberg, *et al., J. Clin. Invest.* **78,** 1673 (1986).

[10] C. L. Lucore and B. E. Sobel, *Circulation* **77,** 660 (1988).

[11] C. Korninger, J. M. Stassen, and D. Collen, *Thromb. Haemostasis* **46,** 658 (1981).

saccharide attached to Asn-117.[12,13] Another, present on hepatic parenchymal cells, is a specific t-PA receptor that appears to recognize a determinant near the amino terminus of the enzyme.[14] The third receptor, originally described by Schartz and co-workers, does not bind free t-PA but specifically recognizes the complex between t-PA and PAI-1.[15-17] Due to the combined action of these receptor systems, the half-life of t-PA introduced into the circulation of patients with myocardial infarctions is only 6 min.[18-20]

t-PA is synthesized and secreted by vascular endothelial cells.[21-24] The amino acid sequence of the human t-PA precursor, originally deduced from the nucleotide sequence of cloned cDNAs, consists of 563 amino acids.[25,26] As this precursor is transported along the secretory pathway, oligosaccharide groups are added at three positions (Asn-117, Asn-184, and Asn-448), and a hydrophobic signal sequence and a hydrophilic prosequence are both removed by endopeptidic cleavage.[27-29] The resulting

[12] A. Hotchkiss, C. J. Refino, C. K. Leonard, J. V. O'Connor, C. Crowley, J. McCabe, K. Tate, G. Nakamura, D. Powers, A. Levinson, M. Mohler, and M. W. Spellman, *Thromb. Haemostasis* **60**, 225 (1988).

[13] C. L. Lucore, E. T. A. Fry, D. A. Nachowiak, and B. E. Sobel, *Circulation* **77**, 906 (1988).

[14] J. Krause, *Fibrinolysis* **2**, 133 (1988).

[15] D. A. Owensby, B. E. Sobel, and A. L. Schartz, *J. Biol. Chem.* **263**, 10587 (1988).

[16] P. A. Morton, D. A. Owensby, B. E. Sobel, and A. L. Schwartz, *J. Biol. Chem.* **264**, 7228 (1989).

[17] P. A. Morton, D. A. Owensby, T. C. Wun, J. J. Billadello, and A. L. Schwartz, *J. Biol. Chem.* **265**, 14093 (1990).

[18] M. Verstraete, H. Bounameaux, F. De Cock, F. Van de Werf, and D. Collen, *J. Pharmacol. Exp. Ther.* **235**, 506 (1985).

[19] T. Nilsson, P. Wallen, and G. Mellbring, *Scand. J. Haematol.* **33**, 49 (1984).

[20] E. J. P. Brommer, F. H. M. Derkx, M. A. D. H. Shcalekamp, G. Dooijewaard, and M. M. V. D. Klaauw, *Thromb. Haemostasis* **59**, 404 (1988).

[21] R. A. Allen and D. S. Pepper *Thromb. Haemostasis* **45**, 43 (1981).

[22] E. Levin, *Proc. Natl. Acad. Sci. U.S.A.* **80**, 6804 (1983).

[23] E. G. Levin and D. J. Loskutoff, *J. Cell Biol.* **94**, 631 (1982).

[24] A. J. van Zonneveld, G. T. G. Chang, J. van den Berg, *et al.*, *Biochem. J.* **235**, 385 (1986).

[25] D. Pennica, W. E. Holmes, W. J. Kohr, *et al.*, *Nature (London)* **301**, 214 (1983).

[26] T. J. R. Harris, T. Patel, F. A. O. Martson, S. Little, and J. S. Emtage, *Mol. Biol. Med.* **3**, 279 (1986).

[27] G. Pohl, M. Kallström, N. Bergsdorf, P. Wallen, and H. Jornvall, *Biochemistry* **23**, 3701 (1984).

[28] G. A. Vehar, W. J. Kohr, W. F. Bennett, *et al.*, *Bio Technology* **2**, 1051 (1984).

[29] J. Sambrook, D. Hanahan, and M. J. Gething, *Mol. Biol. Med.* **3**, 459 (1986).

527-amino acid glycoprotein is secreted as a single-chain enzyme that is subsequently cleaved (between Arg-275 and Ile-276) into a two-chain form.[30,31] This two-chain enzyme consists of a 34-kDa amino-terminal heavy chain and a 31-kDa carboxyl-terminal light chain that are connected by a disulfide bond between Cys-264 and Cys-395.[30-32] It is noteworthy that both single- and two-chain t-PAs are catalytically active.[6,31,33-37] Unlike precursors of other serine proteases of the hemostatic/fibrinolytic system, single-chain t-PA is, therefore, not a true zymogen. Both single- and two-chain t-PA bind to and are stimulated by fibrin, are rapidly inhibited by PAI-1,[38] and are rapidly cleared from the circulation by the liver.[39] Two-chain t-PA, however, has a k_{cat} for hydrolysis of peptide substrates that is approximately threefold higher than that of the single-chain enzyme.[37]

Mature t-PA is a complex protein that is organized by 16 disulfide bridges into a series of five discrete, putative structural domains that are homologous to regions of other secreted and cell surface proteins, including several serine proteases present in plasma[32,40] (Fig. 1). Residues 4–50 of the mature protein form a "finger domain" that is closely related to regions of fibronectin involved in binding to fibrin.[41] Residues 51–87 share

[30] D. C. Rijken and D. Collen, *J. Biol. Chem.* **256,** 7035 (1981).

[31] P. Wallen, N. Bergsdorf, and M. Ranby, *Biochim. Biophys. Acta* **719,** 318 (1982).

[32] T. Ny, F. Elgh, and B. Lund, *Proc. Natl. Acad. Sci. U.S.A.* **81,** 5355 (1984).

[33] M. Ranby, *Biochim. Biophys. Acta* **704,** 461 (1982).

[34] D. L. Higgins and G. A. Vehar, *Biochemistry* **26,** 7786 (1987).

[35] K. M. Tate, D. L. Higgins, W. E. Holmes, M. E. Winkler, H. L. Heyneker, and G. Vehar, *Biochemistry* **26,** 338 (1987).

[36] L. C. Petersen, M. Johannessen, D. Foster, A. Kumar, and E. Mulvihill, *Biochim. Biophys. Acta* **952,** 245 (1988).

[37] J. A. Boose, E. Kuismanen, R. Gerard, J. Sambrook, and M. J. Gething, *Biochemistry* **28,** 635 (1988).

[38] C. Hekman and D. J. Loskutoff, *Arch. Biochem. Biophys.* **262,** 199 (1988).

[39] H. Bounameaux, M. Verstraete, and D. Collen, *Thromb. Haemostasis* **54,** 61 (1985).

[40] L. Patthy, *Cell (Cambridge, Mass.)* **41,** 657 (1985).

[41] L. Banjai, A. Varadi, and L. Patthy, *FEBS Lett.* **163,** 37 (1983).

FIG. 1. Schematic representation of the amino acid sequence[25] of the precursor of human t-PA, including the signal sequence (S) and the prosegment (P). The solid lines indicate the predicted disulfide bridges, and the site of proteolytic cleavage into heavy and catalytic light chains is shown by the bold arrow. The small arrows marked with letters B–M indicate the map positions of individual exons in the protein.[32] The locations of the finger (F), epidermal growth factor (E), kringle 1 (K_1), and kringle 2 (K_2) domains of the heavy chain are shown, together with notation (roman numerals) of the exon(s) that encode each domain. This diagram is modified from Ny *et al.*[32]

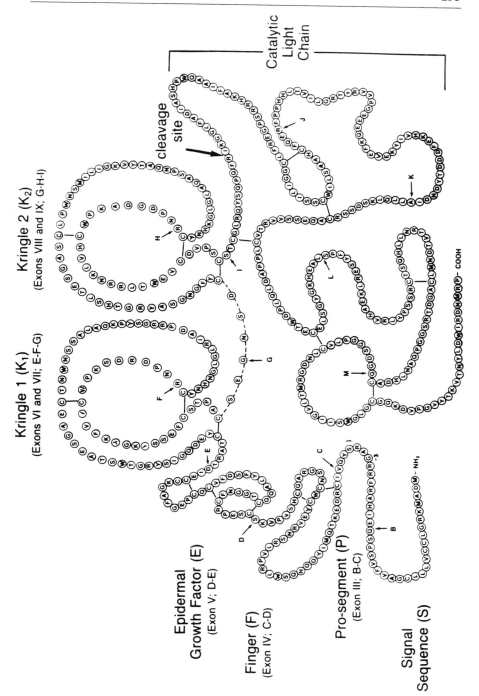

homology with epidermal growth factor[42] and with similar domains in a variety of other proteins, including urokinase,[43] protein C,[44] coagulation factors IX and X,[45,46] the receptor for low density lipoprotein,[47] and differentiation-specific proteins encoded by *Drosophila melanogaster* and *Caenorhabditis elegans*.[48,49] Residues 88–175 and 176–263 form two sequential "kringle" domains (K1 and K2), each built around three characteristic intradomain disulfide bonds.[50–52] Domains of similar structure are found in plasminogen, prothrombin, urokinase, and lipoprotein (a).[40,53,54] The remainder of the molecule (residues 276–527) constitutes the catalytic domain, which is homologous to trypsin and other members of the chymotrypsin family of serine proteases and contains a characteristic catalytic triad of amino acids (His-322, Asp-371, and Ser-478).[6,25,55]

II. Selection of Sites for Mutagenesis

A. Deletion of Putative Structural Domains of t-PA

Isolation and characterization of the 29-kb gene encoding human t-PA revealed that there was a very striking correspondence between the end points of the putative structural domains of t-PA, defined by homology to domains in other proteins, and the location of intron–exon boundaries in the gene.[32,56] This observation suggested that t-PA has evolved into its present mosaic form by a process of duplication and exchange of exons or groups of exons and led to the proposal that the structural domains

[42] R. F. Doolittle, D. F. Feng, and M. S. Johnson, *Nature (London)* **307**, 558 (1984).
[43] P. Verde, M. P. Stoppelli, P. Galeffi, P. D. Nocera, and F. Blasi, *Proc. Natl. Acad. Sci. U.S.A.* **81**, 4727 (1984).
[44] D. Foster and E. W. Davie, *Proc. Natl. Acad. Sci. U.S.A.* **81**, 4766 (1984).
[45] D. S. Anson, K. H. Choo, D. J. G. Rees, *et al.*, *EMBO J.* **3**, 1053 (1984).
[46] S. P. Leytus, D. C. Chung, W. Kisiel, K. Kurachi, and E. W. Davie, *Proc. Natl. Acad. Sci. U.S.A.* **81**, 3699 (1984).
[47] D. W. Russell, W. J. Schneider, T. Yamamoto, K. L. Luskey, M. S. Brown, and J. L. Goldstein, *Cell (Cambridge, Mass.)* **37**, 577 (1984).
[48] K. A. Wharton, K. M. Johansen, T. Xu, and S. Artavanis-Tsakonas, *Cell (Cambridge, Mass.)* **43**, 567 (1985).
[49] I. Greenwald, *Cell (Cambridge, Mass.)* **43**, 583 (1985).
[50] G. Olsson, L. Andersen, O. Lindqvist, *et al.*, *FEBS Lett.* **145**, 317 (1982).
[51] S. K. Holland, K. Harlos, and C. C. F. Blake, *EMBO J.* **6**, 1875 (1987).
[52] C. H. Park and A. Tulinsky, *Biochemistry* **25**, 3977 (1986).
[53] L. Patthy, M. Trexler, Z. Vali, L. Banyai, and A. Varadi, *FEBS Lett.* **171**, 131 (1984).
[54] J. W. McLean, J. E. Tomlinson, W. J. Kuang, *et al.*, *Nature (London)* **330**, 132 (1987).
[55] P. Wallen, G. Pohl, N. Bergsdorf, M. Ranby, T. Ny, and H. Jornvall, *Eur. J. Biochem.* **132**, 681 (1983).
[56] S. J. F. Friezner-Degen, B. Rajput, and E. Reich, *J. Biol. Chem.* **261**, 6972 (1986).

of t-PA might contribute autonomous functions to the enzyme.[40] This hypothesis that the individual domains of t-PA could be viewed as autonomous structural and functional modules became the focus of many of the initial structure–function studies of t-PA. Experiments reported by several investigators tested aspects of this hypothesis by analyzing the properties of mutated proteins that lack various segments of the wild-type enzyme that lie between naturally occurring restriction sites within t-PA cDNA.[57–60] While many of these sites map near the borders of putative structural domains of the protein, others lie some distance away. This raised the possibility that the loss of function exhibited by certain of the mutants is due to malfolding rather than the absence of a particular structural domain from the protein. A more rigorous experimental design, therefore, is to utilize oligonucleotide-directed site-specific mutagenesis to delete precisely the exons that encode the individual domains of t-PA.[61–64] (Detailed protocols for oligonucleotide-mediated mutagenesis are presented in Section III.) Characterization of the resulting mutated enzymes led to the following views regarding the function of individual domains of t-PA: (1) The role of the signal sequence is to assure the entry of the enzyme into the secretory pathway. (2) No function has yet been assigned to the pro segment of t-PA; however, preliminary evidence from our laboratory suggests that this region may modulate both the rate of folding of the enzyme and its rate of transport along the secretory pathway. (3) The finger domain has been implicated in the binding of t-PA to intact fibrin and may also be involved in the clearance of t-PA by hepatic receptors. (4) The growth factor domain appears to be involved in the recognition of t-PA by hepatic, parachymal receptors. (5) The kringle I domain contains the mannose-rich oligosaccharide that has been implicated in clearance of t-PA by receptors on hepatic endothelial cells. In addition, our laboratory has demonstrated that the kringle I domain can mediate stimulation of the enzyme by large fragments of soluble fibrin. (6) We and others have

---

[57] H. Kagitani, M. Tagawa, K. Hatanaka, *et al.*, *FEBS Lett.* **189**, 145 (1985).

[58] A. J. van Zonneveld, H. Veerman, M. E. MacDonald, J. A. van Mourik, and H. Pannekoek, *J. Cell. Biochem.* **32**, 169 (1986).

[59] A. J. van Zonneveld, H. Veerman, and H. Pannekoek, *Proc. Natl. Acad. Sci. U.S.A.* **83**, 4670 (1986).

[60] J. H. Verheijen, M. P. M. Caspers, G. T. G. Chang, G. A. W. de Munk, P. H. Pouwels, and B. E. Enger-Valk, *EMBO J.* **5**, 3525 (1986).

[61] G. R. Larsen, K. Henson, and Y. Blue, *J. Biol. Chem.* **263**, 1023 (1988).

[62] M. J. Browne, J. E. Carey, C. G. Chapman, *et al.*, *J. Biol. Chem.* **263**, 1599 (1988).

[63] N. K. Kalyan, S. G. Lee, J. Wilhelm, *et al.*, *J. Biol. Chem.* **263**, 3971 (1988).

[64] M. J. Gething, B. Alder, J. A. Boose, R. D. Gerard, E. L. Madison, D. McGookey, R. S. Meidell, L. M. Roman, and J. F. Sambrook, *EMBO J.* **7**, 2731 (1988).

demonstrated that the kringle II domain can mediate the stimulation of t-PA by both large and small fragments of fibrin or fibrinogen. Others have shown that the kringle II domain also mediates the binding of t-PA to partially degraded fibrin. (7) The light chain or protease domain contains the active site of the enzyme and is therefore directly involved in the catalytic cleavage of plasminogen and the inhibition by serpins.

B. Homology Modeling

1. Design of Point Mutations. Under the most favorable circumstances, structure–function studies of an enzyme would be performed by proposing a role(s) for a particular residue(s) based on careful scrutiny of the known structure, site-specific mutagenesis of the targeted residue, and an extensive characterization of the mutated protein including enzyme kinetics and X-ray crystallography. Unfortunately, however, because of the absence of structural data, this comprehensive approach is not yet possible with any of the multidomain serine proteases that participate in coagulation, fibrinolysis, and complement activation. For many of these enzymes, however, including t-PA, the structures of related domains from homologous proteins are available, and these structures can be used to predict the unknown structure of interest. The resulting model then serves as the basis for hypotheses that can be tested by site-specific mutagenesis. A study of the interaction of t-PA and its primary endogenous inhibitor, PAI-1, provides an interesting example of this approach to structure–function studies.[65,66] In this study, use of the known crystal structure of the complex between bovine trypsin and bovine pancreatic trypsin inhibitor to model interactions between t-PA and PAI-1 facilitated the design and construction of point mutants of t-PA that were resistant to inhibition by the suicide substrate PAI-1 but maintained normal reactivity toward the natural substrate plasminogen. A similar approach was then adopted to design and construct a variant of PAI-1 that could rapidly inhibit one of the serpin-resistant variants of t-PA.[67]

2. Design of Chimeric Domains. In cases where homologous domains perform different functions (e.g., the kringle domains of t-PA and urokinase), the construction of chimeric domains can sometimes be used to elucidate the molecular basis of the specific function of each domain.

[65] E. L. Madison, E. J. Goldsmith, R. D. Gerard, M. J. Gething, and J. F. Sambrook, *Nature (London)* **336**, 721 (1989).

[66] E. L. Madison, E. J. Goldsmith, R. D. Gerard, M. J. Gething, J. F. Sambrook, and R. S. Bassel-Duby, *Proc. Natl. Acad. Sci. U.S.A.* **87**, 3530 (1990).

[67] E. L. Madison, E. J. Goldsmith, M. J. Gething, J. F. Sambrook, and R. D. Gerard, *J. Biol. Chem.* **265**, 21423 (1990).

While this approach can be adopted without the benefit of any structural information, it becomes much more robust when it is based on homology modeling as well as primary sequence alignment. For example, kringle structures contain a highly conserved hydrophobic core from which four surface loops emerge.[52] The primary sequences of these surface loops show very little similarity among known kringles; it therefore seems quite likely that functions unique to a particular kringle will map to one or more of these surface loops. Consequently, mutagenesis that accomplishes the switching of corresponding surface loops, both singly and in combination, among different kringles might prove particularly informative. For example, loop switches between the kringles of t-PA and the single kringle of urokinase have shed light on the structural basis of the stimulation of the activity of t-PA by fibrin.[68,69] Loop switches between the kringles of t-PA, on the other hand, have helped explain the differential response of these two kringles to partial cyanogen bromide digests of fibrinogen.[68,69] In addition, data from these loop switch experiments facilitate the subsequent design of appropriate point mutations. Similar strategies are also likely to be useful in the elucidation of structure–function relationships of the finger, growth factor, and protease domains of t-PA.

3. Carbohydrate Shielding Mutagenesis. The addition of supernumerary consensus Asn-linked glycosylation sites is another strategy that can be used to examine the role of specific surface loops of a protein that have been identified by homology modeling. This "carbohydrate shielding" mutagenesis has been successfully used to identify a region in the growth factor domain of t-PA that interacts with specific, hepatic receptors.[70] Following the implication of a particular surface loop by carbohydrate shielding mutagenesis, a series of point mutations within the loop is normally prepared to examine the role of individual residues.

4. Design of Small Deletion and Insertion Mutations. The chymotrypsin family of serine proteases has evolved to have members functioning in the same milieu that are exclusive in their reactivity toward both substrates and inhibitors. It has long been appreciated that mutations in the specificity pocket of a serine protease can alter the substrate selectivity of the enzyme. Comparison of the primary sequences of the catalytic domains of a number of serine proteases, however, suggested additional mechanisms by which this selectivity could evolve.[65–67] In comparison to trypsin, the catalytic domain of t-PA has acquired five amino acid inser-

---

[68] E. L. Madison, M. J. Gething, and J. K. Sambrook, unpublished observations.

[69] E. L. Madison, Ph.D. Dissertation, University of Texas, Southwestern Medical Center (1990).

[70] R. S. Bassel-Duby, N. J. Jiang, T. Bittick, E. L. Madison, D. McGookey, K. Orth, R. Shohet, J. F. Sambrook, and M. J. Gething, (1991).

tions of four or more residues; each insertion appears to have occurred in a surface loop, and two of these loops are located near the active site. It has now become clear that, at least in some cases, insertions of this type mediate interactions with substrates or inhibitors that alter the enzyme's specificity. For example, a structural report has revealed that a nine-residue insertion into the 60-loop (chymotrypsin numbering) of thrombin projects into the substrate-binding cleft and forms part of the S1 and S2 subsites.[71] Furthermore, we have shown that the precise deletion of one of these insertions in t-PA (i.e., residues 296–302) dramatically modulates the interaction of the enzyme with its cognate serpin, PAI-1, without significantly affecting reactivity toward its normal substrate, plasminogen.[65]

C. Charged-to-Alanine Scanning Mutagenesis

Another approach that has been used to map functional determinants of t-PA that does not rely on homology modeling is known as "charged-to-alanine" scanning.[72,73] In this scheme charged residues that are clustered within a stretch of four to eight contiguous amino acids are identified and, for each cluster, a variant is constructed in which all charged residues within the cluster are converted to alanine residues. Because mutations made by this strategy are very likely to occur on the surface of the protein, charged-to-alanine scanning is appropriate for the analysis of functions (such as protein–protein interactions) that are expected to involve surface-exposed residues.

III. Protocols for Oligonucleotide-Mediated Mutagenesis

A. Procedure and General Comments

Protocols for oligonucleotide-mediated mutagenesis[74–78] as modfied by Kunkel[79,80] involve the following steps (see Fig. 2).

[71] W. Bode, I. Mayr, U. Baumann, R. Huber, S. R. Stone, and J. Hofsteenge, *EMBO J.* **8,** 3467 (1989).
[72] S. Bass, M. Mulkerin, and J. A. Wells, *Proc. Natl. Acad. Sci. U.S.A.* **88,** 4498 (1991).
[73] W. F. Bennett, N. F. Paoni, B. A. Keyt, D. Botstein, A. J. S. Jones, L. Presta, F. M. Wurm, and M. J. Zoller, *J. Biol. Chem.* **266,** 5191 (1991).
[74] K. F. Norris, F. Norris, L. Christiansen, and N. Fiil, *Nucleic Acids Res.* **11,** 5103 (1983).
[75] M. J. Zoller and M. Smith, *Nucleic Acids Res.* **10,** 6487 (1982).
[76] M. J. Zoller and M. Smith, this series, Vol 100, p. 468.
[77] M. J. Zoller and M. Smith, *DNA* **3,** 479 (1984).
[78] M. J. Zoller and M. Smith, this series, Vol. 154, p. 329.
[79] T. A. Kunkel, *Proc. Natl. Acad. Sci. U.S.A.* **82,** 488 (1985).
[80] T. A. Kunkel, J. D. Roberts, and R. A. Zakour, this series, Vol. 154, p. 367.

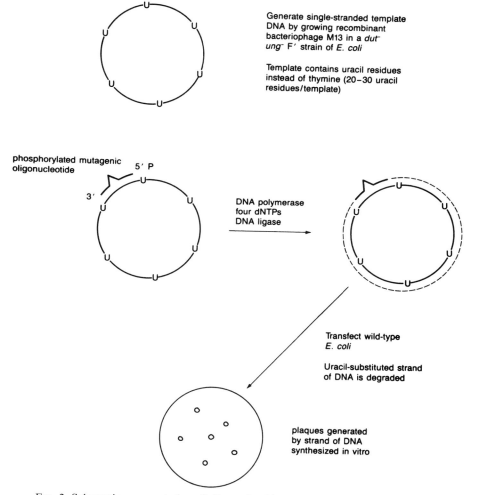

FIG. 2. Schematic representation of oligonucleotide-mediated, site-specific mutagenesis using the Kunkel method (see text for details).

Step 1: Cloning of Appropriate Fragment of DNA into Bacteriophage M13 Vector. Step 1 requires only techniques that are standard in any molecular biology laboratory and have been well described.[81] The DNA fragment cloned into the phage (or phagemid) vector should be as small as possible for the following reasons: (i) Large segments of DNA can be

---

[81] J. F. Sambrook, E. F. Fritsch, and T. Maniatis, "Molecular Cloning: A Laboratory Manual." Cold Spring Harbor Lab. Cold Spring Harbor, NY:, 1989.

unstable in M13 vectors and are prone to suffer spontaneous deletions during propagation. (ii) The chance that the mutagenic oligonucleotide will hybridize to an inappropriate site rather than the desired sequence increases as the size of the target DNA increases. (iii) To ensure that the mutagenized target DNA contains the desired mutation and no other, it is essential to sequence the entire fragment after oligonucleotide-mediated mutagenesis has been completed. The shorter the target DNA, the easier the task of determining its entire sequence.

Step 2: Preparation of Single-Stranded DNA from Recombinant Bacteriophage M13. Standard oligonucleotide-directed site-specific mutagenesis[74–78] as modified by Kunkel[79,80] uses a single-stranded template DNA in which a small number of thymine residues have been replaced by uracil. These templates are produced in strains of *Escherichia coli* carrying the *dut⁻* mutation, which causes a deficiency in dUTPase.[82,83] The intracellular pool of dUTP is greatly increased because of the inability of the cells to convert dUTP to dUMP, and some of the dUTP is incorporated into DNA at sites normally occupied by thymine. *Escherichia coli* normally synthesizes an enzyme (uracil *N*-glycosylase) that removes uracil residues that have been incorporated into DNA.[84,85] However, in *ung⁻* strains, uracil is not removed, and a small fraction of the thymine residues in the bacterial DNA are therefore replaced by uracil. This proportion is increased in the *dut⁻ ung⁻* strains of *E. coli* to the point where M13 bacteriophages grown in such strains contain between 20 and 30 uracil residues in their DNA. When these bacteriophages are used to infect an *ung⁺* strain of bacteria, the uracil is rapidly removed, generating sites that (1) block DNA synthesis and (2) are susceptible to cleavage by specific nucleases.[86] The destruction of the viral (+) strand results in a reduction of infectivity of approximately five orders of magnitude.

The Kunkel method of site-directed mutagenesis takes advantage of this strong selection against uracil-substituted DNA. The template DNA is first prepared by growth of an appropriate bacteriophage M13 recombinant in a strain of *E. coli* that is *dut⁻ ung⁻* F'. The resulting uracil-containing single-stranded DNA is used as the template in a standard mutagenesis procedure to generate a heteroduplex molecule with uracil in the template strand and thymine in the strand synthesized in the *in vitro* reaction. Transformation of this DNA into a *ung⁺* strain results in

[82] E. B. Konrad and I. R. Lehman, *Proc. Natl. Acad. Sci. U.S.A.* **72**, 2150 (1975).
[83] S. J. Hochhauser and B. Weiss, *J. Bacteriol.* **134**, 157 (1978).
[84] T. Lindahl, *Proc. Natl. Acad. Sci. U.S.A.* **71**, 3649 (1974).
[85] B. K. Duncan, P. A. Rockstroh, and H. R. Warner, *J. Bacteriol.* **134**, 1039 (1978).
[86] T. Lindahl, *Annu. Rev. Biochem.* **51**, 61 (1982).

destruction of the template strand, with the concomitant suppression of the production of wild-type bacteriophages. A large proportion (greater than 90%) of the progeny bacteriophages are therefore derived from replication of the tranfected uracil-free ($-$) strand. Because the synthesis of this strand is primed by the mutagenic oligonucleotide, many of the progeny bacteriophages carry the desired mutation.

Step 3: Design and Synthesis of Mutagenic Oligonucleotides. The design of an appropriate mutagenic oligonucleotide for the substitution, addition, or deletion of single nucleotides is guided by three major considerations. First, approximately 8–10 perfectly matched nucleotides are required to suppress displacement of the oligonucleotide by the Klenow fragment of *E. coli* DNA polymerase I during the primer extension reaction. Second, between 7 and 9 perfectly matched nucleotides are required at the 3' terminus of the mutagenic oligonucleotide to ensure the formation of a stable complex that is required both to assure efficient priming and to prevent exonucleolytic degradation during primer extension. Finally, the longer the oligonucleotide, the smaller the difference in thermal stability between a perfectly matched hybrid and one containing a single mismatched base pair. The aim, therefore, is to use the shortest mutagenic oligonucleotide that, under the conditions used for primer extension, will form stable hybrids both upstream and downstream from the mismatch. Under normal circumstances, the mutagenic oligonucleotide should be between 17 and 19 nucleotides in length, with the mismatch centrally located.

Oligonucleotides 25 or more nucleotides in length are used to insert, delete, or substitute two or more bases. Optimally, there should be 12–15 perfectly matched nucleotides on either side of the central looped-out region.

Step 4: Hybridization of Mutagenic Oligonucleotides to Target DNA. A phosphorylated mutagenic oligonucleotide is mixed in a 2- to 10-fold molar excess with the single-stranded template DNA in a small volume of buffer. The template and oligonucleotide are heated briefly to 20° above the estimated T_m, to denature any regions of secondary structure, and then cooled slowly to room temperature. Hybrids form as the temperature of the reaction mixture drops below the relevant T_m. The stoichiometry of the reagents in the mixture ensures that virtually all of the single-stranded template DNA is driven into hybrids with the oligonucleotide. This protocol works well for oligonucleotides of a wide variety of lengths and base compositions. However, annealing mixtures containing a mutagenic oligonucleotide that is exceptionally rich in A + T may need to be cooled to lower temperatures (12°–16°) to stabilize the hybrids.

Step 5: Extension of Hybridized Oligonucleotide by DNA Polymerase.
After the annealing reaction is complete, a mixture containing a DNA
polymerase, dNTPs, DNA ligase, and ATP is added, and the primer
extension is allowed to proceed for 2–15 hr at the appropriate temperature
for the polymerase being used. Any of several DNA polymerases may
be used in the extension reaction. Until recently, most workers relied
exclusively on the Klenow fragment of DNA polymerase I, which lacks
5′ exonucleolytic activity and is therefore incapable of degrading the mutagenic oligonucleotide. While the Klenow fragment is relatively inexpensive
and generally gives yields of mutants that are more than adequate, there
are certain advantages to using other polymerases. By contrast to the
Klenow fragment, neither T4 DNA polymerase[79,87,88] nor T7 DNA polymerase[89] is readily able to displace the mutagenic oligonucleotide during
the primer extension; moreover, both of these enzymes allow the use of
significantly shorter extension/ligation reaction times than does the
Klenow fragment. Finally, the efficiency of mutatagenesis is reported
to be higher with these two DNA polymerases than with the Klenow
fragment.[90–92]

DNA synthesis is initiated at the 3′ terminus of the mutagenic oligonucleotide and the oligonucleotide is extended by DNA polymerase until
the 3′ end of the newly synthesized strand encounters the 5′ terminus of
the phosphorylated mutagenic oligonucleotide. The two segments of DNA
then become ligated by the action of bacteriophage T4 DNA ligase. This
sealing of the phosphodiester bond protects the 5′ terminus of the mutagenic oligonucleotide from displacement by the newly growing strand and
prevents 5′ exonucleolytic editing of mismatched nucleotides after the
DNA is transfected into *E. coli*. A successful extension–ligation reaction
results in the production of covalently closed heteroduplexes that contain
mismatched nucleotides at the site(s) of the desired mutation(s).

Step 6: Transfection of Competent Bacteria. Transfection of competent
bacteria is performed by standard protocols[81]; any of several standard
laboratory strains of *E. coli* (e.g., T.G.-1, XL1 Blue, or MV 1190) can be
used as the host.

*Step 7: Preparation of Single-Stranded DNA from Mutagenized Re-
combinant Bacteriophage and Confirmation by Sequencing That Muta-*

[87] Y. Massamune and C. C. Richardson, *J. Biol. Chem.* **246,** 2692 (1971).
[88] J. G. Nossal, *J. Biol. Chem.* **249,** 5668 (1974).
[89] R. L. Lechner, M. J. Engler, and C. C. Richardson, *J. Biol. Chem.* **258,** 11174 (1983).
[90] J. Geisselsoder, R. Witney, and P. Yuckenberg, *BioTechniques* **5,** 786 (1987).
[91] K. Bebeneck and T. A. Kunkel, *Nucleic Acids Res.* **17,** 5408 (1989).
[92] M. Schena, *Comments (U.S. Biochem. Corp)* **15,** 23 (1989).

genized Bacteriophage M13 DNA Carries Desired Mutation and No Other. Transfection of the DNA synthesized in the *in vitro* extension/ligation reaction into competent *E coli* results in the production of a large number of recombinant M13 plaques. When mutagenesis is performed by the Kunkel method, a large proportion of these plaques (often greater than 80%) contain bacteriophages that carry the desired mutation. It is therefore not necessary to screen these plaques by hybridization with the mutagenic oligonucleotide. Instead, bacteriophages in the plaques are used to infect *E. coli* and single-stranded DNA is isolated from the resulting progeny bacteriophages. It is essential to determine the sequence of the entire region of DNA that was placed into the M13 vector prior to mutagenesis to ensure that the mutagenized, bacteriophage DNA contains the desired mutation and no other.

By contrast to the Kunkel method, earlier protocols for oligonucleotide-mediated mutagenesis did not utilize a selection against the wild-type strand of DNA used as a template in the synthesis reaction. Consequently, the majority of plaques obtained after transfection of the products of the synthesis reaction into competent *E. coli* contain wild-type recombinant bacteriophage. In addition, some plaques that do contain mutated bacteriophage will also contain the wild-type recombinant M13. Identification of plaques that contain only the mutated bacteriophage by direct sequencing would therefore be both tedious and inefficient. Plaques that contain mutated DNA can be easily identified, however, by first transferring the phage DNA onto a nitrocellulose filter and then hybridizing the filter with the mutagenic oligonucleotide. Hybridization is usually carried out under conditions that allow the radiolabeled, mutagenic oligonucleotide to form stable hybrids with both mutant and wild-type DNA. By progressively increasing the stringency of the subsequent washes, it is almost always possible to find conditions that (1) cause dissociation of mismatched hybrids, such as those formed between the mutagenic oligonucleotide and wild-type DNA, and (2) do not dissociate the perfect hybrids formed by the oligonucleotide and the desired mutant. Bacteriophages are eluted from plaques that hybridize to the mutagenic oligonucleotide under stringent conditions and are used to prepare ssDNA for analysis by sequencing.

Step 8: Recovery of Mutated Fragment of DNA from Double-Stranded Replicative Form of Recombinant Bacteriophage M13. Mutated, recombinant bacteriophages that have the desired DNA sequence are used to infect *E. coli*, and the double-stranded replicative form of the viral DNA is isolated by standard protocols.[81] The mutated variant of the DNA fragment that was originally ligated into the bacteriophage vector is recovered by digestion of the double-stranded M13 DNA by the appropriate restriction enzyme(s) followed by preparative gel electrophoresis.

Step 9: Substitution of Mutagenized Fragment for Corresponding Segment of Wild-Type DNA. The restriction fragment carrying the desired mutation is used to replace the corresponding wild-type DNA fragment by standard recombinant methods.[81] The resulting full-length cDNA, encoding the mutated protein of interest, is then ligated into an appropriate expression vector.

B. Detailed Protocols for Oligonucleotide-Mediated Mutagenesis

Step 1: Preparation of Uracil-Containing Template DNA

1. Transfer a single plaque produced by the appropriate bacteriophage M13 recombinant to a microfuge tube containing 1 ml of 2× YT medium. Incubate the tube at 60° for 5 min to kill bacterial cells. Vortex the tube vigorously for 30 sec to release the bacteriophages trapped in the top agar and then centrifuge the tube at 12,000 g for 2 min at 4° in a microfuge. It is important to kill the bacterial cells to prevent them from continuing to produce thymine-containing viral DNA during the next round of bacteriophage growth.

2. Transfer 50 μl of the supernatant to a 500-ml flask containing 50 ml of 2× YT medium with 0.25 μg/ml uridine. Add 5 ml of a mid-log-phase culture of *E. coli* strain CJ236 (*dut⁻ ung⁻* F′).[80] Incubate the culture with vigorous shaking (300 cycles/min on a rotary shaker) for 6 hr at 37°. The bacteriophage suspension used as an inoculum typically contains between 10^9 and 10^{10} plaque-forming units (pfu)/ml. The low multiplicity of viral infection (0.02–0.2 pfu/cell) ensures that the vast majority of the bacteriophages recovered from the culture have been generated in the *dut⁻ ung⁻* F′ strain of *E. coli*. There is no need to supplement the medium with thymidine or adenosine as originally described by Kunkel.[79] Efficient aeration is critical for efficient growth of bacteriophages.

4. Remove the cells by centrifugation at 5000 g for 30 min at 4°. Transfer the bacteriophage-containing supernatant to a 250-ml centrifuge bottle that will fit into a rotor such as a Sorvall GS3.

5. Determine the relative titer of the bacteriophage suspension of *E. coli* strain CJ236 (*dut⁻ ung⁻* F′) and an *ung⁺* strain such as JM109 or TG1. The titer on strain CJ236 should be four to five orders of magnitude greater than on the *ung⁺* strain used. To save time, most workers proceed with the purification of bacteriophage particles before the results of the titration are available. If the purification is postponed, the crude bacteriophage suspension should be stored in ice. The yield of bacteriophages varies from recombinant to recombinant. Typically, the titer of virus particles in the supernatant, determined on a *dut⁻ ung⁻* F′ strain (CJ236), is 5×10^{10} pfu/ml. If the yield of single-stranded DNA is inadequate,

there are two possible remedies: (i) Determine the titer of the bacterio-phage stock used to infect the dut^- ung^- F' strain of *E. coli*. Adjust the volume of the inoculum so as to achieve a multiplicity of infection of 0.1 pfu/bacterial cell. Grow the infected culture for 6 hr as described. (ii) Incubate the infected cultures for 12 hr rather than 6 hr. It is possible, however, that deleted variants will outgrow the original recombinant dur-ing extended periods of incubation. It is therefore advisable to verify that the majority of the single-stranded DNA used as a template is of the correct size. It is also essential to sequence the entire segment of foreign DNA after mutagenesis to ensure that no deletions or other types of mutations have occurred at sites other than the immediate target sequence.

6. Measure the volume of the bacteriophage suspension, and then add 0.25 volume of NaCl/PEG solution [15% (w/v) polyethylene glycol (PEG 8000), 2.5 *M* NaCl]. Mix the contents of the centrifuge bottle by swirling, and then store the bottle on ice for 1 hr.

7. Recover the precipitated bacteriophage particles by centrifugation at 10,000 *g* for 20 min at 4°. Remove the supernatant by aspiration, and then invert the bottle to allow the last traces of supernatant to drain away. Use a pipette attached to a vacuum line to remove any drops of solution adhering to the walls of the bottle.

8. Resuspend the bacteriophage pellet in 4 ml of TE (pH 7.6). Transfer the suspension to a 15-ml Corex centrifuge tube, and wash the walls of the centrifuge bottle with another 2 ml of TE (pH 7.6). Transfer the washing to the Corex tube. Vortex the suspension vigorously for 30 sec, and then store the tube on ice for 1 hr.

9. Vortex the suspension vigorously for 30 sec, and then centrifuge it once more at 5000 *g* for 20 min at 4° in a fixed-angle rotor (e.g., Sorvall SS34).

10. Taking care not to disturb the pellet of bacterial debris, transfer the supernatant to a polypropylene tube. Extract the suspension twice with phenol equilibrated to pH 8.0 and once with phenol:chloroform. Separate the phases by centrifugation at 4000 *g* for 5 min at room tempera-ture. Avoid transferring material from the interface.

11. Transfer the aqueous phase from the final extraction to a glass centrifuge tube (e.g., a 30-ml Corex tube). Measure the volume of the solution, and add 0.1 volume of 3 *M* sodium acetate (pH 5.2), followed by 2 volumes of ethanol at 0°. Mix the contents of the tube thoroughly, and then store the tube on ice for 30 min.

12. Recover the DNA by centrifugation at 5000 *g* for 20 min at 4°. Carefully remove the supernatant. Add 10 ml of 70% ethanol at room temperature, vortex briefly, and recentrifuge.

13. Remove the supernatant and store the inverted tube at room tem-

perature until the last traces of ethanol have evaporated. Redissolve the DNA in 200 μl of TE (pH 7.6).

14. Measure the DNA spectrophotometrically at 260 nm (1 OD = 40 μg/ml). Analyze the size of an aliquot of the DNA (0.5 μg) by gel electrophoresis, using as marker single-stranded DNA of the original bacteriophage M13 recombinant.

Step 2: Phosphorylation of Mutagenic Oligonucleotide

1. Combine:
   Mutagenic oligonucleotide (200 pmol)
   H_2O to 16.5 μl
   $10\times$ T4 polynucleotide kinase buffer (2 μl)
   10 mM ATP (1 μl)
   T4 polynucleotide kinase (4 units)
$10\times$ polynucleotide kinase buffer contains 500 mM Tris (pH 7.6 at 25°), 100 mM MgCl$_2$, 50 mM dithiothreitol (DTT), and 1 mM EDTA

2. Incubate the reaction for 1 hr at 37° and then heat the reaction for 10 min at 70° to inactivate the polynucleotide kinase.

Step 3: Annealing of Primer to Template

1. Combine:
   Uracil-containing ssDNA template (0.1 pmol)
   Phosphorylated, mutagenic oligonucleotide (0.2–1.0 pmol)
   $10\times$ annealing buffer (1 μl)
   H_2O (adjust to 10 μl total reaction volume)
$10\times$ annealing buffer contains 200 mM Tris, pH 7.4 (at 25°), 20 mM MgCl$_2$, and 500 mM NaCl

2. Incubate the reaction at 70–75° for 3 min. Place the reaction in a beaker of 70–75° water and allow to cool slowly (approximately 1°/min). After the reaction has cooled to 30°, place it in an ice–water bath.

Step 4: Synthesis of Complementary DNA Strand

1. While the annealing reaction is still in an ice water bath, add the following reagents:
   10X synthesis Buffer (1 μl)
   T4 DNA ligase [1 μl (2–5 units)]
   T4 DNA polymerase [1 μl (2–3 units)]
$10\times$ synthesis buffer contains 5 mM each of dATP, dCTP, dGTP, and dTTP as well as 100 mM Tris (pH 7.4 at 25°), 10 mM ATP, 50 mM MgCl$_2$, and 20 mM DTT

2. Incubate the synthesis reaction on ice for 5 min, at room temperature for 5 min, and, finally, at 37° for 2 hr. The synthesis reaction is started

at low temperatures to polymerize a small number of nucleotides onto the 3' end of the primer, therefore increasing the thermal stability of the primer–template duplex before shifting the reaction to higher temperatures where the polymerization rate is greatly enhanced.[80]

3. Stop the synthesis reaction by adding 2 μl of 0.5 M EDTA and 85 μl of TE.

Step 5: Transfection of Competent Escherichia coli and Identification of Mutated Bacteriophages

1. Use 10 μl of the completed synthesis reaction to transform 100 μl of competent, ung^+ bacteria by standard protocols.[81]

2. Pick well-isolated recombinant plaques, prepare single-stranded M13 DNA, and sequence the DNA by standard protocols.[81] The DNA sequencing is normally done in two stages. First, the presence of the desired mutation is confirmed, usually by the dideoxy-mediated chain termination method. Second, once the presence of the desired mutation is confirmed, the entire segment of foreign DNA carried in the vector is sequenced to eliminate the possibility of adventitious mutations at ectopic sites.

Step 6: Substitution of Mutagenized Fragment for Corresponding Segment of Wild-Type DNA. Restriction enzymes are used to recover the foreign DNA (or a subfragment carrying the desired mutation) from the double-stranded replicative form of the bacteriophage genome that has been sequenced. The mutated fragment is then used to replace the homologous segment of the wild-type cDNA, the mutated protein is expressed and isolated, and the phenotypic effects of the mutation are measured by the appropriate assays.

IV. Protocol for Expression of Variants of t-PA by Transient Transfection of Cos-1 Cells

A few eukaryotic proteins have been expressed efficiently and inexpensively in prokaryotic hosts. However, many eukaryotic proteins synthesized in bacteria fold incorrectly or inefficiently and, consequently, exhibit low specific activities. In addition, production of authentic, biologically active eukaryotic proteins from cloned DNA frequently requires posttranslational modifications, such as accurate disulfide bond formation, glycosylation, phosphorylation, oligomerization, or specific proteolytic cleavage—processes that are not performed by bacterial cells. This problem is particularly severe when expression of functional membrane or secretory proteins (e.g., the proteases involved in coagulation, fibrinolysis, and complement activation) is required.

Because of these problems, many methods have been developed to introduce cloned eukaryotic DNA into cultured mammalian cells. For the rapid screening of variants of t-PA, we have relied on transient transfection of Cos-1 cells by the DEAE-dextran method.[81,93–95] Ancillary to the transfection procedure, the mutated cDNA is ligated into an expression vector that contains an origin of replication, an enhancer, a promoter, an intron, and polyadenylation signals from the simian virus, SV40.

Transient Transfection Using DEAE-Dextran

1. Prepare a 50 mg/ml solution of DEAE-dextran as follows: Dissolve 100 mg of DEAE-dextran (M_r 500,000; Pharmacia, Piscataway, NJ) in 2 ml of distilled H_2O. Sterilize the solution by autoclaving for 20 min at 15 lb/inch^2 on liquid cycle. (Note: Autoclaving also assists dissolution of the polymer.)

2. At 24 hr prior to transfection, split a confluent plate of Cos-1 cells 1 : 5 or 1 : 6 to give approximately 2×10^5 cells/100-mm plate. Add 10 ml of Dulbecco's modified Eagle's medium (DMEM) + 10% fetal calf serum, and incubate the cultures for 20–24 hr at 37° in a humidified incubator in an atmosphere of 5–7% CO_2. The cells should be approximately 40–60% confluent at the time of transfection. If the cells are grown for less than 12 hr prior to transfection, they will be less well anchored to the substratum and more likely to detach during exposure to DEAE-dextran.

3. Prepare a DNA/DEAE-dextran/DMEM solution (1.5 ml/100-mm plate) by mixing 0.1–4 μg/ml supercoiled or circular DNA and DEAE-dextran (final concentration, 400 μg/ml) in DMEM. The amount of DNA required to achieve maximal levels of transient expression depends on the exact nature of the construct and should be determined in preliminary experiments. If the construct carries a replicon that will function in the transfected cells, 100–200 ng of DNA per 10^5 cells should be sufficient; if no replicon is present, larger amounts of DNA may be required (up to 2 μg per 10^5 cells).

4. Remove the medium by aspiration, and wash the monolayers twice with prewarmed (37°) DMEM + 10 mM HEPES (pH 7.15).

5. Add the DNA/DEAE-dextran/DMEM solution (1.5 ml/100-mm dish). Rock gently to spread the solution evenly across the monolayer. Return the cultures to the incubator for 60–120 min. At 15- to 20-min intervals, remove the dishes from the incubator briefly and swirl them gently.

[93] A. Vaheri and J. S. Pagano, *Virology* **27**, 434 (1965).
[94] J. H. McCutchan and J. S. Pagano, *J. Natl. Cancer Inst. (U.S.)* **41**, 351 (1968).
[95] D. Warden and H. V. Thorne, *J. Gen. Virol.* **3**, 371 (1968).

6. Remove the DNA/DEAE-dextran/DMEM solution by aspiration. Gently wash the monolayers twice with prewarmed (37°) DMEM + 10 mM HEPES (pH 7.15), taking care not to dislodge the transfected cells.

7. Add prewarmed (37°) medium (10 ml/100-mm dish) supplemented with serum and chloroquine diphosphate (100 μM final concentration), and incubate the cultures for 3–5 hr at 37° in a humidified incubator in a atmosphere of 5–7% CO_2. The efficiency of transfection is increased severalfold by treatment with chloroquine, which may act by inhibiting the degradation of the DNA by lysomal hydrolases.[96] Chloroquine diphosphate is stored as a sterile stock solution (100 mM; 60 mg/ml in water) in foil-wrapped tubes at $-20°$. Note, however, that the cytotoxic effects of a combination of DEAE-dextran and chloroquine can be severe. It is therefore important to carry out preliminary experiments to determine the maximum permissible length of exposure to chloroquine after treatment of cells with DEAE-dextran.

8. Remove the medium by aspiration, and wash the monolayers three times with prewarmed (37°) serum-free medium. Add to the cells medium (10 ml/100-mm dish) supplemented with serum, and incubate the cultures for 12–24 hr at 37° in a humidified incubator in an atmosphere of 5–7% CO_2. For assays that involve replicate samples or treatment of transfected cells under multiple conditions or over a time course, it is desirable to avoid dish-to-dish variation in transfection efficiency. In these cases, it is best to transfect large monolayers of cells (100-mm dishes) and then to trypsinize the cells after 24 hr of incubation and distribute them among several smaller dishes.

9. At 12 to 24 hr after the removal of chloroquine, wash the cells three times with serum-free DMEM and add 10 ml of serum-free medium to each plate.

10. Continue incubation of the transfected cells at 37°. Collect the conditioned media between 60 and 72 hr after the removal of chloroquine.

11. Centrifuge the conditioned media at 1000 g for 20 min to remove dead cells and debris, perform radioimmune and enzymatic assays, and snap freeze aliquots of each sample.

V. Assays of Activities of Variants of t-PA

A large number of assays for plasminogen activators have been developed.[97] Depending on the purity of the sample, the enzyme concentration

[96] H. Luthman and G. Magnusson, *Nucleic Acids Res.* **11**, 1295 (1983).

[97] J. H. Verheijen, *in* "Tissue Type Plasminogen Activator (t-PA): Physiological and Clinical Aspect" (C. Kluft, ed.), Vol. 1, p. 123. CRC Press, Boca Raton, FL, 1988.

is usually measured by immunological methods (immunoradiometric or enzyme-linked immunosorbent assays) or by active site titration with radiolabeled reagents (e.g., tritiated diisopropylphosphate). Complete characterization of a preparation, including the fraction of enzyme molecules that are active, requires the results of both assays.

The first assay for plasminogen activator activity to gain wide acceptance, the fibrin plate method, was developed by Astrup and Mullertz.[98] A defined volume of sample is placed onto a thin film of fibrin that contains plasminogen. The plate is then incubated at 37° for 12–24 hr, and the size of the resulting zone of lysis is measured. The activity in the sample is estimated by comparison of the lysis zone formed by the sample to those formed by various dilutions of a standard t-PA preparation. Reliable results with the fibrin plate assay require careful choice (and subsequent standardization) of all assay reagents as well as the preparation of plasminogen-free fibrin plates to be used as a control.

Kinetic analysis of the enzymatic activity of variants of t-PA is usually performed in purified systems utilizing chromogenic or fluorogenic substrates. In spite of the high specificity of t-PA, the enzyme does hydrolyze certain ester or amide bonds in small synthetic molecules.[99] These synthetic substrates consist of a small number of amino acids (or derivatives of amino acids) with the chromogenic or fluorogenic leaving group at the carboxyl terminus. Hydrolysis of synthetic substrates by t-PA (i.e., the "direct" assay) proceeds according to Michaelis–Menten kinetics and can be analyzed by standard methods.[100–102] The activity of variants of t-PA toward the natural substrate plasminogen can be measured in coupled or "indirect" assays that contain t-PA, a fibrin analog, plasminogen, and a plasmin-specific synthetic substrate. The coupled, chromogenic assay does not proceed by Michaelis–Menten kinetics; however, under appropriate conditions, the absorbance change observed is linear with the plasminogen activator concentration and with the square of the reaction time.[97]

Activity assays based on the lysis of blood clots or purified fibrin–plasminogen clots, suspended either in buffer or human plasma, have also been widely reported.[97] The rate of clot lysis can be assessed by measurement of released fibrin degradation products by radioactive or immunological methods. Alternatively, the decrease in turbidity of the

[98] T. Astrup and S. Mullertz, *Arch. Biochem.* **40,** 346 (1952).
[99] P. Friberger, G. Claeson, M. Knos, L. Aurell, S. Arielly, and R. Simonsson, *Prog. Fibrinolysis* **4,** 149 (1979).
[100] D. C. Rijken, G. Wijngaards, M. Zaal-De Jong, and J. Welbergen, *Biochim. Biophys. Acta* **580,** 140 (1979).
[101] M. Ranby, N. Bergsdorf, and T. Nilsson, *Thromb. Res.* **27,** 175 (1982).
[102] P. Wallen, N. Bergsdorf, and M. Ranby, *Biochim. Biophys. Acta* **719,** 318 (1982).

reaction mixture can be used to monitor the progress of clot lysis.[103] In general, clot lysis assays have much lower sensitivity than either the fibrin plate assay or the indirect chromogenic assay and, like the fibrin plate assay, are difficult to describe mathematically. However, these assays are presumably a more accurate reflection of the activity of t-PA *in vivo* than other *in vitro* assays.

VI. Protocols for Chromogenic Assays of t-PA

A. Direct Assay

1. Combine the following reagents into wells of a microtiter plate:
   10× direct assay buffer (10 μl)
   t-PA sample, 10–100 ng t-PA (80 μl)
   10 mM Spec t-PA (10 μl)

Each reaction is performed in duplicate. The 10× direct assay buffer is 0.25 M Tris–imidazole (pH 8.4) and 2.1 M NaCl. Spec t-PA is methylsulfonyl-D-cyclohexyltyrosylglycylarginine *p*-nitroaniline acetate and is purchased from American Diagnostica (Greenwich, CT).

2. Incubate the microtiter plate at 37° for 30–180 min.
3. Read the optical density at 405 nm every 2–5 min.

B. Indirect Assay

1. Combine the following reagents into wells of a microtiter plate:
   t-PA sample, 0.1–1.0 ng (25 μl)
   Lys-plasminogen, final concentration of 0.2 μM (25 μl)
   Desafib, final concentration of 25 μg/ml (25 μl)
   Spec PL, final concentration of 0.5 mM (25 μl)

Each reaction is performed in duplicate. Reaction conditions are 50 mM Tris-HCl (pH 7.5), 100 mM NaCl, 1 mM EDTA, and 0.01% Tween 80. Desafib is a soluble fibrin analog prepared by limited digestion of purified human fibrinogen by batroxobin in the presence of a peptide (Gly-Pro-Arg-Pro) that inhibits polymerization. Desafib is purchased from American Diagnostica (Greenwich, CT). Spec PL is H-D-norleucyl-hexahydrotyrosyllysine *p*-nitroanilide diacetate and is purchased from American Diagnostica (Greenwich, CT).

2. Incubate the microtiter plate at 37° for 30–120 min.
3. Read the optical density at 405 nm every 1–3 min.

[103] A. J. S. Jones and A. M. Meunier, *Thromb. Haemostasis* **64**, 455 (1990).

[17] Lipoprotein (a): Purification and Kinetic Analysis

By Jay M. Edelberg, Young-Joon Lee, Timothy N. Young, and Salvatore V. Pizzo

Introduction

Lipoprotein (a) [Lp(a)] levels in excess of 30 mg/dl are associated with a two- to fivefold increased risk of atherosclerosis.[1-9] Elevated Lp(a) levels also are linked to stenosis of carotid and cerebral arteries and rethrombosis of venous grafts.[10-12] Studies demonstrate that Lp(a) accumulates in the lesions of the affected vessels.[13-15] Lp(a) is a low-density plasma lipoprotein first identified by Berg.[16] It contains a lipid core, an associated apoprotein B (apoB) subunit, and an apoprotein (a) [apo(a)] subunit disulfide linked to apoB (for in depth review, see Scanu and Fless[17]). The apo(a) subunit has extensive homology with the fibrinolytic zymogen plasmino-

[1] J. J. Albers, J. M. Adolphson, and W. D. Hazzard, *J. Lipid Res.* **18**, 331 (1977).
[2] M. M. Frick, G. H. Dahlén, K. Berg, M. Valle, and P. Hekali, *Chest* **73**, 62 (1978).
[3] G. G. Rhoads, G. H. Dahlén, K. Berg, N. E. Morton, and A. L. Dannenberg, *JAMA, J. Am. Med. Assoc.* **256**, 2540 (1986).
[4] G. H. Dahlén, J. R. Guyton, M. Attar, J. A. Farmer, J. A. Kautz, and A. M. Gotto, *Circulation* **74**, 758 (1986).
[5] I. B. Sundell, T. K. Nilsson, G. Hallmans, G. Hellsten, and G. H. Dahlén, *Atherosclerosis* **80**, 9 (1989).
[6] J. A. Hearn, S. J. DeMaio, G. S. Roubin, M. Hammarstöm, and D. Sgoutas, *Am. J. Cardiol.* **66**, 1176 (1990).
[7] A. Rosengren, L. Wilhelmsen, E. Eriksson, B. Risberg, and H. Wedel, *Br. J. Med.* **301**, 1248 (1990).
[8] M. Seed, F. Hopplichler, D. Reaveley, S. McCarthy, G. R. Thompson, E. Boerwinkle, and G. Utermann, *N. Engl. J. Med.* **322**, 1494 (1990).
[9] J. Genest, J. L. Jenner, J. R. McNamara, J. M. Ordovas, S. R. Silberman, P. W. Wilson, and E. J. Schaefer, *Am. J. Cardiol.* **67**, 1039 (1991).
[10] A. Muria, T. Miyahara, N. Fujimoto, M. Matsuda, and M. Kameyama, *Atherosclerosis* **59**, 199 (1986).
[11] G. Zenker, P. Koltringer, G. Bone, K. Niederkorn, K. Pfeiffer, and G. Jurgens, *Stroke* **17**, 942 (1986).
[12] H. F. Hoff, G. J. Beck, M. S. Shribuishi, G. Jurguns, J. O'Neil, J. Kramer, and B. Lyle, *Circulation* **77**, 1238 (1988).
[13] M. Rath, A. Niendorf, T. Reblin, M. Dietel, H. Krebber, and U. Beisiegel, *Arteriosclerosis (Dallas)* **9**, 579 (1989).
[14] E. B. Smith and S. Cochran, *Atherosclerosis* **84**, 173 (1990).
[15] U. Beisiegel, A. Niendorf, K. Wolf, T. Reblin, and M. Rath, *Eur. Heart J.* **11E**, 174 (1990).
[16] K. Berg, *Acta Pathol. Microbiol. Scand., Suppl.* **59**, 166 (1963).
[17] A. M. Scanu and G. M. Fless, *J. Clin. Invest.* **85**, 1709 (1990).

gen (Pg), containing both a variable number of copies of the fibrin-binding domains, termed kringles, and a zymogen domain.[18,19] In addition, although apo(a) has a potential proteinase domain, the subunit lacks a critical Pg cleavage site necessary for zymogen activation,[20] and therefore apo(a) cannot be converted to a proteinase by typical Pg activators.[19] Lp(a) down-regulates fibrinolysis by competing with Pg for various vascular binding sites including cellular binding sites, fibrinogen fragments, and heparin. Pg cellular binding sites on endothelial cells, macrophages, and other cells locally concentrate Pg, which is then activated by adjacently bound Pg activators. Lp(a) binds to these sites with an affinity comparable to Pg[21–24] as a result of the apo(a) kringle-4 domains. Lp(a) displacement of Pg from these cellular binding sites may prevent Pg from interacting with Pg activators such as tissue-type Pg activator (t-PA) and urinary-type Pg activator (u-PA) and may thereby depress fibrinolysis.

Kinetic studies demonstrate that Lp(a) is a competitive inhibitor of t-PA-mediated Pg activation in the presence of fibrinogen fragments and fibrin.[25–27] Lp(a) also inhibits heparin-enhanced fibrinolytic activation pathways. Heparin is located on both the endothelial surface and the basement membrane, and in addition to its well-documented role in anticoagulation (for brief review, see Rosenberg *et al.*[28]), appears to be important in the regulation of fibrinolysis.[29] Heparin enhances the rate of plasmin generation by increasing the catalytic activity of both t-PA and u-PA.[30,31]

[18] J. W. McLean, J. E. Tomlinson, W. Kuang, D. L. Eaton, E. Y. Chen, G. M. Fless, and A. M. Scanu, *Nature (London)* **330,** 132 (1987).

[19] D. L. Eaton, G. M. Fless, W. J. Kohr, J. W. McLean, Q. Xu, C. G. Miller, R. M. Lawn, and A. M. Scanu, *Proc. Natl. Acad. Sci. U.S.A.* **84,** 3224 (1987).

[20] K. C. Robbins, L. Summaria, B. Hsieh, and R. Shah, *J. Biol. Chem.* **247,** 6757 (1967).

[21] M. Gonzalez-Gronow, J. M. Edelberg, and S. V. Pizzo, *Biochemistry* **28,** 2375 (1989).

[22] L. A. Miles, G. M. Fless, E. G. Levine, A. M. Scanu, and E. F. Plow, *Nature (London)* **339,** 301 (1989).

[23] K. A. Hajjar, D. Gavish, J. L. Breslow, and R. L. Nachman, *Nature (London)* **339,** 303 (1989).

[24] T. F. Zioncheck, L. M. Powell, G. C. Rice, D. L. Eaton, and R. M. Lawn, *J. Clin. Invest.* **87,** 767 (1991).

[25] J. M. Edelberg, M. Gonzalez-Gronow, and S. V. Pizzo, *Thromb. Res.* **7,** 155 (1990).

[26] J. Loscalzo, M. Weinfeld, G. M. Fless, and A. M. Scanu, *Arteriosclerosis (Dallas)* **10,** 240 (1990).

[27] D. Rouy, P. Grailhe, F. Nigon, J. Chapman, and E. Angeles-Cano, *Atheroscler. Thromb.* **11,** 629 (1991).

[28] R. D. Rosenberg, K. A. Bauer, and J. A. Marcum, *in* "Reviews in Hematology" (E. Murano, ed.), p. 351. PJD Publications, Westbury, NY, 1986.

[29] F. Markwardt and H. P. Klocking, *Haemostasis* **6,** 370 (1977).

[30] J. M. Edelberg and S. V. Pizzo, *Biochemistry* **29,** 5906 (1990).

[31] J. M. Edelberg, M. Weissler, and S. V. Pizzo, *Biochem. J.* **276,** 758 (1991).

Lp(a) inhibits these heparin-enhanced reactions by competing with Pg for access to the heparin-bound Pg activator.[30,31] Lp(a) does not affect the activity of the catalytic domain of either t-PA or u-PA as determined by amidolytic activity assays, nor does Lp(a) displace the Pg activators from fibrinogen fragments or from heparin, because these interactions would have been reflected in an uncompetitive or noncompetitive inhibition constant.

In addition to suppressing fibrinolytic activation pathways, Lp(a) promotes inhibition of plasmin. In the circulation, α_2-antiplasmin (α_2AP) is the major plasmin inhibitor (for review, see Lijnen and Collen[32]). Fibrin and its degradation products decrease the rate of the plasmin inhibition by α_2AP.[33-35] When Lp(a) is added in the presence of fibrinogen fragments, plasmin is rapidly inhibited, but Lp(a) has no effect on the inhibition reaction in the absence of fibrinogen fragments.[35] Kinetic analysis demonstrates that Lp(a) competes with plasmin for fibrinogen fragments, dissociating plasmin from them, thereby promoting plasmin inhibition by α_2AP.

Unlike those antifibrinolytic effects, Lp(a) also depresses one of the fibrinolytic inhibition pathways. Lp(a) protects t-PA from irreversible enzymatic inhibition by Pg activator inhibitor type 1 (PAI-1). In the circulation, the major irreversible inhibitor of t-PA is PAI-1 (for brief review, see Sprengers and Kluft[36]). Lp(a) decreases the rate of t-PA inhibition by PAI-1 in a manner analogous to Lp(a) inhibition of t-PA-mediated Pg activation.[25,30,37] Moreover, like the Lp(a) affect on Pg activation, the decrease in PAI-1 inhibition is dependent on the presence of either fibrinogen or heparin. Kinetic analysis of the Lp(a) depression in t-PA irreversible inhibition indicates that Lp(a) competes with PAI-1 for access to the template-bound Pg activator. Like the Lp(a) effects on Pg activation, Lp(a) decreases the rate of the inhibition by sterically preventing PAI-1 from interacting with fibrinogen- and heparin-bound t-PA.

Thus in vitro Lp(a) competes with Pg for various cellular binding sites, inhibits plasmin generation by Pg activators, and promotes plasmin inhibition, but protects Pg activators from inhibition. This suggests that Lp(a) may both depress and prolong in vivo fibrinolysis, and that at elevated levels Lp(a) may increase the risk of thrombotic events.

[32] H. R. Lijnen and D. Collen, in "Proteinase Inhibitors" (A. J. Barret and G. Salvesen, eds.), p. 457. Elsevier, Amsterdam, 1986.
[33] B. Wiman, L. Boman, and D. Collen, Eur. J. Biochem. 84, 143 (1978).
[34] P. K. Anonick and S. L. Gonias, Biochem. J. 275, 53 (1991).
[35] J. M. Edelberg, and S. V. Pizzo, Biochem. J. 286, 79 (1992).
[36] E. D. Sprengers and C. Kluft, Blood 69, 381 (1987).
[37] J. M. Edelberg, C. Reilly, and S. V. Pizzo, J. Biol. Chem. 266, 7488 (1991).

Purification

Previous purification techniques for Lp(a) are based on the purification of low-density lipoproteins by ultracentrifugation, followed by the separation of the Lp(a) from low-density lipoprotein.[19,38] Briefly, plasma is isolated from blood in the presence of EDTA (0.15% w/v), NaN$_3$ (0.01% w/v), and soybean trypsin inhibitor (0.4 μM) by low-speed centrifugation, followed by the addition of diisopropyl fluorophosphate (1 mM). The plasma is then prepared for ultracentrifugation by the addition of NaBr (1.21 g/ml) and is centrifuged for 20 hr at 15° in a 60-Ti rotor at 59,000 rpm. Lp(a) is then separated from the other low-density lipoproteins by chromatofocusing on PBE 94 (Pharmacia, Piscataway, NJ) or affinity chromatography on lysine-Sepharose.

A purification employing an initial affinity chromatography step has been reported[39] which takes advantage of the previously reported Lp(a) affinity for heparin.[40,41] This purification requires meticulous care to avoid contamination by fibronectin and potentially other proteins. Most recently, we employed the procedure described below.

Freshly citrated apheresis plasma, with an Lp(a) concentration of 8 mg/dl is obtained from a donor. After adding PMSF (1 mM), EDTA (1 mM), bovine pancreatic trypsin inhibitor (2.4 TIU/ml) to minimize proteolytic degradation of Lp(a) as well as NaN$_3$ (0.02%) to prevent microbial degradation, the plasma is stored at $-70°$ until subjected to purification. All purification steps are carried out at 4° and PMSF is added after every step except for the final one. Thawed plasma (500 ml) is filtered to remove particulate matter and sodium citrate and proteinase inhibitors are added as before. After centrifuging for 20 min at 20000 g to remove chylomicrons, the clear supernatant is gently mixed for 2 hr with 100 ml of gelatin-Sepharose, which is pre-equilibrated in 10 mM potassium phosphate buffer, pH 7.4, containing 0.1 M NaCl, EDTA, NaN$_3$, and bovine pancreatic trypsin inhibitor (standard column buffer). The gelatin-Sepharose-supernatant is then packed in a column (50-mm diameter) and washed with column buffer. This step is added to remove fibronectin which is sometimes found in the preparations.

The buffer-diluted plasma is collected and applied to a lysine-Sepharose column (diameter 1.5 cm, bed volume 100 ml) which is pre-equili-

[38] G. M. Fless, C. A. Rolih, and S. V. Pizzo, *J. Biol. Chem.* **259,** 11470 (1984).
[39] G. Dahlén, C. Erikson, and K. Berg, *Clin. Genet.* **14,** 36 (1978).
[40] M. Bihari-Varga, E. Gruber, M. Rothereder, R. Zechner, and G. M. Kostner, *Arteriosclerosis (Dallas)* **8,** 851 (1988).
[41] J. M. Edelberg, M. Gonzelez-Gronow, and S. V. Pizzo, *Biochemistry* **28,** 2730 (1989).

brated with column buffer in which the NaCl concentration is increased to 0.15 M. In order to remove nonspecifically bound proteins, the column is washed with column buffer in which the NaCl concentration is increased to 0.35 M, until the absorbance at 280 nm has returned to baseline. After re-equilibration with two bed volumes of the column buffer, Lp(a) is eluted with column buffer containing 0.2 M ε-aminocaproic acid, 0.15 M NaCl. The eluate, containing Lp(a) and plasminogen, to which Tween 80 (0.04%) is added and is concentrated to 5 ml of volume against polyethyleneglycol (molecular weight, 8000) using dialysis tubing (molecular weight cut off 1000). The concentrate is applied to a Sephacryl S-400 column (diameter 1.5 cm, bed volume 120 ml), which is pre-equilibrated and washed with column buffer containing 0.15 M NaCl. The Lp(a) is eluted in the first sharp peak, followed by plasminogen in the second peak.

The purity of the Lp(a) preparation is examined by SDS–PAGE gradient gels (2.5%–12%). In order to detect possible contamination by plasminogen and fibronectin, Western blots on a nitrocellulose membrane using goat polyclonal antibodies against plasminogen and fibronectin are performed. At the same time, a monoclonal antibody against apolipoprotein (a) is used on the Western blot in order to confirm the presence of Lp(a). The preparation is also examined for the presence of other apoproteins by SDS–PAGE gradient gels (2.5–20%) and Western blotting using mouse anti-apoprotein B antibody and goat anti-apoprotein C-II antibody.

The Lp(a) preparation consists of two kinds of Lp(a): triglyceride-rich Lp(a) and triglyceride-poor Lp(a). The triglyceride-rich Lp(a), with a 5 : 1 triglyceride : cholesterol ratio and 1 : 1 Lp(a) : cholesterol ratio, contains apoproteins B and C, thus, making it comparable to very low density lipoprotein or its remnant. The triglyceride-poor Lp(a), with a 1 : 2 triglyceride : cholesterol ratio and 2.5 : 1 Lp(a) : cholesterol ratio, contains only apoprotein B, making it comparable to low density lipoprotein.

In order to separate triglyceride-poor and rich fractions of Lp(a), the fraction, to which Tween 80 (0.04%) is added, is adjusted to $d = 1.019$ g/ml with NaCl, layered in a centrifugal tube over 2 ml of buffer solution adjusted to $d = 1.21$ mg/dl with NaCl and NaBr, and centrifuged in a Beckman Type 65 rotor, at 40,000 rpm for 20 hr at 4°. The triglyceride-rich Lp(a) floats at the top, triglyceride-poor Lp(a) is obtained from the midportion, and lipid-free Lp(a) is recovered at the bottom. Further application to a Sephadex G-25 column, equilibrated with phosphate buffer, is employed to desalt the preparation.

Theory of Template-Dependent Competitive Inhibition

Functional studies of the effects of Lp(a) on fibrinolysis require the ability to perform detailed kinetic analysis. The systems under study are generally complex and in some cases have required deriving new kinetic equations. The following discussions present the kinetic models and equations employed to study the effects of Lp(a) on fibrinolysis.

As discussed previously, Lp(a) regulates fibrinolysis by competing with Pg, plasmin, and PAI-1 for binding to templates on which fibrinolysis is regulated. Unlike many other competitive inhibitors, Lp(a) does not interact directly with the enzymes; rather Lp(a) binds to vascular surfaces that modulate these reactions. In this way Lp(a) regulates the interactions between both bound fibrinolytic enzymes and their substrates. Kinetic analysis of the role of Lp(a) in the various fibrinolytic pathways must include a characterization of the effects of the different vascular surfaces on these pathways. This analysis is divided into studies of the activator pathways (Pg activation) and studies of the inhibitor pathways (plasmin and Pg activator inhibition). The kinetic analyses of both sets of pathways are characterized by equations that describe the rates of the activation or inhibition, the general modulation of the reactions, and the inhibition of the modulated reactions.

Description of the effects of the Lp(a) on Pg activation involves the measurement of the initial velocity of Pg activation, calculation of the kinetic effects of modulators on the activation, and determination of the inhibitory effects of Lp(a). Study of the role of Lp(a) in plasmin and Pg activator inhibition requires the measurement of second-order inhibition rate constants, the determination of the effects of modulators of inhibition, and quantification of the effects of Lp(a).

Lipoprotein (a) and Fibrinolytic Activation Pathways

Determination of Initial Rate of Plasminogen Activation

The kinetic studies described here employ a coupled assay to measure the initial rate of Pg activation. The method employed to measure this rate relies on monitoring the activity of the reaction product, plasmin. Plasmin amidolytic activity is in turn monitored by spectrophotometric measurement of absorbance or fluorescence of hydrolyzed plasmin substrates, for example, Val-Leu-Lys-p-nitroanilide. The coupled assay to measure the rate of Pg activation is given in Scheme I. Using a plot of the first-order derivative of either absorbance or fluorescence with respect to time t yields the values of the instantaneous rate of plasmin activity from which the initial rate of Pg activation is calculated. Calculation of the instantaneous rate of plasmin amidolytic activity is dependent on the concentrations of both plasmin $[Pm_t]$ and substrate $[S_t]$, the catalytic rate (k_s) and Michaelis constant (K_s) of substrate hydrolysis, as well as the extinction or luminence coefficient (ε) of the hydrolyzed substrate. Equa-

$$\text{Plasminogen} \xrightarrow{\text{t-PA or u-PA}} \text{plasmin} \qquad \text{[reaction 1]}$$

$$\text{Plasmin substrate} \xrightarrow{\hspace{1cm}} \text{signal} \qquad \text{[reaction 2]}$$

SCHEME I

tion (1), which describes the rate of amidolytic cleavage of a chromogenic substrate, is given as follows[42]:

$$\delta_{\text{absorbance}}/\delta t = k_s[\text{Pm}_t]/\varepsilon(1 + K_s/[\text{S}_t])/k_s \qquad (1)$$

This can be rearranged to solve for Pm_t in Eq. (2):

$$[\text{Pm}_t] = \delta_{\text{absorbance}}/\delta t \varepsilon(1 + K_s/[\text{S}_t])/k_s \qquad (2)$$

During the initial phase of the reaction, when substrate is still greater than 95% of its initial concentration, the rate of plasmin generation ($\delta[\text{Pm}_t]/\delta t$ or $V_{\text{plasmin 0}}$) is constant, and it is determined by treating the concentration of substrate as a constant. Therefore, taking the derivative of both sides of Eq. (2) with respect to time yields

$$\delta[\text{Pm}_t]/\delta t = \delta^2_{\text{absorbance}}/\delta t^2 \varepsilon(1 + K_s/[\text{S}_t])/k_s \qquad (3)$$

Under the initial conditions the velocity of plasmin generation ($V_{\text{plasmin 0}}$) is constant, and therefore the second-order derivative of absorbance with respect to time ($\delta^2_{\text{absorbance}}/\delta t^2$) is equal to the first-order derivative divided by time ($\delta_{\text{absorbance}}/\delta t/\text{time}$). The rate of plasmin generation under these initial conditions will be given by Eq. (4):

$$V_{\text{plasmin 0}}(t) = \delta_{\text{absorbance}}/\delta t\{\varepsilon(1 + K_s/[\text{S}_0])/k_s\} \qquad (4)$$

$V_{\text{plasmin 0}}$ is then determined by calculating the slope of a plot of $\delta_{\text{absorbance}}/\delta t$ versus time multiplied by the value $\{\varepsilon(1 + K_s)/[\text{S}_0]\}/k_s$.

Other methods have been employed to determine the initial rate of Pg activation. However, these methods generally are limited by the difficulty in assuring initial conditions, such as methods that employ stopped-assay measurements of absorbance and plots of absorbance versus t^2 to calculate the velocity of plasmin generation. Unfortunately, by this approach it is difficult to determine whether the initial conditions are still present when the assay is stopped. In the use of plots of absorbance versus t^2 similar problems may be encountered because greater weight is given to measurements at greater times, increasing the possibility that initial conditions may not exist when the velocity is determined.

[42] B. R. Erlarger, N. Kokowsky, and W. Cohen, Arch. Biochem. Biophys. 95, 271 (1961).

General Modulator Analysis of Plasminogen Activation

The rate of Pg activation is modulated by various components of the vasculature, which may also serve as templates for the effects of Lp(a), such as fibrinogen fragments, fibrin monomer, fibrin polymer, heparin, and cell surfaces (for review, see Edelberg and Pizzo[43]). Prior to investigating the role of Lp(a) in Pg activation, it is important to characterize the kinetic effects of these various components on the rate of plasmin generation. A modulated reaction may result in either an increased or decreased rate of product formation. A decreased rate of modulation differs from reversible inhibition, which results in no product formation. Later sections of this chapter will examine the treatment of both reversible and irreversible inhibition. Any enzymatic reaction that obeys Michaelis–Menten steady-state kinetics may be modulated via changes in either or both k_{cat} and K_m. Furthermore, k_{cat} and K_m modulation in the same reaction does not necessarily depend on the same interaction by the modulator. For example, the modulator may alter the k_{cat} of the reaction through one binding site, and affect the K_m via another binding site. Moreover, these different sites may not be on the same component of the reaction; one site may be on the enzyme and the other on the substrate and the affinities of the modulator for these different sites are not necessarily equal.

Kinetic equations presented here describe the general modulation of an enzymatic reaction. These equations allow the independent analysis of changes in both k_{cat} and K_m. Furthermore, these equations employ the calculated values of k_{cat} and K_m in the analysis and do not rely on any particular method of deriving these kinetic constants.

Examples of modulated reactions employ the following abbreviations: E, enzyme; S, substrate; P, product; A, modulator; K_A, modulator dissociation constant; α, K_m modulation constant; and β, k_{cat} modulation constant. An example of the modulated reaction with changes in k_{cat}, where the modulator binds to the enzyme, is given in Scheme II.

Scheme II is equally valid if the modulator is binding to the substrate. An example of the modulated reaction with changes in K_m, where the modulator binds to the enzyme, is given in Scheme III.

Again, Scheme III is equally valid if the modulator binds to the substrate. The general equations describing modulation are derived for changes in k_{cat}, but are equally valid for modulation of K_m by substituting α for β in Eq. (5). The general equation that describes the modulation of k_{cat} must

[43] J. M. Edelberg and S. V. Pizzo, *Fibrinolysis* **5**, 135 (1991).

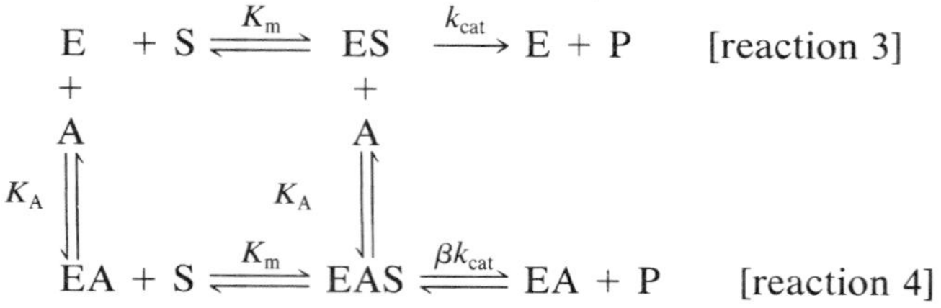

$$E + S \overset{K_m}{\rightleftharpoons} ES \overset{k_{cat}}{\longrightarrow} E + P \qquad \text{[reaction 3]}$$

$$EA + S \overset{K_m}{\rightleftharpoons} EAS \overset{\beta k_{cat}}{\rightleftharpoons} EA + P \qquad \text{[reaction 4]}$$

SCHEME II

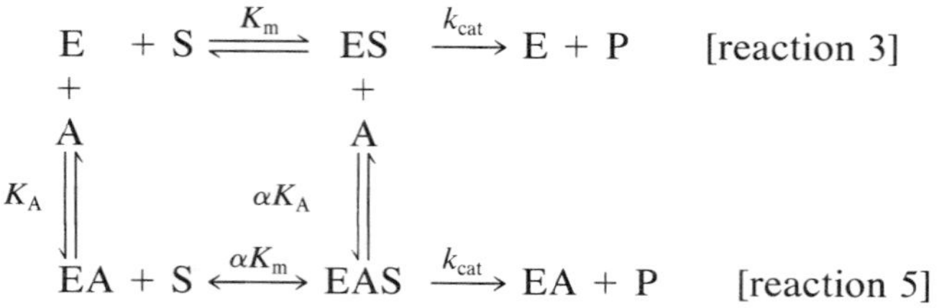

$$E + S \overset{K_m}{\rightleftharpoons} ES \overset{k_{cat}}{\longrightarrow} E + P \qquad \text{[reaction 3]}$$

$$EA + S \overset{\alpha K_m}{\longleftarrow} EAS \overset{k_{cat}}{\longrightarrow} EA + P \qquad \text{[reaction 5]}$$

SCHEME III

satisfy three essential conditions: (1) when the modulator is absent, the reaction is via reaction 3 exclusively, and the catalytic rate measured ($k_{cat\ mod}$) equals k_{cat}; (2) when modulator concentration is infinite, the reaction is via reaction 4 exclusively, and the catalytic rate is βk_{cat}; (3) when modulator concentration is equal to the K_A, then the reaction is via reactions 3 and 4 equally, and the catalytic rate is $k_{cat}(\beta + 1)/2$. These equations assume that the concentration of the component interacting with the modulator, either the enzyme or the substrate, is much lower than K_A. These conditions are satisfied by Eq. (5):

$$k_{cat}/(k_{cat\ mod} - k_{cat}) = \{(K_A)/(\beta - 1)\}/[A] + 1/(\beta - 1) \qquad (5)$$

The values of β and K_A are then determined by a plot of $k_{cat}/(k_{cat\ mod} - k_{cat})$ versus reciprocal [A]. The ordinate equals $(\beta - 1)^{-1}$ and the slope equals $K_A/\beta - 1$. For changes in K_m a similar plot of $K_m/(K_{m\ mod} - K_m)$ versus reciprocal [A] will yield values for α and K_A for K_m changes.

A simplified example of heparin modulation of t-PA-mediated Pg activation is shown in Fig. 1.[30] The double-reciprocal plot of Pg activation versus Pg concentration demonstrates that increasing concentrations of heparin increase the rate of the reaction via increases in k_{cat}. The replot of the data yields the kinetic constants of the modulation.

If the calculated value of β is 0 and/or α is infinite, then the affector component is not a modulator but rather it is a true inhibitor, and the

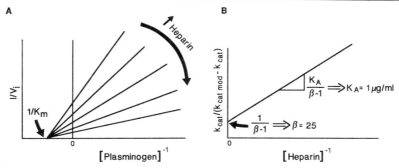

FIG. 1. Kinetic effects of heparin on tissue-type plasminogen activator-mediated plasminogen activation. Data taken from Edelberg and Pizzo.[30] (A) A double-reciprocal plot of the rate of plasminogen activation versus the concentration of plasminogen in the presence of various concentrations of heparin demonstrates that heparin increases the catalytic rate of the activation, but does not alter the K_m of the reaction. (B) From a double-reciprocal plot of $k_{cat}/(k_{cat\ mod} - k_{cat})$ versus heparin concentration the maximum increase in the k_{cat} (β) and that of the binding constant of heparin in the reaction (K_A) are calculated.

reaction cannot yield a product when this component is bound. A further evaluation of the mechanism of inhibition can be obtained by Dixon inhibition analysis described in the next section.

Dixon Inhibition Analysis of Plasminogen Activation

The effects of Lp(a) on the rate of Pg activation on various vascular templates can be kinetically analyzed. To study these effects, increasing concentrations of Lp(a) are added to various concentrations of Pg in the presence of a constant concentration of a Pg activator, plasmin substrate, and a modulator template in the reaction.

The kinetic effects of Lp(a) may be determined by Dixon inhibition analysis.[44–46] Dixon inhibition analysis allows the determination of the mode of inhibition. There are two basic modes of simple reversible inhibition: competitive inhibition, where the inhibitor competes with the substrate for access to the enzyme, and uncompetitive inhibition, where the inhibitor binds the enzyme–substrate complex and prevents it from generating product. In addition, the inhibitor may act as both a competitive and an uncompetitive inhibitor in the same reaction. If the inhibitor constants for both types of inhibition are the same, this is termed noncompeti-

[44] M. Dixon, *Biochem. J.* **55**, 170 (1953).
[45] M. Dixon and E. C. Webb, eds., "Enzymes," p. 332. Academic Press, New York, 1979.
[46] C. G. Knight, *in* "Proteinase Inhibitors" (A. J. Barret and G. Salvesen, eds.), p. 23. Elsevier, Amsterdam, 1986.

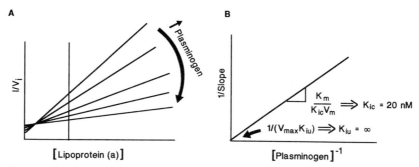

Fig. 2. Kinetic effects of lipoprotein (a) on tissue-type plasminogen activator-mediated plasminogen activation in the presence of fibrinogen fragments. Data taken from Edelberg and Pizzo.[25] (A) A Dixon plot of the reciprocal rate of plasminogen activation versus the concentration of lipoprotein (a) for various concentrations of plasminogen demonstrates that lipoprotein (a) decreases the rate of the activation with a greater effect on the lower than the higher concentrations of plasminogen. (B) A replot of the slope from the Dixon plot versus the reciprocal concentration of plasminogen demonstrates that lipoprotein (a) is a competitive inhibitor of the reaction, and that there is no mixed inhibition.

tive inhibition, and if the constants differ, it is mixed inhibition. Equation (6) is for the integrated Dixon inhibition analysis,

$$1/V_i = (K_m + [S])/V_m[S] + \{(K_m/K_{ic}) + ([S]/K_{iu})\}[I]/V_m[S] \qquad (6)$$

where K_{ic} and K_{iu} are the inhibition constants of competitive and uncompetitive inhibition, respectively.[45] The inverse initial rate of Pg activation is plotted versus inhibitor concentration for various concentrations of substrate. The slope of each of these lines is determined for each substrate concentration and is replotted verus inverse substrate concentration. The abscissa intercept of this plot equals $1(V_m K_{iu})$ and the slope is $K_m/(K_{ic} V_m)$.

A simplified example of Lp(a) competitive inhibition of fibrinogen fragment-enhanced t-PA activation of Pg is shown in Fig. 2.[25] The plot of reciprocal Pg activation versus Lp(a) concentration demonstrates that increasing the concentration of Lp(a) decreases the rate of the activation. The replot of the slopes of this plot versus reciprocal Pg concentration reveals that the Lp(a) is a competitive inhibitor of the activation.

Lipoprotein (a) and Fibrinolytic Inhibition Pathways

Determination of Second-Order Irreversible Inhibition Rate Constant

Determination of the rate of inhibition of either plasmin or Pg activators is based on the measurement of the rate of inhibition of enzymatic activity,

$$E + I \xrightleftharpoons{K_1} EI_{reversible} \qquad [\text{reaction 6}]$$
$$EI_{reversible} \xrightarrow{k_i} EI_{irreversible} \qquad [\text{reaction 7}]$$

SCHEME IV

not on the rate of enzyme–inhibitor complex formation. An irreversible inhibitor inhibits an enzyme in a two-step reaction, first forming a reversible complex with the enzyme, which is then followed by the transformation to a functionally irreversible complex.[47] This is modeled by Scheme IV, where an irreversible inhibitor (I) inhibits an enzyme (E) and the second-order inhibition rate constant (k_2) is the quotient of the kinetic constants K_i and K_1. Kinetically the enzyme is inhibited in both steps of the reaction. Therefore, to measure accurately the rate of this inhibition it is necessary that the activity of the enzyme be measured under conditions that do not dissociate the reversible enzyme–inhibitor complex. These conditions require that the concentrations of the components in the reaction are not significantly diluted during the assay of enzymatic activity. This can be done by continuously measuring enzymatic activity with a substrate (see previous section on Pg activation rate). If the substrate is at a concentration well below its K_m, it will not alter the concentration of free enzyme. If a higher concentration of substrate is required, the substrate will act as a modulator of the inhibition (see next section). The concentration of free enzyme then can be determined by a comparison of the relative velocities of the inhibited reaction versus the uninhibited reaction. The rate of enzymatic inhibition is given by Eq. (7)[48]:

$$\delta[E_t]/\delta t = -k_2[E_t][I_t] \tag{7}$$

The calculation is greatly simplified in the case that equal concentrations of enzyme and inhibitor are employed in the reactions. The rate of inhibition is described by Eq. (8):

$$1/[E_t] = 1/[E_0] + k_2 t \tag{8}$$

A plot of the inverse free enzyme concentration versus time will then yield the second-order inhibition rate constant of the reaction.

General Modulator Analysis of Irreversible Enzymatic Inhibition

The inhibition of Pg activators is effected by various components of the vasculature, which, as in Pg activation, may also serve as templates

---

[47] J. Travis and G. S. Salvesen, *Annu. Rev. Biochem.* **254**, 655 (1983).
[48] A. Fersht, "Enzyme Structure and Mechanism," p. 176. Freeman, New York, 1984.

$$E + I \xrightarrow{k_2} EI_{inactive} \qquad \text{[reaction 8]}$$

$$+$$

$$Mo$$

$$K_{Mo} \Big\updownarrow$$

$$MoE + I \xrightarrow{\alpha k_2} MoEI_{inactive} \qquad \text{[reaction 9]}$$

SCHEME V

for the effects of Lp(a). The kinetic characterization of the modulated inhibition is critical to the analysis of the role of Lp(a) in these systems. Like the effects observed on Pg activation, modulators may either increase or decrease the rate of the inhibition, but, unlike the kinetics of activation, there is only one kinetic measurement that varies, k_2. The modulation of irreversible inhibition in which the modulator (Mo) interacts with the enzyme is described in Scheme V.

The equations describing the modulated inhibition are similar to those for modulated activation in both description and treatment of the data. The equation describing these reactions must satisfy three essential conditions: (1) when Mo is absent, the inhibition is via reaction 8 exclusively, and the second-order inhibition constant measured ($k_{2\,mod}$) is equal to k_2; (2) when the concentration of Mo is infinite, the inhibition is via reaction 9 exclusively, and $k_{2\,mod}$ equals αk_2; (3) when the concentration of Mo is equal to the K_{Mo}, the inhibition is via reactions 8 and 9 equally, and the $k_{2\,mod}$ equals $1/2(k_2 + \alpha k_2)$. This assumes that the concentration of the component interacting with the modulator is much lower than K_{Mo}. These conditions are satisfied by Eq. (9)[49]:

$$k_2/(k_{2\,mod} - k_2) = (K_{Mo}/[Mo] + 1)/(\alpha - 1) \qquad (9)$$

The modulator constants are determined from a plot of $k_2/(k_{2\,mod} - k_2)$ versus the reciprocal concentration of the modulator. The ordinate equals $1/(\alpha - 1)$ and the slope equals $K_{Mo}/(\alpha - 1)$.

Figure 3 illustrates the use of this equation for a simplified example of a modulator that decreases the rate of irreversible enzymatic inhibition, where increasing concentrations of fibrinogen fragments decrease the rate of plasmin inhibition by $\alpha_2 AP$.[35]

[49] J. M. Edelberg and S. V. Pizzo, J. Biol. Chem. **266**, 7494 (1991).

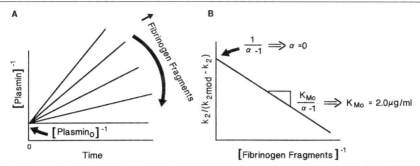

FIG. 3. Kinetic effects of fibrinogen fragments on plasmin irreversible enzymatic inhibition by an equimolar concentration of α_2-antiplasmin. Data taken from Edelberg and Pizzo.[35] (A) A plot of the reciprocal concentration of free plasmin versus time for various concentrations of fibrinogen fragments demonstrates that these fragments decrease the rate of plasmin inhibition. (B) From a plot of $k_2/(k_{2\,mod} - k_2)$ versus reciprocal fibrinogen concentration the maximum change in the rate of inhibition (α) and the binding constant (K_{Mo}) of the fibrinogen fragments are calculated.

General Suppressor Analysis of Modulated Plasminogen Activator Inhibition

Analogous to the effects of Lp(a) on Pg activation, Lp(a) has no effect on the inhibition of Pg activators. Rather, it decreases the rate of the inhibition in the presence of the modulators fibrinogen and heparin. The kinetics of the effects of Lp(a) on the inhibition rate may be mediated by one of two kinds of suppression: Lp(a) may take the modulator out of the inhibition by competing with either the enzyme or the inhibitor for the modulator, or the Lp(a) may bind to either the enzyme- or inhibitor-bound modulator and prevent interaction between the enzyme and inhibitor on the modulator. The former suppression, in which the suppressor competes for, and removes, the modulator from the reaction, is termed competitive suppression of modulated irreversible inhibition. The latter case, when the suppressor binds to the modulator and prevents the inhibition of the enzyme, is termed uncompetitive suppression of modulated irreversible inhibition.

The competitive suppression of modulated inhibition is described by Scheme VI. The equation describing these reactions must satisfy three essential conditions: (1) when the suppressor (Su) is absent, the inhibition is via reactions 8 and 9, and the second-order inhibition constant measured ($k_{2\,sup}$) is described by Eq. (9); (2) when the concentration of Su is infinite, the modulator (Mo) is bound in reaction 10 exclusively, thus all E and I are free and inhibition is via reaction 8, and the $k_{2\,sup}$ is k_2; (3) when the

$$E + I \xrightarrow{\ k_2\ } EI_{inactive} \qquad [reaction\ 8]$$

$$+$$

$$Mo + Su \underset{}{\overset{K_{Su}}{\rightleftharpoons}} MoSu \qquad [reaction\ 10]$$

$$K_{Mo} \Big\Updownarrow$$

$$MoE + I \xrightarrow{\ \alpha k_2\ } MoEI_{inactive} \qquad [reaction\ 9]$$

SCHEME VI

concentration of Su is equal to K_{Su}, the concentration of free Mo is decreased by half, and therefore the inhibition via reaction 9 is decreased by half. This assumes that the concentration of the component interacting with the modulator is much lower than K_{Mo}. These conditions are satisfied by Eq. (10)[49]:

$$k_{2\ sup} = k_2\{1 + \alpha([Mo]/K_{Mo})/(1 + [Su]/K_{Su})\}/$$
$$\{1 + ([Mo]/K_{Mo})/(1 + [Su]/K_{Su})\} \qquad (10)$$

Equation (10) must be solved empirically because the concentrations of free modulator and suppressor are interdependently related. To determine the competitive suppressor constant, the rate of the inhibition should be measured for increasing concentrations of suppressor at a constant concentration of modulator. These rates should then be plotted versus the concentration of suppressor added to the reaction. From this plot the concentration of suppressor needed to decrease the modulated reaction by 50%, or to bind half of the free modulator, can be determined. Subtracting half of the original concentration of the modulator from the concentration of the suppressor needed for a 50% reduction in the modulated reaction yields the K_{Su} for competitive suppression. Figure 4 demonstrates a simplified example of a competitive suppressor of irreversible enzymatic inhibition; increasing the concentration of Lp(a) increases the rate of plasmin inhibition by $\alpha_2 AP$ in the presence of downward modulatory fibrinogen fragments.[35]

The other model of suppression of modulated inhibition involves a suppressor that binds to the modulator with either the enzyme or the inhibitor, and thereby blocks the inhibition. The kinetic model of competitive suppression of modulated inhibition is shown in Scheme VII.

The equation describing these reactions must satisfy three essential conditions: (1) when Su is absent, the inhibition is via reactions 8 and 9, and $k_{2\ sup}$ is described by Eq. (9); (2) when the concentration of Su is infinite, the Mo is bound in reaction 11 exclusively, thus the E and I bound to Mo cannot inhibit via reaction 9, and the remaining free E is inhibited via reaction 8, and $k_{2\ sup}$ is described by Eq. (9) minus the contribution of the inhibition via reaction 9, or $k_{2\ sup} = k_{2\ mod}/(1 + [Mo]/K_{Mo})$; (3) when

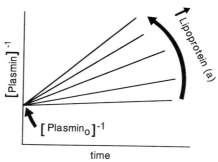

FIG. 4. Kinetic analysis of the effects of lipoprotein (a) on plasmin irreversible enzymatic inhibition by an equimolar concentration of α_2-antiplasmin in the presence of fibrinogen fragments. Data taken from Edelberg and Pizzo.[35] A plot of the reciprocal concentration of free plasmin versus time for various concentrations of lipoprotein (a) demonstrates that lipoprotein (a) increases the rate of the inhibition. The binding constant of lipoprotein (a) for the fibrinogen fragments ($K_{Su} = 4$ nM) is calculated from an empiric solution to Eq. (10).

the concentration of Su is equal to K_{Su}, the Mo bound in reactions 9 and 11 is equal, and therefore the inhibition via reaction 9 is decreased by half. This assumes that the concentration of the component interacting with the modulator is much lower than K_{Mo}. These conditions are satisfied by Eq. (11).[49]

$$\{k_{2\,sup}(1 + [Mo]/K_{Mo})/\alpha k_2([Mo]K_{Mo}) - k_2/\alpha k_2([Mo]/K_{Mo})\}^{-1} = 1 + [Su]/K_{Su}$$

$$(11)$$

The suppression constants are determined by a plot of $\{k_{2\,sup}(1 + [Mo]/K_{Mo})/\alpha k_2([Mo]/K_{Mo}) - k_2/\alpha k_2([Mo]/K_{Mo})\}^{-1}$ versus the concentra-

$$E + I \xrightarrow{\ k_2\ } EI_{inactive} \qquad [reaction\ 8]$$

$$+$$

$$Mo$$

$$K_{Mo} \updownarrow$$

$$MoE + I \xrightarrow{\ \alpha k_2\ } Mo(EI_{inactive}) \qquad [reaction\ 9]$$

$$+$$

$$Su$$

$$K_{Su} \updownarrow \qquad\qquad [reaction\ 11]$$

$$SuMoE$$

SCHEME VII

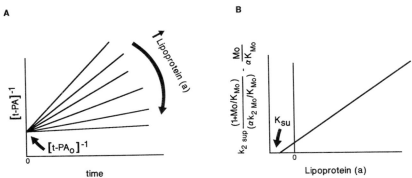

F[IG]. 5. Kinetic analysis of the effects of lipoprotein (a) on tissue-type plasminogen activator irreversible enzymatic inhibition by an equimolar concentration of plasminogen activator inhibitor type 1 in the presence of heparin. Data taken from Edelberg *et al.*[37] (A) A plot of the reciprocal concentration of free tissue-type plasminogen activator versus time for various concentrations of lipoprotein (a) demonstrates that lipoprotein (a) decreases the rate of the inhibition. (B) From a replot of the data versus lipoprotein (a) concentration the binding constant of lipoprotein (a) for heparin (K_{su}) is calculated.

tion of the suppressor. The binding constant of the suppressor for the modulator, K_{Su}, is the slope of the line.

Figure 5 illustrates a simplified example of an uncompetitive irreversible inhibition, where increasing concentrations of Lp(a) decrease the rate of t-PA inhibition by PAI-1 in the presence of heparin.[37]

Conclusion

Lp(a) is an important etiologic factor in the development of atherosclerotic disease. Several *in vitro* investigations suggest that this association is due to an inhibition of fibrinolysis by elevated levels of Lp(a). We hope that the methods described in this chapter will assist in future investigations into the pathophysiologic mechanisms of Lp(a).

Section III

Nonmammalian Blood Coagulation Factors and Inhibitors

[18] Purification and Characterization of Inhibitors of Blood Coagulation Factor Xa from Hematophagous Organisms

By CHRISTOPHER T. DUNWIDDIE, LLOYD WAXMAN, GEORGE P. VLASUK, and PAUL A. FRIEDMAN

Introduction

Hematophagous organisms such as leeches and ticks possess in their salivas an extensive repertoire of biochemical factors that act in concert to maintain blood in a liquid state, both during the ingestion of a blood meal and during its prolonged storage in the gut. As such, inhibitors of blood coagulation[1-5] and of platelet aggregation[6,7] have been identified in the salivas of these animals, as have stimulators of fibrinolysis.[8-10]

The therapeutic potential of leeches or leech-derived substances has been recognized since antiquity. The most extensively studied leech-derived substance, hirudin, the extremely potent anticoagulant found in the saliva of the leech *Hirudo medicinalis,* is a selective, stoichiometric inhibitor of thrombin. Activation of coagulation through either the intrinsic or extrinsic pathway results in the formation of factor Xa (fXa), which catalyzes the conversion of prothrombin to thrombin. Because fXa is the only known physiologically relevant activator of prothrombin, a potent, selective, stoichiometric inhibitor of fXa should also be a very effective anticoagulant. Thus, from a teleological standpoint, the existence of fXa inhibitors of this type in salivas from hematophagous animals would seem likely. As discussed below, this is in fact true.

[1] F. Markwardt, this series, Vol. 19, p. 924.

[2] F. Markwardt and H. Landmann, *Naturwissenschaften* **45,** 398 (1958).

[3] F. Markwardt and H. Landmann, *Naturwissenschaften* **48,** 433 (1961).

[4] K. Hellmann and R. I. Hawkins, *Thromb. Diath. Haemorrh.* **18,** 617 (1967).

[5] J. M. C. Ribeiro, G. T. Makoul, J. Levine, D. R. Robinson, and A. Spielman, *J. Exp. Med.* **161,** 332 (1985).

[6] M. Rigbi, H. Levy, A. Eldor, F. Iraqi, M. Teitelbaum, M. Orevi, A. Horovitz, and R. Galun, *Comp. Biochem. Physiol. C* **88,** 95 (1987).

[7] J. L. Seymour, W. J. Henzel, B. Nevins, J. T. Stults, and R. A. Lazarus, *J. Biol. Chem.* **265,** 10143 (1990).

[8] A. Z. Budzynski, S. A. Olexa, B. S. Brizuela, R. T. Sawyer, and G. S. Stent, *Proc. Soc. Exp. Biol. Med.* **168,** 266 (1981).

[9] A. Z. Budzynski, S. A. Olexa, and R. T. Sawyer, *Proc. Soc. Exp. Biol. Med.* **168,** 259 (1981).

[10] E. M. A. Kelen and G. Rosenfeld, *Haemostasis* **4,** 51 (1975).

Salivary gland extracts from the Mexican leech *Haementeria officinalis* have been shown to contain a potent antimetastatic and anticoagulant activity.[11] The protein responsible for both of these activities has been purified and named antistasin (ATS).[12,13] The anticoagulant properties of ATS result from the selective and tight-binding inhibition of fXa.[14] More recently, a second novel and highly selective inhibitor of fXa has been purified from the soft tick, *Ornithodoros moubata*, and has been designated tick anticoagulant peptide (TAP).[15] These molecules represent the most potent and selective inhibitors of fXa described to date and thus serve as valuable tools for the critical evaluation of the role played by fXa in coagulation. In this report we will review the methods utilized for the purification of ATS and TAP and summarize their physical and inhibitory properties.

Assay Methods

Inhibition of the enzymatic activity of free, solution-phase fXa was measured using a chromogenic substrate-based assay, and inhibition of fXa preassembled in the prothrombinase complex (fXa, factor Va, calcium, and phospholipid) was determined either using reconstituted purified components or standard clotting-time assays. The inhibition of thrombin by interaction at either the active site or the anion-binding "exo site" was evaluated using the chromogen S2238 or fibrinogen as the substrate, respectively.

Solution-Phase Enzyme Assays

Reagents

TBSA buffer: 0.05 M Tris-HCl, 0.15 M NaCl, 0.1% (w/v) bovine serum albumin (BSA), pH 7.5

Human factor Xa and human α-thrombin (Enzyme Research Labs, Southbend, IN)

Methoxycarbonyl-D-cyclohexylglycylglycylarginine *p*-nitroanilide acetate (Spectrozyme FXa, American Diagnostica, Inc., Greenwich,

---

[11] G. J. Gasič, A. Iwakawa, T. B. Gasič, E. D. Viner, and L. Milaś, *Cancer Res.* **44**, 5670 (1984).

[12] G. P. Tuszynski, T. B. Gasič, and G. J. Gasič, *J. Biol. Chem.* **262**, 9718 (1987).

[13] E. Nutt, T. Gasič, J. Rodkey, G. J. Gasič, J. W. Jacobs, P. A. Friedman, and E. Simpson, *J. Biol. Chem.* **263**, 10162 (1988).

[14] C. T. Dunwiddie, N. A. Thornberry, H. G. Bull, M. Sardana, P. A. Friedman, J. W. Jacobs, and E. Simpson, *J. Biol. Chem.* **264**, 16694 (1989).

[15] L. Waxman, D. E. Smith, K. E. Arcuri, and G. P. Vlasuk, *Science* **248**, 593 (1990).

CT), made as a stock solution of 6.67 mM in water and diluted to a working stock of 1.11 mM in TBSA

H-D-phenylalanyl-L-pipecolyl-L-arginine p-nitroanilide dihydrochloride (S2238, Kabi-Vitrum, Stockholm, Sweden), made as a stock solution of 2.0 mM in water and diluted to a working stock of 367 μM in TBSA

Inhibitor solutions represented either crude extracts, column fractions, or purified samples diluted in TBSA buffer

Procedure. Purified human fXa and α-thrombin are quantified by active site titration as described,[16,17] diluted into prechilled TBSA just prior to use, and maintained on ice. The enzymatic activity of solution-phase human fXa (fXa in the absence of phospholipid, factor Va, or calcium cofactors) or human α-thrombin is determined at 22° with a continuous chromogenic assay using the p-nitroanilide substrates Spectrozyme FXa or S2238, respectively. Reactions are measured in a 96-well microtiter plate (Dynatech, Chantilly, VA) using a Vmax kinetic microplate reader (Molecular Devices, Palo Alto, CA) by monitoring the increase in absorbance at 405 nm. All reactions are performed in a final volume of 220 μl TBSA buffer with a final concentration of 0.5 nM enzyme. Inhibited, steady-state velocities are generated by preincubating the enzyme with increasing concentrations of inhibitor for 30 min prior to the addition of substrate (300 μM Spectrozyme FXa, 4.3× K_m or 100 μM S2238, 11× K_m). Under these conditions equilibrium is attained between the enzyme and inhibitor over a wide range of inhibitor concentrations. Residual enzyme activity is determined by measuring the initial velocity of the reaction over a 5-min period. During this time course less than 5% of the substrate is consumed.

Factor Xa Activity in Prothrombinase Complex

Reagents

Buffer A: 20 mM Tris-HCl, 0.15 M NaCl, 0.1% (w/v) bovine serum albumin, 1 mM CaCl$_2$, pH 7.4

Buffer B: 50 mM Tris-HCl, 0.15 M NaCl, 10 mM EDTA, 0.1% bovine serum albumin, pH 8.3

Stop buffer: 100 mM EDTA, 0.1% bovine serum albumin, pH 8.3

Human factor V, human α-thrombin, and human factor Xa, (Enzyme Research Labs)

---

[16] P. E. Bock, P. A. Craig, S. T. Olson, and P. Singh, *Arch. Biochem. Biophys.* **273**, 375 (1989).

[17] G. W. Jameson, D. V. Roberts, R. W. Adams, W. S. A. Kyle, and D. T. Elmore, *Biochem. J.* **131**, 107 (1973).

Hirudin, recombinant desulfatohirudin variant HV-2 (Sigma, St. Louis, MO) with a specific activity of 8000 ATU/mg, made as a stock solution of 1.0 U/ml in TBSA buffer

Rabbit brain cephalin (Sigma), reconstituted with 10 ml 0.85% NaCl and stored frozen at −20°

Bovine prothrombin (Sigma), prepared as a stock solution of 10 U/ml in TBSA buffer

S2238 (Kabi-Vitrum), prepared as a 2.0 mM stock solution in water

Procedure. The inhibition of fXa in the prothrombinase complex is evaluated using an *in vitro* system consisting of reconstituted purified components. In this two-step, coupled assay, thrombin is generated from prothrombin by the reconstituted prothrombinase complex in the first step and the relative amount of thrombin generated is quantitated with a chromogenic substrate-based assay in the second step. Human factor V is converted to factor Va (fVa) by incubation with human α-thrombin at a molar ratio of 100 : 1 in TBSA buffer at 37° for 10 min. Following activation, the thrombin is neutralized by the addition of a titrated amount of hirudin. Reconstituted prothrombinase complex is prepared by combining rabbit brain cephalin (1 : 60 final dilution) with human fVa and human fXa in reaction buffer A to yield final concentrations of 12.5 and 0.167 nM, respectively. Aliquots (10 μl) of various dilutions of the inhibitor are mixed with 10 μl of the preformed prothrombinase complex and 20 μl of buffer A and then preincubated 60 min at 37° to allow equilibration between the inhibitor and the enzyme. The reaction is initiated by the addition of 10 μl (0.1 U) of prothrombin and allowed to incubate for 10 min at 37°. The reaction is rapidly terminated by the addition of 10 μl of stop buffer and then diluted into 1 ml of buffer B and stored on ice until assay. The relative thrombin activity is determined by adding 10 μl of the diluted reaction sample to the wells of a prewarmed 96-well microtiter plate containing 73 μM S2238 (8 × K_m) in a total volume of 210 μl of buffer B. Reaction velocities are determined at 37° by monitoring the increase in absorbance at 405 nm using a Thermo-Max kinetic microplate reader (Molecular Devices). FXa-catalyzed prothrombin activation in this assay is absolutely dependent on added fVa and calcium. Prothrombinase activity is expressed as a percent of the control rate determined in the absence of any inhibitor.

Clotting-Time Assays

Reagents

Thromboplastin C and actin-activated cephaloplastin reagent (Baxter-Dade Division, Miami, FL)

Citrated, pooled, normal human plasma (George King Biomedical Overland Park, KS)

Rabbit, dog, and rhesus monkey plasma samples, prepared from freshly drawn whole blood collected in 0.38% sodium citrate; Plasma, prepared by centrifugation at 2000 g for 10 min

Procedure. Prothrombin times (PT) and activated partial thromboplastin times (APTT) are measured on an Electra 800 automated clot timer (Medical Laboratory Automation, Mt. Vernon, NY) using commercially available reagents. Dilutions of inhibitors are made in TBSA buffer and stored on ice. A 10-μl aliquot of each inhibitor dilution is added to 100 μl of freshly thawed citrated plasma and assayed. Clotting times are measured in seconds and clot formation is determined spectrophotometrically.

Thrombin-Mediated Fibrin Formation Assay

Reagents

Buffer A: 21 mM imidazole, 3.75 mM CaCl$_2$, 75 mM NaCl, 7.5% (w/v) gum acacia (Sigma), pH 7.2

TBP buffer: 20 mM Tris-HCl, 0.1% (w/v) polyethylene glycol (PEG) 8000, pH 8.3

Human α-thrombin and fibrinogen (Enzyme Research Labs)

Hirudin, recombinant desulfatohirudin variant HV-1 (American Diagnostica Inc.)

Procedure. The following reagents are added, in order, to microtiter plate wells: 100 μl buffer A; 25 μl of the appropriate inhibitor diluted in TBP buffer; 25 μl of α-thrombin diluted in TBP buffer (final concentration 0.5 nM). The inhibitor is preincubated with α-thrombin for 30 min at 25° prior to initiation of the reaction by the addition of 100 μl of fibrinogen (final concentration 1.2 μM). The reaction is allowed to proceed for 20 min at 37°, at which time the increase in turbidity is measured by the absorbance at 650 nm using a Thermo-Max kinetic microplate reader. The increase in turbidity is proportional to the amount of fibrin polymerization that transpires following the cleavage of fibrinogen to fibrin by thrombin.

Purification of ATS from Leech Salivary Glands

Crude salivary gland extract (SGE) is prepared according to the procedure of Gasič *et al.*[11] Anterior and posterier salivary glands are dissected from the Mexican leech *H. officinalis,* and homogenized at 4° in 20 mM HEPES, 10 mM CaCl$_2$, pH 7.8. Following two cycles of sonication, the material is centrifuged at 8500 g for 20 min at 4° and the supernatant is collected. The pellet is extracted twice with the same buffer and the supernatants are pooled and centrifuged at 100,000 g for 1 hr at 4°. The final supernatant is passed through a 0.45-μm filter and stored at $-70°$.

The initial purification of ATS from crude SGE involves a two-step chromatographic procedure using the Pharmacia (Piscataway, NJ) fast protein liquid chromatography (FPLC) system described by Tuszynski *et al.*[12] A volume of 25 ml of SGE is loaded onto a 1 × 4 cm heparin-agarose column preequilibrated with 20 mM Tris-HCl, pH 8.7, at a flow rate of 0.5 ml/min. The column is washed until the OD$_{280}$ returns to baseline and is then eluted with a combination of linear and step gradients produced by the gradient maker using equilibration buffer containing 1 M NaCl. As each protein peak begins eluting, the gradient is held manually so that proteins elute under isocratic conditions. After each protein peak elutes the gradient is restarted. Then 1-ml fractions are collected and assayed for anticoagulant activity using a one-stage prothrombin time clotting assay. A single peak of anticoagulant activity elutes at approximately 0.55 M NaCl (Fig. 1). The active fractions are pooled, concentrated by Amicon (Danvers, MA) filtration (12,000 molecular weight exclusion limit), and desalted on Sephadex G-25 columns. This material is applied to a Mono

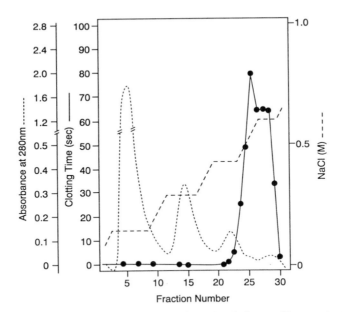

FIG. 1. Heparin-agarose column elution profile. The elution profile was obtained after adsorption and NaCl elution of 25 ml of salivary gland extract from a heparin-agarose column as described (see text). Fractions (1-ml) were collected and clotting-time measurements were performed on 5-μl aliquots of the corresponding column fractions using the prothrombin time assay. Reproduced from Tuszynski *et al.*[12]

Q column equilibrated in 20 mM Tris-HCl, pH 8.7, and eluted in a manner similar to that used for the heparin-agarose column. The anticoagulant activity elutes at approximately 0.25 M NaCl. ATS purified according to this protocol appears homogeneous as assayed by SDS–PAGE; however, comparison of the amino acid composition derived on this material[12] with the subsequently determined amino acid compositions of several highly purified ATS isoforms (Fig. 2) suggests that this material is only partially pure.[13,18] This purification scheme typically yields 200–300 μg of purified protein from 150 mg of SGE.

Two modifications have been made to this original purification protocol.[13] The first modification is a streamlined heparin-agarose column in which adsorbed proteins are eluted stepwise with a 0.35 M NaCl step and a 0.55 M NaCl step rather than a combination linear and step gradient. Anticoagulant activity elutes in the latter fraction as observed previously. The second modification is a substitution of the Mono Q column with an analytical Vydac C_4 RP-HPLC column to increase the resolution of the final step. The concentrated and desalted post-heparin-agarose material is loaded onto a C_4 column equilibrated in aqueous 0.1% CF_3COOH containing 15% isopropanol at a flow rate of 0.5 ml/min. The proteins are resolved with a linear 15–30% isopropanol gradient over 30 min and the eluant is monitored at 220 nm. Anticoagulant activity eluting from this column spans a series of five to six peaks, suggesting the presence of multiple structural ATS isoforms (Fig. 3A). The two predominant peaks in this mixture are purified by repeated RP-HPLC runs and their primary amino acid sequences are determined.[13] These results confirm that at least the two primary HPLC peaks are in fact structural isoforms of the same molecule differing in amino acid sequence at positions 5 and 35 (Fig. 2, isoforms A and B).

Utilizing a similar purification protocol, consisting of anion-exchange chromatography, heparin-agarose chromatography, and C_{18} RP-HPLC, a highly homologous ATS isoform has been purified from salivary gland extracts of the related proboscis leech, *Haementeria ghilianii* (Fig. 2).[19–21]

The complex compositions of the ATS isoforms present in leech SGE create an obstacle in obtaining sufficient quantities of a single isoform of acceptable purity for biochemical characterization. Therefore, a recombi-

[18] E. Nutt, D. Jain, A. B. Lenny, L. Schaffer, P. K. Siegl, and C. T. Dunwiddie, *Arch. Biochem. Biophys.* **285,** 37 (1991).

[19] C. Condra, E. M. Nutt, C. J. Petroski, E. Simpson, P. A. Friedman, and J. W. Jacobs, *Thromb. Haemostasis* **61,** 437 (1989).

[20] D. T. Blankenship, R. G. Brankamp, G. D. Manley, and A. D. Cardin, *Biochem. Biophys. Res. Commun.* **166,** 1384 (1990).

[21] R. G. Brankamp, D. T. Blankenship, P. S. Sunkara, and A. D. Cardin, *J. Lab. Clin. Med.* **115,** 89 (1990).

```
                           5                    10                   15
Isoform A:    |Glu| Gly-Pro-Phe-Gly-Pro-Gly-Cys-Glu-Glu-Ala-Gly-Cys-Pro-Glu
Isoform B:                       Arg
Isoform C:
Isoform D:
H.g. Isoform:


                          20                   25                   30
Isoform A:    Gly-Ser-Ala-Cys-Asn-Ile-Ile-Thr-Asp-Arg-Cys-Thr-Cys-Ser-Gly
Isoform B:
Isoform C:
Isoform D:                                                            Glu
H.g. Isoform:                                                     Pro-Glu


                          35                   40                   45
Isoform A:    Val-Arg-Cys-Arg-Val-His-Cys-Pro-His-Gly-Phe-Gln-Arg-Ser-Arg
Isoform B:             Met
Isoform C:             Met
Isoform D:
H.g. Isoform:              Tyr       Ser


                          50                   55                   60
Isoform A:    Tyr-Gly-Cys-Glu-Phe-Cys-Lys-Cys-Arg-Leu-Glu-Pro-Met-Lys-Ala
Isoform B:
Isoform C:                                   Ile
Isoform D:
H.g. Isoform:              Val       Arg           Thr


                          65                   70                   75
Isoform A:    Thr-Cys-Asp-Ile-Ser-Glu-Cys-Pro-Glu-Gly-Met-Met-Cys-Ser-Arg
Isoform B:
Isoform C:
Isoform D:
H.g. Isoform:


                          80                   85                   90
Isoform A:    Leu-Thr-Asn-Lys-Cys-Asp-Cys-Lys-Ile-Asp-Ile-Asn-Cys-Arg-Lys
Isoform B:
Isoform C:
Isoform D:
H.g. Isoform:


                          95                   100                  105
Isoform A:    Thr-Cys-Pro-Asn-Gly-Leu-Lys-Arg-Asp-Lys-Leu-Gly-Cys-Glu-Tyr
Isoform B:
Isoform C:
Isoform D:
H.g. Isoform:


                          110                  115                  119
Isoform A:    Cys-Glu-Cys-Arg-Pro-Lys-Arg-Lys-Leu-Ile-Pro-Arg-Leu-Ser-OH
Isoform B:
Isoform C:
Isoform D:
H.g.Isoform:              Lys                       Val
```

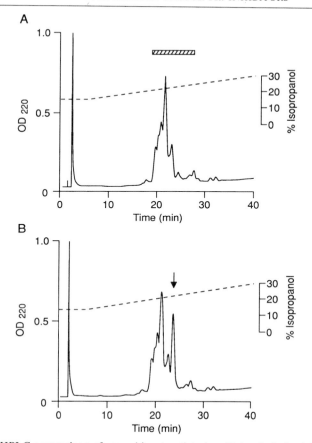

FIG. 3. HPLC comparison of recombinant antistasin with leech-derived ATS isoforms. (A) A partially purified mixture of leech ATS isoforms (38 μg total) was fractionated by HPLC on a C$_4$ column using a 15–30% isopropanol gradient. Those fractions that exhibited fXa inhibitory activity are indicated by a hatched bar. (B) An equivalent amount of the leech ATS isoform mixture was coinjected with 12 μg of purified rATS and chromatographed as in A. The arrow indicates the retention time of rATS run alone on the same gradient. Reproduced from Nutt et al.[18]

FIG. 2. Amino acid sequence comparison of leech antistasin isoforms. The amino acid sequences of four *Haementeria officinalis* ATS isoforms (isoforms A–D) and one *Haementeria ghilianii* (H.g.) ATS isoform are aligned and compared. Isoforms A and B and the *H. ghilianii* isoform[20] were determined from direct amino acid sequence analysis of purified proteins, and the amino acid sequences of isoforms C and D were inferred from the cDNA sequences of two independent ATS clones. The blocked amino-terminal pyroglutamic acid residue is indicated by a box and amino acid differences are shown relative to isoform A.

nant DNA methodology is employed to generate pure preparations of a single ATS isoform. Two structurally distinct cDNA clones (isoforms C and D in Fig. 2) are isolated from a leech salivary gland cDNA library and used for the production of a single recombinant ATS (rATS) isoform.[22]

Purification of Recombinant Antistasin

rATS is produced using an insect cell–baculovirus expression system. Recombinant *Autographa californica* nuclear polyhedrosis virus containing the cDNA from clone 5C-4 in Ref. 22 is used as the viral stock (isoform D in Fig. 2). *Spodoptera frugiperda* (Sf9) cells are infected with the recombinant baculovirus and grown in gassed, 4-liter spinner flasks in serum-free Excell-400 medium as described by Jain *et al.*[23,24] The purification protocol outlined below has been described previously[18] and utilizes S-Sepharose fast flow resin (Pharmacia) instead of heparin-agarose as the initial capture step, because it allows the attainment of considerably higher flow rates, eliminates the batch-to-batch variability associated with heparin-agarose preparations, and is less expensive. Raw growth medium (2 liters), which generally contains 10–14 mg rATS, is harvested and the particulate material and residual cellular debris are removed by centrifugation at 8000 g for 20 min. The supernatant is collected and the pH is raised from 6.4 to 8.2 by the addition of NaOH. Under these conditions a large precipitate forms (presumably residual virus particles) and is removed by further centrifugation at 10,000 g for 20 min. The pH of the clarified medium is reduced to 6.8 with HCl and is loaded at 4° directly onto a 1.6 × 20 cm column containing 38 ml of S-Sepharose equilibrated in 20 mM Tris-HCl, pH 6.8. The column is loaded at a flow rate of 150 ml/hr and the flow through is monitored at 280 nm by an on-line detector. Following loading, the column is washed with 20 mM Tris-HCl, pH 6.8, until a baseline absorbance is attained. Bound proteins are initially eluted with a 0.25 M NaCl step gradient followed by a 250-ml linear gradient to 1.0 M NaCl, and 3.5-ml fractions are collected. Fractions are assayed for fXa inhibitory activity utilizing the chromogenic substrate-based assay, and the peak fractions, which elute as the final protein peak between 0.5 and 0.6 M NaCl, are pooled and desalted by repeated dialysis steps against

[22] J. H. Han, S. W. Law, P. M. Keller, P. J. Kniskern, M. Silberklang, J. Tung, T. B. Gasič, G. J. Gasič, P. A. Friedman, and R. W. Ellis, *Gene* **75**, 47 (1989).
[23] D. Jain, K. Ramasubramanyan, S. Gould, C. Seamans, S. Wang, A. Lenny, and M. Silberklang, *ACS Symp. Ser.* **477**, 91 (1991).
[24] D. Jain, K. Ramasubramanyan, S. Gould, A. Lenny, M. Candelore, M. Tota, C. Strader, K. Alves, G. Cuca, J. Tung, G. Hunt, B. Junker, B. C. Buckland, and M. Silberklang, *Proc. 10th Eur. Soc. Animal Cell Technol. Meet.*, 345 (1991).

double-distilled water using a 3500 molecular weight cutoff membrane. This column typically yields material that is between 60 and 80% pure (Fig. 4B, inset lane 2) with a 75–80% recovery of the total fXa inhibitory activity loaded. Following dialysis, the material is concentrated by lyophilization and loaded directly onto a 2.2×25 cm Vydac preparative C_{18}

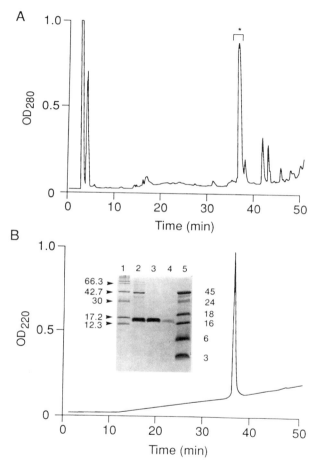

FIG. 4. Analytical analysis of purified rATS. (A) The post-S-Sepharose pooled rATS was purified by preparative C_{18} RP-HPLC. FXa inhibitory activity was localized to a single major peak, as indicated by the starred bracket. (B) Analysis of purified rATS by analytical C_{18} RP-HPLC (10 μg) and reducing/tricine SDS–PAGE (inset). Lanes 1 and 5: molecular weight standards with their associated M_r values $\times 10^{-3}$; lane 2, 10 μg of S-Sepharose-purified rATS; lane 3, 10 μg of HPLC-purified rATS; lane 4, 1 μg of leech-derived ATS. Reproduced from Nutt et al.[18]

column equilibrated with aqueous 0.1% (v/v) CF_3COOH at a flow rate of 8 ml/min. The column is eluted with a linear gradient of acetonitrile in aqueous CF_3COOH from 0 to 50% over 50 min and the eluate is monitored at 280 nm. A symmetrical peak of fXa inhibitory activity representing the major protein species elutes at approximately 38 min and is collected and lyophilized (Fig. 4A). The yield of the preparative HPLC step is 70–75%, providing an overall purification yield of 50%. The purified rATS is resuspended in filtered, HPLC-grade water and stored at 4°.

Properties

Purity. As mentioned above, the complex mixture of ATS isoforms that copurified through multiple purification steps made it difficult to obtain a highly purified preparation of a single isoform for characterization. Therefore, to alleviate the possibility of isoform contamination in a final preparation, the properties described below primarily relate to the purified rATS isoform. The homogeneity of the purified rATS was initially assessed by quantitative amino acid analysis that indicated a homogeneous preparation with a composition consistent with that predicted from the cDNA sequence.[18,22] Analytical C_{18} RP-HPLC analysis demonstrated that the purified rATS eluted as a single symmetrical peak of absorbance with a purity of at least 95% (Fig. 4B). Electrophoretic analysis by tricine/SDS–PAGE under reducing conditions revealed a homogeneous protein species with an electrophoretic mobility indistinguishable from that of leech-derived ATS (Fig. 4B, inset). Coinjection of purified rATS with a mixture of partially purified, leech-derived ATS isoforms, followed by RP-HPLC fractionation, demonstrated that the rATS isoform constituted only a minor component of the leech isoform mixture (Fig. 3B, arrow), thus explaining why this particular isoform was not easily purified from leech SGE.

Physical Properties. The primary structure of rATS is summarized in Fig. 5. It is a 119-amino acid, cysteine-rich, nonglycosylated single-chain polypeptide.[12,13,22] The pI values of both leech-derived ATS (mixed isoforms) and rATS are 9.5–10.0, reflecting the high content of basic amino acids (21 Lys/Arg residues).[12,18] The amino terminus is blocked by a pyroglutamate residue in both the leech-derived protein and the recombinant protein, suggesting that this modification occurs *in vivo* and is not a fermentation or purification artifact.[12,13,18] The ATS molecule exhibits a twofold internal sequence homology between amino acid residues 1–55 and 56–110 with almost perfect alignment of all cysteine residues. The homology is 40% at the amino acid level and 50% at the DNA level, suggesting that ATS may have evolved by a tandem gene duplication

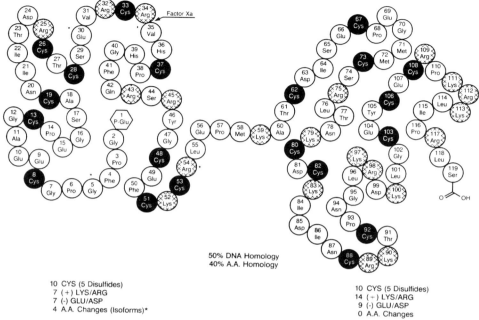

FIG. 5. Primary structure of rATS. The primary sequence of rATS is shown with the postulated bi-domain structure. Cysteine residues are in black and Lys/Arg residues are cross-hatched. The blocked amino-terminal pyroglutamate residue is indicated by a hexagon and the fXa cleavage site is indicated by an arrow. The positions of the amino acid differences constituting the multiple *H. officinalis* isoforms are indicated by asterisks.

event.[13,22] The twofold structural homology of ATS suggests that the molecule is arranged into separate amino- and carboxy-terminal domains, with each domain containing 10 homologous cysteines (Fig. 5). However, the specific disulfide assignments are currently unknown. The identification of the reactive site of ATS at Arg[34] demonstrates that the amino-terminal domain provides the fXa inhibitory function of ATS.[14,25] The high proportion of positively charged residues in the carboxy-terminal domain (residues 58–119) and the shared amino acid sequence homology between residues in this domain with other heparin-binding proteins[20,26,27] imply that the carboxy-terminal domain may provide the heparin-binding characteristics of ATS. Only 2 of the 12 amino acid differences between the four ATS isoforms identified from *H. officinalis* and the single ATS isoform

[25] C. T. Dunwiddie, G. P. Vlasuk, and E. M. Nutt, *Arch. Biochem. Biophys.* **294,** 647 (1992).
[26] G. D. Holt, H. C. Krivan, G. J. Gasič, and V. Ginsburg, *J. Biol. Chem.* **264,** 12138 (1989).
[27] G. D. Holt, M. K. Pangburn, and V. Ginsburg, *J. Biol. Chem.* **265,** 2852 (1990).

purified from *H. ghilianii* (Figs. 2 and 5, asterisks) are located in the carboxy-terminal domain, suggesting that this portion of the molecule is evolutionarily conserved and further implies an important role for this domain.

Inhibitory Characteristics. Utilizing a chromogenic substrate-based assay, ATS was shown to be a reversible, competitive inhibitor of fXa.[14] The selectivity of ATS for fXa was demonstrated by its inability to inhibit several other serine proteases, including thrombin, factor VIIa, chymotrypsin, and elastase.[14] The inhibitory effects of ATS on thrombin were further investigated using the purified rATS isoform. rATS exhibited no inhibitory effects on the amidolytic activity of thrombin at ratios of inhibitor to enzyme as high as 4000 : 1, confirming the results obtained with ATS purified from leech salivary glands. In addition, rATS failed to inhibit thrombin-mediated fibrin formation (see Assay Methods) at molar ratios of inhibitor to enzyme as high as 2000 : 1. These results demonstrate that rATS does not interact with either the active site or the "exo site" of thrombin. rATS, however, does inhibit trypsin with approximately a 100-fold lower potency than fXa, exhibiting a K_i of 21 nM.[28]

Kinetic analysis of ATS revealed that it is a slow, tight-binding inhibitor of fXa (Fig. 6A). ATS interacts with fXa according to a two-step mechanism that postulates the formation of an initial, rapid encounter complex between the inhibitor and the enzyme, which slowly converts to a stable tight-binding complex (Fig. 6B). The equilibrium dissociation constants describing formation of the stable enzyme–inhibitor complex between fXa and the major leech-derived ATS isoform or rATS are 410 ± 30 and 43.4 ± 1.4 pM, respectively.[14,25] The difference in potencies exhibited by these two isoforms probably reflects the amino acid differences near the reactive site at residues 30 and 35 (see below).

rATS inhibits fXa[25] according to a "standard mechanism" of serine protease inhibitors as outlined below[29]:

$$E + I \rightleftharpoons [EI] \rightleftharpoons C \rightleftharpoons [EI]^* \rightleftharpoons E + I^*$$

where E is the enzyme, I is the inhibitor, I* is the inhibitor hydrolyzed at the reactive site peptide bond, [EI] and [EI]* are loose, noncovalent complexes, and C is the stable enzyme–inhibitor complex. According to this mechanism, the inhibitor provides a conformationally constrained, substrate-like bait region, called the reactive site, which interacts with the active site of the enzyme in the manner of a limited proteolytic substrate. The reactive site peptide bond of rATS is located at residues

[28] C. T. Dunwiddie, unpublished data (1991).
[29] M. Laskowski, Jr. and I. Kato, *Annu. Rev. Biochem.* **49,** 593 (1980).

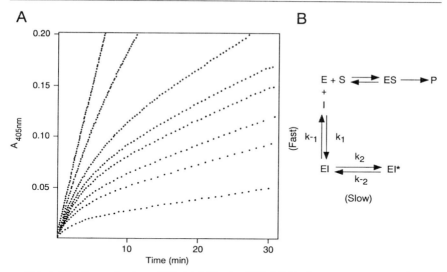

FIG. 6. Kinetic mechanism of fXa inhibition by ATS. (A) Product progress curves demonstrating time-dependent inhibition of fXa by ATS. Reactions were initiated by the addition of fXa to solutions containing substrate and increasing concentrations of ATS. (B) Slow-binding mechanism of inhibition characterized by the rapid formation of an encounter complex (EI), which slowly converts to the stable enzyme–inhibitor complex (EI*).

Arg^{34}-Val^{35} (Fig. 5), and this peptide bond remains intact in the stable enzyme–inhibitor complex.[25]

In addition to inhibiting solution-phase fXa, rATS is a potent and rapid inhibitor of fXa in the prothrombinase complex when assayed either utilizing reconstituted purified components[14,30] or standard, plasma-based clotting assays (Fig. 7).

Purification of Tick Anticoagulant Peptide from Whole Ticks

The soft tick *Ornithodoros moubata* is indigenous to South Africa and is purchased from Antibody Associates, Ltd. (Bedford, TX). This tick is a vector for swine fever and a special permit from the United States Department of Agriculture may be required in order to import it. Alternatively, there are colonies of this tick and related species maintained in the United States. A typical TAP preparation is initiated with 400–500 whole ticks (maintained alive or stored at −70°), constituting about 12 g of material. The ticks are homogenized in batches of 50 ticks with a

---

[30] C. T. Dunwiddie, D. E. Smith, E. M. Nutt, and G. P. Vlasuk, *Thromb. Res.* **64,** 787 (1991).

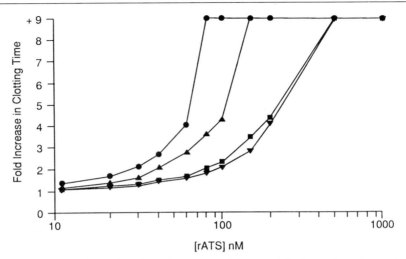

FIG. 7. *In vitro* plasma-based clotting assays with rATS. Dilutions of rATS (10 nM to 1 μM) were tested for their ability to prolong the standard APTT clotting assay using various plasmas. The following plasma samples were tested: ●, human; ▲, rhesus monkey; ■, rabbit; ▼, dog. Reproduced from Nutt *et al.*[18]

Polytron (Brinkmann Instrument Co., Westbury, NY) in 10 ml of 20 mM Bis-Tris-HCl, pH 7.0, containing 0.15 M NaCl and the protease inhibitors E-64, pepstatin, chymostatin, and leupeptin (50 μM each, Sigma). All operations are performed at 4° using plasticware. The homogenate is centrifuged at 100,000 g for 20 min and the resulting pellets are reextracted in 3–5 ml of the same buffer and recentrifuged. The combined supernatants (~800 mg protein) are diluted twofold with water and applied at 1 ml/min to a 50-ml column of Fast Q-Sepharose anion-exchange resin (Pharmacia) equilibrated with 20 mM Bis-Tris-HCl, pH 7.0. The column is washed with five volumes of the same buffer and bound proteins are eluted at 0.5 ml/min with a 500-ml gradient from 0 to 1.0 M NaCl. Then 5-ml fractions are collected and aliquots from each fraction are diluted 1 : 1000 and assayed for fXa inhibitory activity using the chromogenic substrate-based assay. The active fractions, which elute between 0.25 and 0.3 M NaCl, are pooled and yield approximately 60 mg protein in a total volume of 70 ml. This pool is diluted to 1 liter with 50 mM sodium acetate, pH 3.5, and then loaded at 0.5 ml/min onto a 30-ml column of Fast S-Sepharose cation-exchange resin (Pharmacia) preequilibrated with 50 mM sodium acetate, pH 3.5. Following loading, the column is washed with the same buffer until the absorbance at 280 nm returns to baseline. The inhibitor is then eluted with a 250-ml linear gradient of NaCl from 0 to 1.0 M at 0.5 ml/min and 2.5-ml fractions are collected and assayed for fXa inhibitory activity. The inhibitory activity, which elutes between 0.2 and 0.3 M NaCl, is pooled and lyophilized.

The dried material is dissolved in 3 ml of water and the vessel is rinsed with an additional 1 ml of water. The solution is filtered through a 0.45-μm filter and loaded onto a 1 × 25 cm Vydac C_{18} column equilibrated with aqueous 0.1% CF_3COOH at room temperature. The inhibitor is eluted with a linear gradient of acetonitrile from 0 to 60% in 0.1% CF_3COOH over 60 min at a flow rate of 3 ml/min. Then 3-ml fractions are collected and each fraction is assayed for fXa inhibitory activity. The active fractions, which elute at approximately 40 min (Fig. 8A, starred bracket), are pooled and lyophilized to remove organic solvents. The lyophilized material is dissolved in 1 ml of water and rechromatographed on HPLC as above with the exception that the TAP-containing peak is collected by hand. The final material is lyophilized, resuspended in 0.5 ml of water, and analyzed for purity by analytical C_{18} RP-HPLC (0.46 × 15 cm column) as shown in Fig. 8B. The amount and concentration of purified TAP are determined by quantitative amino acid analysis. The typical yield from 400 ticks is approximately 270 μg.

Following our original TAP isolation,[15] we purified TAP from a colony of *O. moubata* ticks maintained on site (Dr. Richard Endris, Merck and Co., Rahway, NJ). Although these ticks were determined to be taxonomically identical to those obtained through Antibody Associates, LTD. (Dr. Richard Endris, personal communication, 1990), the inhibitor purified from these ticks (TAP-2) differs in amino acid sequence at four positions from the TAP preparation described above (TAP-1), as summarized in Table I. These results demonstrate the existence of at least two TAP isoforms. Nonetheless, both inhibitor isoforms are indistinguishable with respect to their inhibitory potencies toward fXa.

TABLE I

AMINO ACID SEQUENCES OF FACTOR Xa INHIBITOR ISOFORMS FROM *O. moubata*

Type	Sequence[a]

	1									10										20									30
TAP-1	Y	N	R	L	C	I	K	P	*R*	D	W	I	D	E	C	D	S	N	E	G	G	E	R	A	*Y*	F	R	N	*G*	K
TAP-2	Y	N	R	L	C	I	K	P	*Q*	D	W	I	D	E	C	D	S	N	E	G	G	E	R	A	*F*	F	R	N	*D*	K

					40										50									60
TAP-1	G	G	C	D	S	F	W	I	C	P	E	D	H	T	G	A	D	Y	Y	S	S	Y	*R*	D	C	F	N	A	C	I
TAP-2	G	G	C	D	S	F	W	I	C	P	E	D	H	T	G	A	D	Y	Y	S	S	Y	*Q*	D	C	F	N	A	C	I

[a] Differences are denoted by italics.

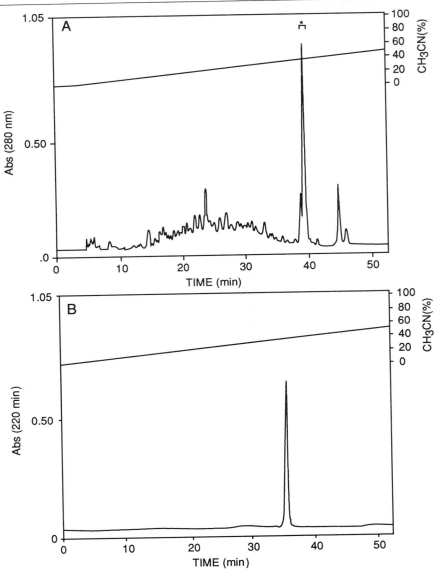

FIG. 8. RP-HPLC purification and analytical analysis of TAP. (A) Following chromatography on Fast Q and Fast S, TAP was chromatographed on a C_{18} column. Those fractions exhibiting fXa inhibitory activity are indicated by a starred bracket. (B) The final purified preparation of TAP was analyzed by analytical C_{18} RP-HPLC.

We have detected fXa inhibitory activity in extracts from several other species of ticks, including *Ornithodoros coriaceus, Rhipicephalus sanguineus* (brown dog tick), and *Amblyomma americanum* (Lone Star tick); however, the levels are considerably lower than those from *O. moubata*.

Properties

Purity. Both isoforms of TAP purified from *O. moubata* were homogeneous as assessed by analytical C_{18} RP-HPLC (Fig. 8B) and revealed a single band (M_r 6000) on SDS–PAGE.[15] Moreover, direct amino acid sequence analysis revealed only one sequence for each peptide (Table I), with yields that were consistent with the amount of protein determined by quantitative amino acid analysis.

Physical Properties. TAP is an acidic polypeptide (pI 4.5) of 60 amino acids (M_r 6985),[15] and on gel filtration under nondenaturing conditions it elutes as a monomer. Its amino acid sequence has limited homology to the Kunitz-type inhibitors such as bovine pancreatic trypsin inhibitor (BPTI), with 15 or 16 homologous amino acids, depending on how the sequences are aligned.[15,31]

Like the Kunitz inhibitors, TAP contains six cysteine residues arranged in three disulfides. The pairing has been determined on the recombinant version of TAP, isoform-1 (rTAP): Cys-5 to Cys-59, Cys-15 to Cys-39, and Cys-33 to Cys-55.[31,32] This topological arrangement is the same as that in the Kunitz-type inhibitors. The invariant spacing of the Cys residues, however, is not maintained; only Cys-55 and Cys-59 have the same spacing as that exhibited by the Kunitz inhibitors.[33] In addition, although treatment of BPTI with dithiothreitol rapidly and selectively reduces one disulfide bond (Cys-14 to Cys-38), the reduction of disulfides in rTAP proceeds at a slower rate and appears to be nonspecific.[31] Thus, in spite of sequence homology with the Kunitz inhibitors, TAP probably differs significantly in its tertiary structure.

The majority of native (tick-derived) TAP appears to be covalently modified, possibly by phosphorylation.[34] However, because the inhibitory properties of native TAP are indistinguishable from those of rTAP, this modification is not essential for activity.

[31] M. Sardana, V. Sardana, J. Rodkey, T. Wood, A. Ng, G. P. Vlasuk, and L. Waxman, *J. Biol. Chem.* **266,** 13560 (1991).
[32] M. Neeper, L. Waxman, D. E. Smith, C. Schulman, M. Sardana, R. W. Ellis, L. Schaffer, P. K. S. Siegl, and G. P. Vlasuk, *J. Biol. Chem.* **265,** 17746 (1990).
[33] T. C. C. Wun, K. K. Kretzmer, T. J. Girard, J. P. Miletich, and G. J. Broze, Jr., *J. Biol. Chem.* **263,** 6001 (1988).
[34] G. P. Vlasuk and L. Waxman, unpublished data (1991).

Inhibitory Characteristics. TAP is a unique protease inhibitor due to its high degree of specificity for blood coagulation fXa. The specificity of native TAP was assessed by observing the effects of this inhibitor on the amidolytic activity of various serine proteases using the appropriate peptidyl chromogenic substrates.[15] There was no inhibition of blood coagulation factor XIa, thrombin, trypsin, chymotrypsin, elastase, kallikrein, urokinase, plasmin, tissue-type plasminogen activator, or *Staphylococcus aureus* V8 protease, using a 300-fold molar excess of TAP over the enzyme. In addition, TAP exhibited no inhibition of human coagulation factor VIIa activity measured by the release of tritiated activation peptide from human factor X in the presence of thromboplastin and calcium.[15] TAP also had no effect on the proteases of the alternative complement pathway at concentrations of the inhibitor (rTAP) up to 5 μM (Comp-Quick CH50, Sigma).

The results discussed above demonstrate that TAP exhibits no inhibition of the catalytic activity of a broad spectrum of serine proteases, including thrombin. To evaluate the inhibitory capacity of TAP toward the "exo site" of thrombin, the effects of TAP on thrombin-mediated fibrin formation were measured *in vitro* utilizing the recombinant form of the inhibitor[32] (see Methods). As a positive control, recombinant desulfatohirudin (rHIR) partially inhibited thrombin-mediated fibrin formation in this assay at 0.5 nM, with complete inhibition occurring at 50 and 500 nM (Table II). In contrast, rTAP only modestly inhibited thrombin-mediated

TABLE II
INHIBITION OF THROMBIN-MEDIATED
FIBRIN FORMATION[a]

[Inhibitor] (μM)	Inhibition %	
	rTAP	rHIR
10	43.1 ± 4.1	ND
5	35.6 ± 2.4	ND
2	20.6 ± 1.6	ND
1	7.8 ± 1.9	ND
0.5	0	100
0.05	0	100
0.005	0	91.7 ± 0.99
0.0005	0	48.3 ± 3.1

[a] Thrombin-mediated fibrin formation was measured as described under "Assay Methods." ND, Not done. Each value represents the mean ± SEM of three determinations.

fibrin formation at concentrations of the inhibitor that were at least 20,000-fold greater than those required for rHIR. Taken together, these results provide staunch evidence of the exquisite specificity of TAP for fXa.

A comparison with the Kunitz inhibitors provides an explanation for the inability of TAP to inhibit other serine proteases. BPTI and other members of this family form a complex with trypsinlike enzymes in which a Lys or Arg residue positioned adjacent to the second Cys in the sequence is inserted into the active site of the protease, where it interacts with the Asp residue of the catalytic triad of the enzyme. Two potent Kunitz-type inhibitors of factor Xa, the peptide trypstatin[35] and the second domain of the lipoprotein-associated coagulation inhibitor (LACI)[36] present in plasma, both have an Arg residue after the second Cys. In contrast, the corresponding amino acid in TAP is an Asp, whose negatively charged side chain should not be readily accommodated by proteases that hydrolyze proteins on the carboxy-terminal side of basic residues. Because TAP contains no Cys residues that are followed by an Arg, it is not possible to assign a reactive site by analogy to members of the Kunitz family.

A detailed kinetic analysis of the interaction of native TAP with fXa was not possible due to the limited quantities of the tick-derived material. Nonetheless, the intrinsic K_i for native TAP was determined from steady-state velocities and yielded a value of 0.280 nM (Fig. 9). Utilizing the recombinant version of the inhibitor,[32] we were able to show that rTAP is a reversible, slow, tight-binding inhibitor of factor Xa, displaying a competitive type of inhibition.[37] The binding is stoichiometric and exhibits a K_i of 0.180 ± 0.002 nM, a value in close agreement with that obtained for native TAP. The calculated association and dissociation rate constants were 2.85 ± 0.07 × 10^6 M^{-1} sec^{-1} and 5.54 ± 1.78 × 10^{-4} sec^{-1}, respectively.[37] The active site of fXa is required for tight binding of TAP because ^{35}S-labeled rTAP bound only to fXa and not to diisopropyl fluorophosphate-inactivated fXa or to zymogen factor X.[37] Most serine protease inhibitors, including ATS, share a common mechanistic pathway consisting of binding of the protease through its active site to the reactive site of the inhibitor followed by slow hydrolysis of the inhibitors reactive site peptide bond.[29] However, we have been unable to demonstrate evidence for peptide bond cleavage of TAP by fXa.[34] In this regard, TAP resembles the binding of hirudin to thrombin.[38]

[35] H. Kido, Y. Yokogoshi, and N. Katunuma, *J. Biol. Chem.* **263**, 18104 (1988).
[36] T. G. Girard, L. A. Warren, W. F. Novotny, K. M. Likert, S. G. Brown, J. P. Miletich, and G. P. Broze, Jr., *Nature (London)* **338**, 518 (1989).
[37] S. P. Jordan, L. Waxman, D. E. Smith, and G. P. Vlasuk, *Biochemistry* **29**, 11095 (1990).
[38] S. R. Stone and J. Hofsteenge, *Biochemistry* **25**, 4622 (1986).

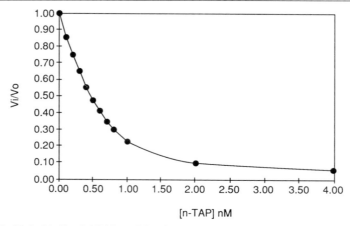

[n-TAP] nM

FIG. 9. Tight-binding inhibition of fXa by TAP. FXa (0.5 nM) was preincubated for 30 min with increasing concentrations of purified TAP and the residual enzyme activity was determined by measuring the initial velocity of the reaction following the addition of substrate at a concentration equal to 4.3 times its K_m.

In addition to its inhibitory effects on the amidolytic activity of fXa, TAP was also active in several human plasma-based clotting assays. Prothrombin clotting times, activated partial thromboplastin times, and Stypven times were all prolonged in a concentration-dependent manner.[15]

[19] Hirudin and Hirudin-Based Peptides

By STUART R. STONE and JOHN M. MARAGANORE

The anticoagulant properties of leech saliva were first described over 100 years ago by Haycraft,[1] and hirudin was isolated as the active anticoagulant ingredient from the salivary glands of the European medicinal leech *Hirudo medicinalis* by Markwardt in 1957.[2] About 30 years later, the first recombinant expression of hirudin was reported.[3-7] Hirudin is a polypep-

[1] J. B. Haycraft, *Proc. R. Soc. London, Ser. B* **36**, 478 (1884).
[2] F. Markwardt, *Hoppe-Seyler's Z. Physiol. Chem.* **308**, 147 (1957).
[3] J. Dodt, T. Schmitz, T. Schäfer, and C. Bergmann, *FEBS Lett.* **202**, 373 (1986).
[4] C. Bergmann, J. Dodt, S. Köhler, E. Fink, and H. G. Gassen, *Biol. Chem. Hoppe-Seyler* **367**, 731 (1986).

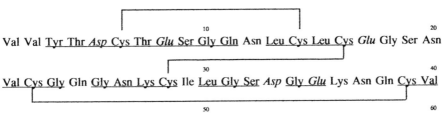

Val Val <u>Tyr Thr</u> *Asp* <u>Cys Thr</u> *Glu* <u>Ser Gly Gln</u> Asn <u>Leu Cys Leu Cys</u> *Glu* Gly Ser Asn

Val <u>Cys Gly</u> Gln <u>Gly Asn Lys Cys</u> Ile <u>Leu Gly Ser</u> *Asp* <u>Gly</u> *Glu* Lys Asn Gln <u>Cys</u> Val

<u>Thr Gly</u> *Glu* <u>Gly Thr Pro Lys Pro</u> Gln <u>Ser</u> His Asn *Asp* <u>Gly</u> *Asp* <u>Phe</u> *Glu* *Glu* Ile <u>Pro</u>

Glu *Glu* <u>Tyr</u> Leu Gln

FIG. 1. Sequence of hirudin variant 1. The sequence given in the three-letter code is that determined by Bagdy *et al.*[8] and Dodt *et al.*[9] The disulfide bridges determined by Dodt *et al.*[11] are also given. Acidic residues are indicated by italics and residues invariant in other hirudin sequences[14] are underlined.

tide of 65 residues (Fig. 1)[8,9] that contains three disulfide bridges: $Cys^{6'}$-$Cys^{14'}$,[10] $Cys^{16'}$-$Cys^{28'}$;, and $Cys^{22'}$-$Cys^{39'}$.[11] The amino acid sequence is particularly rich in acidic residues; 6 of the last 13 amino acids are acidic (Fig. 1). In natural hirudin isolated from the leech, Tyr-63' is sulfated, which further increases the number of negatively charged residues in this region. This posttranslational modification is not, however, found in recombinant molecules expressed in *Escherichia coli* or yeast. More than 20 different isoforms of hirudin from *H. medicinalis* have been isolated and sequenced[12-14] and the conserved residues are noted in Fig. 1. Most

[5] E. Fortkamp, M. Rieger, G. Heisterberg-Moutses, S. Schweitzer, and R. Sommer, *DNA* **5**, 511 (1986).

[6] B. Meyhack, J. Heim, H. Rink, W. Zimmerman, and W. Märki, *Thromb. Res., Suppl.* **7**, 33 (1987).

[7] G. Loison, A. Findeli, S. Bernard, M. Nguyen-Juilleret, M. Marquet, N. Riehl-Bellon, D. Carvallo, L. Guerra-Santos, S. W. Brown, M. Courtney, C. Roitsch, and Y. Lemoine, *Bio/Technology* **6**, 72 (1988).

[8] D. Bagdy, E. Barabas, L. Graf, T. E. Peterson, and S. Magnusson, this series, Vol. 45, p. 669.

[9] J. Dodt, H. Müller, U. Seemüller, and J.-Y. Chang, *FEBS Lett.* **165**, 180 (1984).

[10] Amino acid residues in the sequence of hirudin are designated with a prime following the number to distinguish them from amino acid residues in the sequence of thrombin.

[11] J. Dodt, U. Seemüller, R. Maschler, and H. Fritz, *Biol. Chem. Hoppe-Seyler* **366**, 379 (1985).

[12] J. Dodt, N. Machleidt, U. Seemüller, R. Maschler, and H. Fritz, *Biol. Chem. Hoppe-Seyler* **367**, 803 (1986).

[13] R. P. Harvey, E. Degryse, L. Stefani, F. Schamber, J.-P. Cazenave, M. Courtney, P. Tolstoshev, and J.-P. Lecocq, *Proc. Natl. Acad. Sci. U.S.A.* **83**, 1084 (1986).

[14] M. Scharf, J. Engel, and D. Tripier, *FEBS Lett.* **255**, 105 (1989).

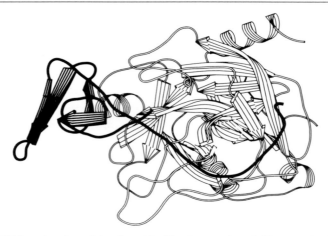

FIG. 2. Ribbon drawing of the thrombin–hirudin complex. β Sheets and α helices are represented as arrows and coils, respectively. The hirudin structure is darker and is found in front of thrombin. It is composed of four short β sheets linked by loops and a long C-terminal tail that wraps around the thrombin molecule and binds to the anion-binding exosite. The N-terminus of hirudin binds in the active site of thrombin. The thrombin molecule consists of an A and B chain. The A chain is composed of two helical portions and is found at the back of the B chain while the B chain consists mainly of β sheets. The structure represented is that of Grütter et al.[17] and the plot was made using the program of Priestle.[17a]

notably, the acidic nature of the C-terminal region and the positions of the disulfide bridges are conserved between isoforms. However, other residues are also conserved and many of these residues are involved in the binding of hirudin to thrombin.[15–17]

The structure of the thrombin–hirudin complex presented in Fig. 2 shows the novel mechanism by which hirudin inhibits thrombin. The N-terminal three residues of hirudin are bound in the active site cleft of thrombin, but the orientation of the polypeptide chain is opposite to that observed for other protein inhibitors of serine proteinases. Whereas the polypeptide chain of other inhibitors runs from N-terminal to C-terminal into the active site and forms an antiparallel β sheet with residues 214–218 of thrombin, the polypeptide chain of the first three amino acids of hirudin runs from N-terminal to C-terminal out of the active site and forms a

[15] T. J. Rydel, K. G. Kavichandran, A. Tulinsky, W. Bode, R. Huber, C. Roitsch, and J. W. Fenton, II, Science 249, 277 (1990).
[16] T. J. Rydel, A. Tulinsky, W. Bode, and R. Huber, J. Mol. Biol. 221, 583 (1991).
[17] M. G. Grütter, J. P. Priestle, J. Rahuel, H. Grossenbacher, W. Bode, J. Hofsteenge, and S. R. Stone, EMBO J. 9, 2361 (1990).
[17a] J. P. Priestle, J. Appl. Crystallogr. 21, 572 (1988).

parallel β sheet with these residues in thrombin.[15-17] The remainder of the N-terminal core region of hirudin closes off the active site of thrombin. The C-terminal tail of hirudin extends 35 Å across the surface of thrombin and is bound to a surface groove (the anion-binding exosite) that is rich in basic amino acids. The acidic residues in the C-terminal region of hirudin make several electrostatic interactions with the anion-binding exosite. In addition, numerous hydrophobic contacts occur between residues of the C-terminal tail and the anion-binding exosite. The structure of the N-terminal core of hirudin in the complex corresponds to those determined by two-dimensional NMR for hirudin in solution.[18-20] In solution, however, the C-terminal tail (residues 44–65) is disordered.

Peptide fragments corresponding to either the C-terminal tail[21-23] or the N-terminal core[24-26] of hirudin have been produced by synthetic or recombinant methods, respectively. C-Terminal tail fragments inhibit competitively thrombin cleavages of fibrinogen and show anticoagulant properties in plasma. However, as these fragments bind to a locus quite distant from the thrombin catalytic site,[27-29] the peptides do not inhibit thrombin reactivity toward small synthetic substrates nor thrombin interactions with antithrombin III and the antithrombin III–heparin complex.[26,30] Although C-terminal tail peptides do not inhibit the cleavage of small synthetic substrates, a conformational change resulting from their binding to the anion-binding exosite[31] appears to either accelerate or decel-

[18] G. M. Clore, D. K. Sukumaran, M. Nilges, J. Zarbock, and A. M. Gronenborn, *EMBO J.* **6,** 529 (1987).

[19] D. K. Sukumaran, G. M. Clore, A. Preuss, J. Zarbock, and A. M. Gronenborn, *Biochemistry* **26,** 333 (1987).

[20] H. Haruyama and K. Wüthrich, *Biochemistry* **28,** 4301 (1989).

[21] S. Bajusz, I. Favszt, E. Barabas, M. Dioszegi, and D. Bagdy, *in* "Peptides 1984" (U. Ragnarson, ed.), p. 473. Almqvist & Wiksell, Stockholm, 1984.

[22] J. L. Krstenansky and S. J. T. Mao, *FEBS Lett.* **211,** 10 (1987).

[23] J. M. Maraganore, B. Chao, M. L. Joseph, J. Jablonski, and K. L. Ramachandran, *J. Biol. Chem.* **264,** 8692 (1989).

[24] J. Dodt, S. Köhler, T. Schmitz, and B. Wilhelm, *J. Biol. Chem.* **265,** 713 (1990).

[25] J.-Y. Chang, J.-M. Schläppi, and S. R. Stone, *FEBS Lett.* **260,** 209 (1990).

[26] S. Dennis, A. Wallace, J. Hofsteenge, and S. R. Stone, *Eur. J. Biochem.* **188,** 61 (1990).

[27] P. Bourdon, J. W. Fenton, II, and J. M. Maraganore, *Biochemistry* **29,** 6379 (1990).

[28] J.-Y. Chang, P. K. Ngai, H. Rink, S. Dennis, and J.-M. Schlaeppi, *FEBS Lett.* **261,** 287 (1990).

[29] E. Skrzypczak-Jankun, V. E. Carperos, K. G. Ravichandran, A. Tulinsky, M. Westbrook, and J. M. Maraganore, *J. Mol. Biol.* **221,** 1379 (1991).

[30] M. C. Naski, J. W. Fenton, II, J. M. Maraganore, S. T. Olson, and J. A. Shafer, *J. Biol. Chem.* **265,** 13484 (1990).

[31] S. Konno, J. W. Fenton, II, and G. B. Villanueva, *Arch. Biochem. Biophys.* **267,** 158 (1988).

erate the rate for substrate cleavage.[26,30,32–34] N-Terminal core peptides produced by recombinant DNA techniques inhibit thrombin cleavage of small synthetic substrates by binding to the active site of thrombin with an affinity markedly reduced compared to hirudin.[24–26,33] Affinity of N-terminal core fragments for thrombin is not increased significantly in the presence of C-terminal tail peptides, indicating the absence of cooperativity for interactions of individual hirudin domains with thrombin.[26]

Recombinant Expression of Hirudin and Construction of Mutants

Several systems have been developed for the recombinant expression of hirudin. Expression in *Saccharomyces cerevisiae* is most suited for the commercial production of recombinant hirudin because of the high yields that can be obtained in yeast.[6,7] For structure–function studies, however, recombinant expression in *E. coli* has been most widely used. In comparison with yeast expression, the time required to produce a mutant protein in *E. coli* is substantially less. In *E. coli,* hirudin has been expressed in the periplasmic space of the bacterium[3,35] and as a fusion protein in the cytoplasm.[36] Recombinant expression in the periplasmic space of *E. coli* results in an easy purification of the recombinant hirudin and the system used in the laboratory of one of the authors (S.R.S.) is described briefly below.

Recombinant expression of hirudin is achieved by using the pIN-III vector developed by Inouye and co-workers.[37] This vector utilizes the signal sequence of the ompA protein to direct expression to the periplasmic space of the bacterium. The gene encoding recombinant hirudin (rhir) was synthesized[38] and ligated into the pIN-III–ompA$_2$ plasmid. In the resultant plasmid (pIN-III–ompA$_2$/hir), the sequence encoding the ompA signal peptide is followed immediately by that of hirudin and both of these sequences are contained within an *Xba*I/*Bam*HI restriction fragment.[35] For site-directed mutagenesis, this fragment is subcloned into the replicative form of the bacteriophage M13. Site-specific mutations can be introduced into rhir by using one of the standard methods of site-directed mutagenesis, such as that of Kunkel.[39] After identification of the mutants

[32] L.-W. Liu, T.-K. Vu, C. T. Esmon, and S. R. Coughlin, *J. Biol. Chem.* **266**, 16977 (1991).

[33] T. Schmitz, M. Rothe, and J. Dodt, *Eur. J. Biochem.* **195**, 251 (1991).

[34] G. L. Hortin and B. L. Trimpe, *J. Biol. Chem.* **266**, 6866 (1991).

[35] P. J. Braun, S. Dennis, J. Hofsteenge, and S. R. Stone, *Biochemistry* **27**, 6517 (1988).

[36] J. B. Lazar, R. C. Winant, and P. H. Johnson, *J. Biol. Chem.* **266**, 685 (1991).

[37] J. Ghrayeb, H. Kimua, M. Takahara, H. Hsiung, Y. Masui, and M. Inouye, *EMBO J.* **3**, 2437 (1984).

[38] H. Rink, M. Liersch, P. Sieber, and F. Meyer, *Nucleic Acids Res.* **12**, 6369 (1984).

[39] R. A. Kunkel, *Proc. Natl. Acad. Sci. U.S.A.* **82**, 488 (1985).

by dideoxy sequencing, the *Xba*I/*Bam*HI restriction fragment from the replicative form of M13 is recloned into pIN-III–ompA$_2$. The resultant mutant pIN-III–ompA$_2$/hir plasmid is used to transform *E. coli* JM109. Recombinant hirudins are prepared from 1-liter cultures of *E. coli* JM109 in LB medium containing 100 μg of ampicillin/liter.

Purification of Recombinant Hirudin

Reagents

Cold (4°) TE buffer: 10 m*M* Tris-HCl, pH 8.1, containing 1 m*M* EDTA
50 m*M* Bis-Tris-HCl, pH 6.5
50 m*M* Bis-Tris-HCl, pH 6.5, containing 0.5 *M* NaCl
0.1% (v/v) Trifluoroacetic acid
0.1% (v/v) Trifluoroacetic acid in 70% acetonitrile (HPLC buffer B)

After harvesting the bacteria from a 1-liter culture by centrifugation, the rhir is released from the periplasmic space by suspending the pellet in cold (4°) 10 m*M* TE buffer (50 ml). After centrifugation to remove the bacteria, the supernatant is filtered and applied to a Mono Q fast protein liquid chromatography (FPLC) column (Pharmacia, Uppsala, Sweden) that has been equilibrated with 50 m*M* Bis-Tris-HCl buffer, pH 6.5. All FPLC steps are performed with a flow rate of 1.0 ml/min. FPLC and HPLC steps are performed at room temperature. The Mono Q column is subsequently developed by using a 45-ml gradient of 0–0.3 *M* NaCl in the equilibration buffer. Fractions are assayed for their thrombin inhibitory activity by using the assay for the determination of the concentration of hirudin described below. A typical elution profile is shown in Fig. 3. Fractions containing inhibitory activity are pooled and lyophilized. The lyophilizate is redissolved in 0.1% trifluoroacetic acid and applied to a 20 × 250 mm C$_{18}$ reversed-phase HPLC column equilibrated in the same solvent. The flow rate for the HPLC purification is 10 ml/min. The column is washed for 5 min with 10% buffer B (70% acetonitrile in 0.1% trifluoro-acetic acid) before eluting the rhir with a gradient of 10–40% buffer B over 35 min; the percentage of buffer B is then increased to 90% over the next 10 min. A typical elution profile is shown in Fig. 4.

By using the above method for expression and purification, yields of 1 mg of pure rhir per liter of culture are routinely obtained.

Kinetic Analysis of the Inhibition of Thrombin by Hirudin

Hirudin is a competitive inhibitor of thrombin with respect to tripepti-dyl *p*-nitroanilide substrates.[40–42] With low (picomolar) concentrations of

---

[40] S. R. Stone and J. Hofsteenge, *Biochemistry* **25**, 4622 (1986).

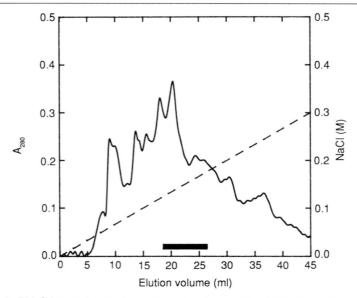

FIG. 3. FPLC ion-exchange chromatography of recombinant hirudin on Mono Q. The chromatogram shown is that for the mutant E62Q. The filtered extract (50 ml) obtained after releasing the contents of the periplasmic space of the bacteria by osmotic shock was loaded onto the Mono Q column. The majority of the protein did not bind to the column. The column was eluted with a 45-ml gradient of 0.0 to 0.3 M NaCl (---) in 50 mM Bis–Tris–HCl buffer, pH 6.5, at a flow rate of 1.0 ml/min. The bar indicates the fractions in which the majority of the thrombin inhibitory activity was found.

hirudin, the kinetic mechanism for the inhibition of thrombin in the presence of substrate can be represented by Scheme I, where E, I, and S represent thrombin, hirudin, and the substrate, respectively; k_1 is the observed association rate constant for the interaction under the conditions of the assay and k_2 is the dissociation rate constant. These parameters are related to the inhibition constant (K_I) by the expression given in Scheme I. The equation describing the progress curve of product (P) generation for this mechanism has been derived by Cha[43] and analysis of data for the inhibition of thrombin by hirudin according to this equation yields estimates for k_1, k_2, and K_I.[35] It should be noted, however, that there is evidence that the mechanism for the formation of the thrombin–hirudin complex involves more than one step. Studies using picomolar concentra-

[41] S. R. Stone and J. Hofsteenge, *Protein Eng.* **4**, 295 (1991).
[42] S. R. Stone, P. J. Braun, and J. Hofsteenge, *Biochemistry* **26**, 4617 (1987).
[43] S. Cha, *Biochem. Pharmacol.* **25**, 2695 (1976).

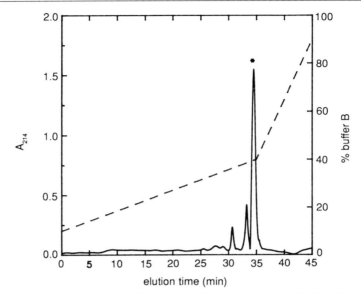

Fig. 4. HPLC reversed-phase chromatography of recombinant hirudin. The fractions indicated by the bar in Fig. 3 were pooled and lyophilized. The lyophilized pool was redissolved in 0.1% trifluoroacetic acid and loaded onto a 20 × 250 mm C_{18} column at a flow rate of 10 ml/min; this flow rate was used for all subsequent steps. The column was washed for 5 min with 10% buffer B (70% acetonitrile in 0.1% trifluoroacetic acid) before the column was developed with a gradient of 10–40% buffer B over 35 min, followed by an increase in the percentage of buffer B to 90% over the next 10 min (---). The absorbance was monitored at 214 nm and the peaks collected by hand. The peak marked with an asterisk contained the thrombin inhibitory activity.

tions of hirudin and a conventional recording spectrophotometer indicate the existence of at least two steps. The first step is ionic strength dependent and does not involve the active site of thrombin; the rate of the association of hirudin with thrombin was found to be independent of the substrate concentration.[40–42] In a subsequent step, hirudin binds to the active site.

$$
\begin{array}{c}
\text{E} + \text{S} \underset{}{\overset{K_m}{\rightleftharpoons}} \text{ES} \overset{k_{cat}}{\longrightarrow} \text{E} + \text{P} \\
+ \\
\text{I} \\
k_1 \Big\Updownarrow k_2 \qquad K_I = k_2/k_1 \\
\text{EI}
\end{array}
$$

SCHEME I

Using higher concentrations of rhir and a stopped-flow spectrophotometer, a third step was detected.[44] There is good evidence that the first step involves the binding of the C-terminal region of hirudin to the anion-binding exosite of thrombin.[35,40,44,45] This step is rate limiting with low hirudin concentrations; thus, the mechanism presented in Scheme I can be used to analyze the data. However, considering the complexity of the mechanism, it is apparent that the parameters k_1 and k_2 determined from the analysis of the data are not rate constants for a particular step. The parameter k_1 represents the effective rate constant for the association of hirudin with thrombin with low (picomolar) hirudin concentrations. Values of 1.4×10^8 and 4.7×10^8 M^{-1} sec^{-1} were obtained for this parameter for rhir and native hirudin, respectively.[35] The corresponding values for the inhibition constant (K_I) are 0.23 and 0.020 pM.

Estimates of the K_I value for hirudin found in the literature vary by over three orders of magnitude. There are at least three possible reasons for this variation:

1. *Reaction conditions*. The K_I value for hirudin increases markedly as the ionic strength is increased.[45] The pH of the assay will also affect the value of K_I; a pH optimum between 7.5 and 8.0 is observed.[46]

2. *Quality of thrombin*. γ-Thrombin, a degraded form of thrombin, can arise through autolysis[47,48] and has a much reduced affinity for hirudin.[42,49,50] The presence of γ-thrombin in preparations used for kinetic studies can lead to higher estimates for K_I and may lead to data suggesting noncompetitive inhibition.[41] Simulations indicated that a contamination with γ-thrombin of less than 5% can significantly affect the data obtained. The thrombin used can be prepared by using one of the published methods.[47,51] The purity of the preparation used should be assessed by gel electrophoresis and it should contain less than 2% γ-thrombin. Labeling with [^{3}H]diisopropyl fluorophosphate followed by autoradiography can

---

[44] M. P. Jackman, M. A. A. Parry, J. Hofsteenge, and S. R. Stone, *J. Biol. Chem.* (1992) (in press).

[45] S. R. Stone, S. Dennis, and J. Hofsteenge, *Biochemistry* **28**, 6857 (1989).

[46] A. Betz, J. Hofsteenge, and S. R. Stone, *Biochemistry* **31**, 1168 (1992).

[47] J. W. Fenton, II, M. J. Fasco, A. B. Stackrow, D. L. Aronson, A. M. Young, and J. S. Finlayson, *J. Biol. Chem.* **252**, 3587 (1977).

[48] J. W. Fenton, II, B. H. Landis, D. A. Walz, and J. S. Finlayson, *in* "Chemistry and Biology of Thrombin" (R. L. Lundblad, J. W. Fenton, II, and K. G. Mann, eds.), p. 43. Ann Arbor Sci. Publ., Ann Arbor, MI, 1977.

[49] B. H. Landis, M. P. Zabinski, G. J. M. Lafleur, D. H. Bing, and J. W. Fenton, II, *Fed. Proc., Fed. Am. Soc. Exp. Biol.* **37**, 1445 (1978).

[50] S. R. Stone and J. Hofsteenge, *Biochemistry* **30**, 3950 (1991).

[51] R. L. Lundblad, *Biochemistry* **10**, 2501 (1971).

be used to assess the extent of contamination with γ-thrombin.[52] The concentration of active molecules in the preparation should be determined by active site titration with MUGB[53] or NPGB.[54] Good preparations are usually greater than 90% active.

3. *Experimental protocols*. Because hirudin is a tight-binding inhibitor of thrombin, protocols appropriate for this type of inhibition must be used. The concentrations of thrombin and hirudin used must be as close as possible to the apparent K_I value in order to obtain an accurate estimate of this parameter. The concentration of active hirudin molecules should be determined by titration with thrombin. Nonlinear regression analysis should be used to fit the data to the appropriate equations.

All groups presently working on the inhibition of thrombin by hirudin are paying careful attention to the above points with the result that all estimates for K_I values that have appeared in the literature over the last several years have been of the same order of magnitude.[24,35,36,41,55,56] All values published have been in the low picomolar range and it seems that previous higher estimates for the K_I were in error. The reaction conditions and methods of data analysis used in one of our laboratories (S.R.S.) is presented below. Similar procedures are used in other laboratories working in the field.

Procedures for Determination of Kinetic Parameters
for Hirudin Inhibition

Reagents

66.7 mM Tris-HCl buffer, pH 8.0, containing 133 mM NaCl and 0.27% (w/v) polyethylene glycol (M_r 6000)

Chromogenic substrate: 2.0 mM S2238 (D-Phe-pipecolyl-Arg p-nitroaniline) or 2.0 mM S2266 (D-Val-Leu-Arg p-nitroaniline) (Kabi-Pharmacia, Molndal, Sweden)

Human α-thrombin: 200 to 2 nM in the above buffer

Assay for Determination of Hirudin Concentration. The assay mixture for this assay contains 750 μl of Tris-HCl buffer, 100 μl of 2.0 mM S2266,

[52] D. L. Bing, M. Cory, and J. W. Fenton, II, *J. Biol. Chem.* **252,** 8027 (1977).
[53] G. W. Jameson, D. V. Roberts, R. W. Adams, W. S. A. Kyle, and D. T. Elmore, *Biochem. J.* **131,** 107 (1973).
[54] T. Chase and E. Shaw, *Biochem. Biophys. Res. Commun.* **29,** 508 (1967).
[55] J. Dodt, S. Kohler, and A. Baici, *FEBS Lett.* **229,** 87 (1988).
[56] J. I. Witting, D. V. Brezniak, J. R. Hayes, and J. W. Fenton, II, *Thromb. Res.* **63,** 473 (1991).

and the rhir sample and water to a final concentration of 990 μl. This assay mixture is warmed to 37°. The pH of the assay mixture at this temperature will be 7.8 if the pH of the buffer is adjusted at room temperature and the final concentration of the buffer will be 50 mM Tris-HCl containing 100 mM NaCl and 0.2% (w/v) polyethylene glycol. The reaction is started by the addition of 10 μl of 200 nM thrombin (2 nM final concentration in the assay). The generation of p-nitroaniline is followed at 405 nm in a spectrophotometer as described by Lottenberg et al.[57] and the steady-state velocity of the reaction is measured. The reactions are usually followed for 5 to 15 min, during which time the reaction rates are linear.

Assay for Determination of Kinetic Parameters for Hirudin Inhibition of Thrombin. The assay mixture for this assay contains 750 μl of Tris-HCl buffer (described above), 50 μl of 2.0 mM S2238, and the rhir sample and water to a final concentration of 990 μl. This assay mixture is warmed to 37° and the reaction is started by the addition of 10 μl of 2–5 nM thrombin (20–50 pM final concentration in the assay). The progress curve for the formation of p-nitroaniline is followed as described above for 20 to 30 min and the concentration of p-nitroaniline is recorded at time intervals of 15 to 60 sec. Data obtained when greater than 10% of substrate is utilized are not used in the data analysis.

Data Analysis

Determination of Hirudin Concentration

The concentration of active hirudin molecules that form a tight complex with thrombin can be determined by titration of 2.0 nM thrombin in the presence of 200 μM D-Val-Leu-Arg-pNA. The dependence of the steady-state velocity on the amount of hirudin present in the assay can be described by Eq. (1)[40]:

$$v_s = \frac{v_0}{2E_t} \{ sqrt[(K_{I'} + \alpha I_\alpha - E_t)^2 + 4K_{I'}E_t] - (K_{I'} + \alpha I_\alpha - E_t) \} \quad (1)$$

where v_s is the observed steady-state velocity, v_0 is the steady-state velocity in the absence of hirudin, E_t is the total molar enzyme concentration, $K_{I'}$ is the apparent inhibition constant, I_α is hirudin (μl/ml) added, and α is the molar concentration of 1.0 μl/ml of hirudin.[58] Steady-state velocities are determined in the absence of hirudin and with four to five concentra-

---

[57] R. Lottenberg, U. Christensen, C. M. Jackson, and P. L. Coleman, this series, Vol. 80, p. 341.

[58] J. W. Williams and J. F. Morrison, this series, Vol. 63, p. 437.

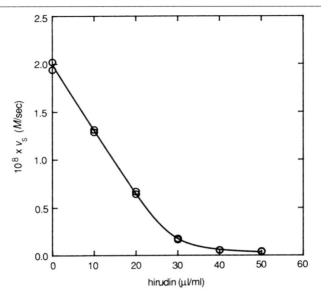

FIG. 5. Determination of recombinant hirudin concentration by titration against thrombin. The effect of the indicated volumes of a stock solution of the mutant V56L on the steady-state velocity (v_s) of thrombin was measured under the conditions indicated in the text with 200 μM S2266. The data were fitted to Eq. (3) by nonlinear regression and the fit of the data to this equation is shown. The analysis yielded the following estimates for the parameters: $v_0 = (9.97 \pm 0.19) \times 10^4\ M^{-1}\ \text{sec}^{-1}$, $K_{I'} = 27.4 \pm 4.0$ pM, and $\alpha = (7.02 \pm 0.17) \times 10^{-2}$ nmol/liter.

tions of hirudin below that of thrombin. In addition, at least one rhir concentration greater than the thrombin concentration is used. A typical set of data is shown in Fig. 5. Analysis of the data by nonlinear regression yields values for α that can be used to calculate the concentrations of hirudin in stock solutions. The determination of an accurate estimate for α is facilitated if the concentration of hirudin is high relative to its apparent inhibition constant ($K_{I'}$). Thus, a substrate with a high Michaelis constant (K_m) is used, because $K_{I'}$ is related to the true inhibition constant (K_I) by Eq. (2),[40]

$$K_{I'} = K_I(1 + S/K_m) \tag{2}$$

where S is the substrate concentration. In addition, with a poor substrate, higher concentrations of both thrombin and hirudin can be used. For the mutants that do not bind tightly to thrombin, the concentration must be obtained by amino acid analysis.

Determination of Inhibition Constants

Three types of inhibition have been observed with hirudin mutants and these correspond to classical, tight-binding, and slow, tight-binding inhibition in the nomenclature of Morrison.[59] It should be noted that the distinction between the three types of inhibition is somewhat arbitrary and will depend on the reaction conditions chosen.[60] Classical inhibition is observed with mutants that exhibit a low affinity for thrombin—i.e., when concentrations of hirudin at least 10 times the concentration of thrombin must be used in order to obtain inhibition. Steady-state velocity data for such mutants can be analyzed according to the Dixon equation.[61] Tight-binding inhibition is observed when the steady-state velocity in the presence of hirudin is achieved fairly rapidly (<5 min), and significant inhibition is observed when the concentration of hirudin is similar to that of thrombin. Data conforming to tight-binding inhibition can be fitted to Eq. (3) by nonlinear regression:

$$v_s = \frac{v_0}{2E_t} \{ \text{sqrt}[(K_{I'} + I_t - E_t)^2 + 4K_{I'}E_t] - (K_{I'} + I_t - E_t) \} \tag{3}$$

where I_t is the molar concentration of hirudin. These analyses yield values for $K_{I'}$ and the expression given in Eq. (2) can be used to calculate K_I. Despite multistep interactions of thrombin with hirudin, K_I is an accurate measure of the overall inhibition. Analysis of tight-binding inhibition data has been used by a number of groups to obtain estimates of K_I values for hirudin.[36,40,56,62] In some cases, it may be possible to determine the concentration of hirudin and its K_I value from the same set of data by using Eq. (1) for the analysis.

The slow, tight-binding inhibition is observed (Fig. 6) when the steady-state velocity in the presence of hirudin is only relatively slowly achieved ($t_{0.5} > 5$ min). In most cases, this slow approach to the steady state can only be observed with relatively low concentrations of rhir (<50 pM). Slow, tight-binding inhibition data can be fitted to Eq. (4) by nonlinear regression,[40]

$$P = v_s t + \frac{(v_0 - v_s)(1 - d)}{(1 - d)} \ln \left[\frac{(1 - de^{-k't})}{(1 - d)} \right] \tag{4}$$

[59] J. F. Morrison, *Trends Biochem. Sci.* **7**, 102 (1982).
[60] J. F. Morrison and C. T. Walsh, *Adv. Enzymol. Relat. Areas Mol. Biol.* **61**, 201 (1987).
[61] I. H. Segel, "Enzyme Kinetics," Chapter 6. Wiley, New York, 1975.
[62] E. Degryse, M. Acker, G. Defreyn, A. Bernat, J. P. Maffrand, C. Roitsch, and M. Courtney, *Protein Eng.* **2**, 459 (1989).

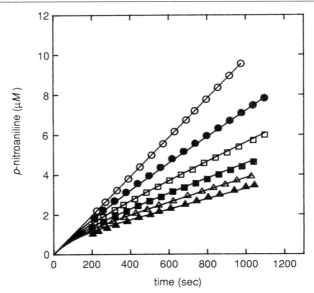

FIG. 6. Slow-binding inhibition of thrombin by recombinant hirudin. The data shown were obtained with 50 pM thrombin and the following concentrations of the mutant V2L: 0 (○), 140 (●), 280 (□), 420 (■), 560 (△), and 700 pM (▲). The substrate S2238 was present at a concentration of 118 μM. The data were fitted by nonlinear regression to Eq. (4) and the lines show the fit of the data to Eq. (4). The estimates obtained for $K_{I'}$ and k_1 were 239 ± 3 pM and $(1.06 \pm 0.02) \times 10^7 M^{-1} sec^{-1}$. Data obtained at times less than 200 sec are not shown and thereafter only every second point is shown.

where P is the amount of p-nitroaniline formed at time t and v_s is defined by Eq. (3). The constants d and k' are defined by Eqs. (5) and (6)[58]:

$$d = \frac{K_{I'} + I_t + E_t - \text{sqrt}[(K_{I'} + I_t + E_t)^2 - 4E_tI_t]}{K_{I'} + I_t + E_t + \text{sqrt}[(K_{I'} + I_t + E_t)^2 - 4E_tI_t]} \qquad (5)$$

$$k' = k_1 \, \text{sqrt}[(K_{I'} + I_t + E_t)^2 - 4E_tI_t] \qquad (6)$$

These analyses yield values for $K_{I'}$ and k_1. The value for K_I can be calculated from that of $K_{I'}$ using Eq. (2) and the value of k_1 is approximately equal that determined from the analysis.[40,41]

Dodt and co-workers have used a slightly different approach to obtain values for K_I and k_1 for hirudin.[24,55] These workers have used a fluorogenic substrate and a higher ionic strength. The higher ionic strength increases the value of K_I and, thus, higher concentrations of hirudin are required to achieve the same degree of inhibition. The rate of formation of the complex is also slower; k_1 is smaller. In addition, lower concentrations of thrombin can be used because of the increased sensitivity of the assay.

The overall result of these different assay conditions is that slow-binding inhibition data can be obtained with concentrations of hirudin that are at least 10-fold greater than the thrombin concentration. Consequently, the data can be analyzed without making the tight-binding assumption.[60]

Structure–Function Relationship Studies with Mutants of Recombinant Hirudin

The effects of over 70 mutations on the stability of the thrombin–hirudin complex are presented in Table I. We have presented only the data for mutants produced by the group of Stone and Hofsteenge. Other groups have produced similar mutants but these data are not included because the reaction conditions used were slightly different. For the effects of these mutations, the reader is referred to the original publications.[24,36,55,62,63] As a result of the efforts of all these groups, together with those of the protein crystallographers working on the thrombin–hirudin complexes,[15–17] we have a good idea of the importance of various interactions to the formation of the thrombin–hirudin complex.

The results presented in Table I indicate the importance of contacts between thrombin and two regions of hirudin. The N-terminal five amino acids of hirudin are bound to the active site cleft of thrombin. Each of these residues makes a contribution to the stability of the complex; interactions with Val-1' and Tyr-3', however, seem to be most important. The C-terminal region of hirudin is bound to the anion-binding exosite of thrombin and both nonpolar and electrostatic interactions play a role in this interaction. Glu-57', Glu-58', and the sulfated tyrosine at position 63 appear to be the most important of the acidic residues in the C-terminal of hirudin, but all of the negatively charged residues in this region make some contribution to binding energy. This is due to their interaction with the positive electrostatic potential created by the anion-binding exosite.[64] The binding of the C-terminal region also has a nonpolar component, with important contacts being made by Phe-56', Pro-60', and Tyr-63'.

Hirudin-Based Peptide Derivatives

Two classes of hirudin-based, C-terminal tail peptides have been described.[22,23,65,66] Class I peptides derived from amino acids 45 to 65 of

[63] R. C. Winant, J. B. Lazar, and P. H. Johnson, *Biochemistry* **30**, 1271 (1991).

[64] A. Karshikov, W. Bode, A. Tulinsky, and S. R. Stone, *Protein Sci.* (in press).

[65] J. M. Maraganore, P. Bourdon, J. Jablonski, K. L. Ramachandran, and J. W. Fenton, II, *Biochemistry* **29**, 7095 (1990).

[66] J. DiMaio, B. Gibbs, D. Munn, J. Lefebvre, F. Ni, and Y. Konishi, *J. Biol. Chem.* **265**, 21698 (1990).

hirudin bind to the anion-binding exosite of thrombin with the consequence of inhibiting competitively thrombin cleavages of fibrinogen.[30] Because these peptides fail to interact with the catalytic site of thrombin, inhibition of thrombin hydrolysis of synthetic p-nitroanilide peptides is not observed. More recently, a new class (Class II) of hirudin-based peptides has been described that contain an N-terminal extension capable of binding to the catalytic site joined by a linker segment to structural determinants for binding to the anion-binding exosite. The bivalent interactions of these peptides, designated "hirulogs," with thrombin results in inhibition of thrombin hydrolysis of chromogenic substrates in addition to inhibition of the procoagulant functions of the enzyme. A related series of bivalent thrombin inhibitor peptides have also been derived using the peptides derived from the structure of the human platelet thrombin receptor.[32,67] X-Ray crystallographic structures of thrombin complexes with both classes of hirudin-based peptides have been determined,[29] serving to confirm observations established through kinetic and chemical modification studies.[22,26–28,30,65]

Hirudin-based peptides are prepared by conventional peptide synthesis procedures, and are purified by reversed-phase HPLC procedures. Although a variety of chromatographic conditions can be applied to purification of crude samples of synthetic hirudin-based peptides, the laboratory of one of the authors (J.M.M.) has employed routinely a C_8 HPLC column equilibrated in 0.1% trifluoroacetic acid and developed with a linear gradient of increasing acetonitrile concentration from 0 to 50% over 45 min at a flow rate of 1.0 ml/min. The effluent stream is monitored at 214 nm. Particularly as applied to the purification of Tyr-sulfated peptides (prepared as described below), simultaneous monitoring of absorbance at 214 and 280 nm is preferred. In this case, the reduced absorbance at 280 nm of the modified tyrosine peptide results in its facile identification.

Sulfation of Tyr-63' in hirudin accounts for a 10-fold increase in the affinity of the hirudin C-terminal tail with the anion-binding exosite of thrombin.[22,35] In X-ray crystallographic studies,[29] tyrosine sulfation was found to stabilize interactions with thrombin through a complex hydrogen-bonding network, including contributions from the backbone amide nitrogen of Ile-82 and the phenolic hydroxyl group of Tyr-76. Although recombinant forms of hirudin expressed in E. coli and S. cerevisiae lack a sulfated Tyr at position 63', peptide fragments of the hirudin C-terminal tail have been prepared with the analogous Tyr modified as the O-sulfate ester by chemical methods.[23,68] One approach for tyrosine sulfation of hirudin

[67] T.-K. Vu, V. I. Wheaton, D. T. Hung, I. Charo, and S. R. Coughlin, Nature (London) 353, 674 (1991).
[68] P. Bourdon, J. Jablonski, B. H. Chao, and J. M. Maraganore, FEBS Lett. 294, 163 (1991).

TABLE I
Effect of Mutations on Inhibition Constant (K_1) of
Recombinant Hirudin*

Mutation	K_1 (pM)	$\Delta\Delta G_b$ (kJ mol^{-1})[a]
rhir[a]	0.23	
Native hirudin[a]	0.020	−6.3
N-Terminal region		
V1L[b]	0.24	0
V1G[c]	2.7	6.2
V1S[c]	7.9	9.1
V1E[b]	295	18.4
V1K[c]	10.1	9.7
V1R[c]	0.32	0.8
V2L[b]	10.3	9.8
V2G[c]	141	16.5
V2S[c]	0.52	2.1
V2E[b]	248	18.0
V2K[c]	0.90	3.5
V2R[c]	0.027	−5.6
V1, 2L[b]	9.9	9.7
V1, 2G[b]	694	20.6
V1, 2S[b]	175	17.1
V1, 2E[b]	67,600	32.4
V1, 2K[b]	152	16.7
V1, 2R[c]	0.18	0.8
V1, 2I[b]	0.099	−2.2
V1, 2F[b]	0.24	0
Y3A[c]	31	12.6
T4A[c]	5.1	7.9
T4S[c]	1.3	4.5
T4W[d]	8.8	9.4
D5A[c]	1.4	4.7
D5E[c]	1.8	5.3
N-Terminal extensions		
M0[b]	5430	25.9
G0[b]	488	19.7
CH$_3$CO-rhirQQR[b]	9340	22.7[e]
CH$_3$NH$_2$+-rhirQQR[b]	24.3	7.4[e]
C-Terminal region		
Acidic residues		
D53A[f]	0.21	−0.3
D55N[f]	0.60	2.4
E57Q[f]	2.3	5.9
E58Q[f]	1.8	5.3
E61Q[a]	0.37	1.2
E62Q[a]	0.55	2.2
E57, 58Q[a]	2.4	6.0
E61, 62Q[f]	0.91	3.5
E57, 58, 62Q[a]	8.6	9.3
E57, 58, 61, 62Q[a]	14.1	10.6
Nonpolar residues		
F56Y[g]	0.19	−0.6
F56W[g]	0.41	1.4
F56A[g]	0.49	1.9
F56S[d]	18.3	11.2

TABLE I (*continued*)

Mutation	K_I (pM)	$\Delta\Delta G_b$ (kJ mol^{-1})[a]
F56V[g]	13.6	10.5
F56T[g]	26.3	12.2
F56L[g]		
F56I[g]	7.0	8.7
P60A[g]	2.1	5.7
P60G[g]	3.0	6.6
Y63F[g]	0.26	0.3
Y63A[g]	0.53	2.1
Y63V[g]	1.7	5.1
Y63L[g]	0.79	3.1
Y63E[g]	0.49	1.9
Core region		
K27Q[a]	0.30	0.6
K36Q[a]	0.22	0
K47Q[a]	2.0	5.5
K27Q/K36Q/K47R (rhirQQR)[b]	1.4	4.6
E17A[d]	0.82	3.2
E35Q[d]	0.76	3.0
S19D[d]	0.37	1.2
S19A[d]	0.21	−0.3
N20A[d]	0.23	0
P46A[d]	0.78	3.1
P48A[d]	0.52	2.1
H51Q[a]	0.24	0
H51R[d]	0.16	−0.9
N52M[h]	0.12	−1.7
Fragments of hirudin		
rhir(1-43)[i]	299,000	36.3
rhir(1-51)[h]	19,600	29.6
rhir(1-52)[j]	24,400	29.8

* All values for K_I were determined at 37° with a pH of 7.8 and an ionic strength of 0.125. The binding energy (ΔG_b) for the formation of the complex between human α-thrombin and rhir is −75 kJ mol^{-1}. The change in binding energy for a particular mutation ($\Delta\Delta G_b$) is calculated relative rhir [$\Delta\Delta G_b = \Delta G_b$(mutant) − ΔG_b(rhir)]. The nomenclature for the mutations shows the residue replaced (in the single-letter code) followed by its number and then the residue used for the replacement; e.g., V1G represents a mutant in which Val-1′ was replaced by glycine.

[a] P. J. Braun, S. Dennis, J. Hofsteenge, and S. R. Stone, *Biochemistry* **27**, 6517 (1988).

[b] A. Wallace, S. Dennis, J. Hofsteenge, and S. R. Stone, *Biochemistry* **28**, 10079 (1989).

[c] A. Betz, J. Hofsteenge, and S. R. Stone, *Biochemistry* (in press).

[d] A. Betz, P. Hopkins, and S. R. Stone, unpublished results.

[e] The $\Delta\Delta G_b$ values for these N-terminal extensions are calculated relative to hirQQR.

[f] A. Betz, J. Hofsteenge, and S. R. Stone, *Biochem. J.* **275**, 801 (1991).

[g] A. Betz, J. Hofsteenge, and S. R. Stone, *Biochemistry* **30**, 9848 (1991).

[h] S. Dennis, A. Wallace, J. Hofsteenge, and S. R. Stone, *Eur. J. Biochem.* **188**, 61 (1990).

[i] J.-Y. Chang, J.-M. Schlappi, and S. R. Stone, *FEBS Lett.* **260**, 209 (1990).

[j] M. P. Jackman, M. A. A. Parry, J. Hofsteenge, and S. R. Stone, unpublished results.

peptides has employed a modification of a method described by Nakahara et al.[69]

Reagents

Cold (2–4°) dimethylformamide, packed under N_2 under a molecular sieve from Pierce Chemical Co. (Rockford, IL)

Concentrated sulfuric acid

Dicyclohexylcarbodiimide (DCC)

The hirudin-based peptide is dissolved in cold dimethylformamide at a final concentration of 13 mM, containing 0.4 M DCC. While cooled in an ice bath, sulfuric acid is added to the peptide solution to a final concentration of 0.18 M. The reaction rapidly proceeds to completion and is terminated by addition of five volumes of deionized water. Following centrifugation or, for larger reactions, filtration over a glass sieve to remove the cyclohexylurea precipitant, the reaction mixture is applied directly to purification on a reversed-phase C_8 HPLC column equilibrated in 0.1% trifluoroacetic acid. The column is developed with a linear gradient of increasing acetonitrile concentration from 0 to 50% over 45 min at a flow rate of 1.0 ml/min in the same trifluoroacetic acid-containing solvent. The effluent stream is monitored at both 214 and 280 nm for identification of the Tyr-sulfated peptide. A typical HPLC chromatogram of crude, Tyr-sulfated, N-acetyl-Hir(53′-64′) (a peptide fragment designated "hirugen"), prepared at a 25-mg scale, is shown in Fig. 7. Complications in obtaining high yields of the Tyr-sulfated peptides arise most often as a consequence of excessively hydrated peptide samples. This can be avoided, however, by storage of peptide samples under desiccant and by performing the chemical modification in a glove box controlled for humidity.

Thrombin Inhibition by Synthetic Hirudin-Based Peptides

Class I hirudin-based peptide fragments bind to the anion-binding exosite of thrombin to inhibit competitively thrombin interactions with fibrinogen and other (but not all) macromolecular thrombin substrates or receptors.[30,67,70,71] As a consequence, approaches for measuring thrombin inhibition by hirudin (as described above) or by Class II peptides (as

[69] T. Nakahara, M. Waki, H. Uchimura, M. Hirano, J. S. Kim, T. Matsumoto, K. Nakamura, K. Ishibashi, H. Hirano, and A. Shiraishi, *Anal. Biochem.* **154,** 194 (1986).

[70] J. M. Maraganore and J. W. Fenton, II, *in* "Fibrinogen, Thrombosis, Coagulation, and Fibrinolysis" (C. Y. Liu and S. Chien, eds.), p. 177. Plenum, New York, 1991.

[71] S. R. Stone and J. M. Maraganore, *in* "Thrombin: Structure and Function" (L. Berliner, ed.). Plenum, New York, 1992.

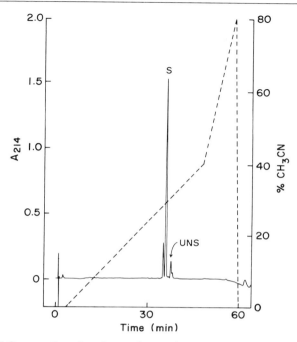

FIG. 7. HPLC separation of crude reaction products from tyrosine sulfation of hirugen [N-acetylhirudin(53-64)]. The reaction product from tyrosine sulfation of N-acetylhirudin (53-64) was applied to reversed-phase HPLC on an Aquapore RP-300 octasilyl column equilibrated in 0.1% TFA/water. An increasing linear gradient of acetonitrile concentration was used to develop the column and to achieve separation of the sulfated product (S) from unreacted peptide (UNS) and other unidentified reaction products.

described below) using chromogenic substrate assays cannot be employed. Inhibition of thrombin cleavages of fibrinogen by Class I peptides is measured by plasma coagulation assays,[23,72] including activated partial thromboplastin time (APTT), prothrombin time (PT), and thrombin time (TT); turbidometric assay for fibrin polymerization[22]; immunoassay for release of fibrinopeptide A[73]; and HPLC assay for generation of fibrinopeptides.[30,33] Interpretation of results from the first two assays listed above is complicated by recent observations from Kaminski and Mossesson[74] that thrombin binding to fibrin monomer I via the anion-binding exosite

[72] J. A. Jakubowski and J. M. Maraganore, *Blood* **75,** 399 (1990).
[73] J. I. Weitz, M. Hudoba, D. Massel, J. M. Maraganore, and J. Hirsh, *J. Clin. Invest.* **86,** 385 (1990).
[74] M. Kaminski, K. R. Siebenlist, and M. Mosesson, *J. Lab. Clin. Med.* **117,** 218 (1991).

accelerates the rate of fibrin polymerization. By measuring fibrin formation, plasma and turbidometric assays are thus poor methods for determining the independent action of Class I hirudin peptides toward thrombin cleavages in fibrinogen. Despite this shortcoming, plasma coagulation assays, namely APTT assays, are routinely used for measurement of thrombin inhibition by Class I hirudin-based peptides.

Reagents

Pooled, normal, citrated human plasma
Activated partial thromboplastin
0.025 M calcium chloride solution
Phosphate-buffered saline (PBS)

APTT assays are performed by a semiautomated method employing a commercially obtained fibrometer (Coag-A-Mate XC, Organon Technicon, Durham, NC). Plasma (100 μl) is added to the well of a 12-test cassette provided by the instrument's manufacturer. The hirudin-based peptide (0–50 μg/ml) is added in a 25-μl aliquot to the plasma sample. The plasma and sample are mixed and then placed in the fibrometer for equilibration to 37° and initiation of the coagulation reaction by addition of activated partial thromboplastin (100 μl) and recalcification with 0.025 M calcium chloride (100 μl). The APTT is recorded and values are expressed as percent control (test samples containing plasma and PBS in the absence of hirudin-based peptide). Provided in Fig. 8 is a dose–response for hirugen [N-acetyl-Hir(53'-64')] prolongation of APTT in human plasma.

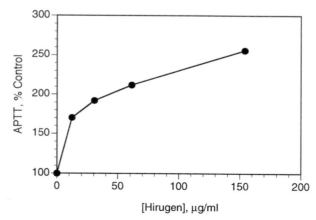

FIG. 8. Anticoagulant activity of hirugen in activated partial thromboplastin time (APTT) assays. The anticoagulant activity of hirugen was measured in APTT assays using citrated normal human plasma. Dose-dependent prolongation of APTT was observed.

The antithrombin actions of hirudin peptides are species specific. Potency toward thrombin from human and subhuman primate sources is maximal, but inhibition of thrombin from other species, including enzymes from canine and bovine sources, is reduced some 5- to 10-fold. Accordingly, although the anticoagulant activity of hirudin peptides can be measured toward nonhuman plasma sources, the use of human plasma is certainly preferred.

Unlike Class I hirudin peptides, which bind to the anion-binding exosite of thrombin, Class II peptides interact additionally with the catalytic site of thrombin. The peptide "hirulog-1" [D-Phe-Pro-Arg-Pro-(Gly)$_4$-Asn-Gly-Asp-Phe-Glu-Glu-Ile-Pro-Glu-Glu-Tyr-Leu] is characteristic of this group and was demonstrated to inhibit thrombin-catalyzed hydrolysis of a chromogenic substrate and to block modification by diisopropyl fluorophosphate of Ser-195 in thrombin.[65] The bivalent interactions of hirulog-1 with thrombin were confirmed in X-ray crystallographic studies.[29]

The kinetics for Class II hirudin-based peptide inhibition of thrombin are readily measured using chromogenic substrates essentially as described above for hirudin. Some important exceptions with respect to hirudin are (1) the absence of tight-binding inhibition and (2) the slow cleavage of some of these inhibitors by thrombin. Accordingly, the kinetics for Class II peptide inhibition of thrombin are described by different equations, for example, those described recently by Witting et al.[75] For routine analyses, the laboratory of one of the authors (J.M.M.) has employed a 96-well plate assay for measuring thrombin inhibition. This method is particularly well-suited to routine analysis of numerous peptide derivatives to define structure–function relationships for thrombin inhibition.

Reagents

Substrate stock: 125 μM stock solution of Spectrozyme TH (American Diagnostica, New York, NY), H-D-hexahydro-tyrosyl-L-alanyl-L-arginine p-nitroanilide, in a 50 mM sodium borate buffer, pH 8.4, 100 mM NaCl (assay buffer)

Enzyme stock: 10 nM human thrombin in 50 mM Tris-HCl buffer, pH 7.4, 100 mM NaCl, 0.1% (w/v) polyethylene glycol (PEG) 8000

Peptide stock: 0.5 to 50 μM peptide in assay buffer

In a 96-well Immunolon 1 microtiter plate (Dynatech, Chantilly, VA), 5 to 50 μl of substrate stock and 0 to 25 μl of peptide stock are mixed and then adjusted to a 200-μl volume with assay buffer. Final concentrations of Spectrozyme TH and peptide inhibitor thus obtained are 2.5 to 25 μM

[75] J. I. Witting, P. Bourdon, D. V. Brezniak, J. M. Maraganore, and J. W. Fenton, II, *Biochem. J.* **283,** 737 (1992).

and 0 to 100 nM, respectively. The microtiter plate is placed on a Thermomax Plate Reader (Molecular Devices, Menlo Park, CA), and the reaction is started by addition of 50 μl of enzyme stock. The reaction at 23° is monitored for absorbance at 405 nm for 1 min with measurements obtained every 6 sec . Initial velocities values (v) in mOD/min, determined using the first six data points, are transformed to units of moles product/second using the extinction coefficient for the p-nitroanilide product. Results for the uninhibited and inhibited reactions are plotted as 1/v versus 1/[substrate] for determination of K_i values, where

$$K_i = \frac{\text{Slope}_{\text{uninhibited}}[\text{Peptide Inhibitor}]}{\text{Slope}_{\text{inhibited}} - \text{Slope}_{\text{uninhibited}}} \tag{7}$$

Listed in Table II are Class II peptides tested by this method and K_i values for thrombin inhibition thus obtained. The results of these studies illustrate the ability to improve the potency of Class II peptides as compared with the prototype peptide in this family, hirulog-1. For example, substitution of the P_3 D-Phe residue with D-cyclohexylalanine results in a 20-fold increase in binding affinity. Similarly, substitutions at the C-terminal end of the peptides, such as tyrosine sulfation, can lead to substantial increases in potency.[68] Further, substitution of the P_1–P_1' dipeptide sequence with noncleavable isosteres has been shown to provide resistance to thrombin proteolysis.[76,77]

Hirudin Peptides as Probes of Thrombin Structure and Function

Because Class I peptides, such as the peptide hirugen, bind to the anion-binding exosite of thrombin, they serve as useful tools to evaluate thrombin activities dependent on exosite interactions. Further, prior to the availability of a structure for the hirudin–thrombin complex,[15–18] these peptides were employed as affinity reagents[27] or in chemical modification experiments[28] to identify amino acid residues within or neighboring the anion-binding exosite.

Studies with hirugen have identified numerous thrombin activities that are exosite dependent, including thrombin interactions with fibrinogen,[30] thrombomodulin,[78] and the cellular thrombin receptor.[79] Yet other func-

[76] T. Kline, C. Hammond, P. Bourdon, and J. M. Maraganore, *Biochem. Biophys. Res. Commun.* **177**, 1049 (1991).

[77] J. DiMaio, F. Ni, B. Gibbs, and Y. Konishi, *FEBS Lett.* **282**, 47 (1991).

[78] M. Tsiang, S. R. Lentz, W. A. Dittman, D. Wen, E. M. Scarpati, and J. E. Sadler, *Biochemistry* **29**, 10602 (1990).

[79] T.-K. Vu, D. T. Hung, V. I. Wheaton, and S. R. Coughlin, *Cell (Cambridge, Mass.)* **64**, 1057 (1991).

TABLE II

INHIBITION OF THROMBIN-CATALYZED HYDROLYSIS OF SYNTHETIC
p-NITROANILIDE SUBSTRATE BY CLASS II HIRUDIN-BASED PEPTIDES[a]

Designation	Sequence	K_i (nM)
Hirulog-1[b]	(D-F)PRPGGGGNGDFEEIPEEYL	1.4
Exosite binding segment modifications		
S-Hirulog-1[b]	(D-F)PRPGGGGNGDFEEIPEE(OSO$_3$Y)L	0.4
dI-Hirulog-1[b]	(D-F)PRPGGGGNGDFEEIPEE(I$_2$Y)L	0.3
Hirulog-β1[b]	(D-F)PRPGGGGDGDFEESLDDIDQ	1.1
Hirulog-β2[b]	(D-F)PRPGGGGDGDFEPIPEEYLQ	5.4
Hirulog-β3[b]	(D-F)PRPGGGGDGDYEPIPEEA(Cha)(D-E)	0.3
Catalytic site binding segment modifications		
Hirulog-B1[c]	(D-Npa)PRPGGGGNGDFEEIPEEYL	4.3
Hirulog-B2[c]	(D-Cha)PRPGGGGNGDFEEIPEEYL	0.1
Hirulog-A1[c]	(D-F)PRGGGGGGGNGDFEEIPEEYL	66.0
Hirulog-A2[c]	(D-F)PR(Src)GGGGGNGDFEEIPEEYL	13.5
Hirulog-A3[c]	(D-F)PR(HPro)GGGGNGDFEEIPEEYL	2.7
Hirulog-2a[d]	(D-F)P(βHar)GGGGGNGDFEEIPEEYL	7.4
Hirulog-2b[d]	(D-F)P(βHar)PGGGGNGDFEEIPEEYL	4.6
Hirulog-2c[d]	(D-F)P(βHar)VGGGGNGDFEEIPEEYL	205
Hirulog-3a[d]	(D-F)PR[CH$_2$NH]GGGGGNGDFEEIPEEYL	20.0
Hirulog-1[c]	acGDFLAEGGGVRPGGGGNGDFEEIPEEYL	3.2
Hirulog-2[c]	(GBA)GGGGGNGDFEEIPEEYL	7000
Linker segment modifications		
Hirulog-L1[c]	(D-F)(PR[TEG-succ]NGDFEEIPEEYL	237
Hirulog-L2[c]	R[TEG-succ]NGDFEEIPEEYL	840

[a] K_i values were determined from initial velocity data obtained at 23°, I 0.15 M, pH 8.4, in a sodium borate buffer. Substrate was Spectrozyme TH (H-D-hexahydrotyrosyl-L-alanyl-L-arginine-p-nitroanilide). (OSO$_3$-Y), O-Sulfatotyrosyl; (I$_2$-Y), diiodotyrosyl; Cha, cyclohexylalanyl; (D-Npa), D-naphthylalanyl; (Src), sarcosine; (HPro), hydroxyproline; (βHar), β-homoarginyl; (GBA), guanidinobenzoyl; [TEG-succ], tetraethylene glycol succinyl.

[b] P. Bourdon, J. Jablonski, B. H. Chao, and J. M. Maraganore, FEBS Lett. **294,** 163 (1991).

[c] C. Hammond, P. Bourdon, J. Jablonski, T. Kline, J. W. Fenton, II, and J. M. Maraganore, unpublished results.

[d] T. Kline, C. Hammond, P. Bourdon, and J. M. Maraganore, Biochem. Biophys. Res. Commun. **177,** 1049 (1991).

tions of thrombin appear to be exosite independent, namely thrombin interactions with factors V and VIII[80] and with heparin and the heparin: antithrombin III complex.[30] These latter findings provide evidence for other thrombin interaction domains, a conclusion supported by the chemi-

[80] F. A. Ofosu, J. W. Fenton, II, J. M. Maraganore, M. A. Blachjman, X. Yang, L. Smith, N. Anvari, M. R. Buchanan, and J. Hirsh, Biochem. J. **283,** 893 (1992).

cal modification studies of Church and co-workers,[81] who identified discontinuous protection of lysine residues from derivatization by pyridoxyl 5'-phosphate in the presence of heparin or fibrinogen.

Conclusions

Hirudin and hirudin-based peptides have served as exemplary tools for studies on thrombin structure–function relationships and on thrombin biological activities. Binding of hirudin and peptide fragments with thrombin serves generally as a model for protein–protein and, in particular, protease–inhibitor interactions. Certainly, thrombin–hirudin interactions represent a unique model for protease inhibition. With the determination of three-dimensional structures of thrombin complexes with these reagents,[16,17,29] studies on hirudin and peptide fragments will continue to provide important insights on thrombin, its interactions with physiologic substrates, and blood coagulation mechanisms.

[81] F. C. Church, C. W. Pratt, C. M. Noyes, T. Kalayanamit, G. B. Sherrill, R. B. Tobin, and J. B. Meade, *J. Biol. Chem.* **264,** 18419 (1989).

[20] *Limulus* Clotting Factor C: Lipopolysaccharide-Sensitive Serine Protease Zymogen

By Tatsushi Muta, Fuminori Tokunaga, Takanori Nakamura, Takashi Morita, and Sadaaki Iwanaga

Introduction

The bacterial endotoxin lipopolysaccharide (LPS)-induced clotting phenomenon of the hemocyte lysate from horseshoe crab is thought to be a defense mechanism that serves to immobilize invading gram-negative bacteria.[1-4] Because of this specific property and its extreme sensitivity to endotoxins, *Limulus* hemolymph has been applied in the last decade

[1] J. Levin and F. B. Bang, *Bull. Johns Hopkins Hosp.* **115,** 265 (1964).
[2] E. H. Muller, J. Levin, and R. Holme, *J. Cell. Physiol.* **816,** 533 (1975).
[3] S. Iwanaga, T. Morita, T. Miyata, T. Nakamura, and J. Aketagawa, *J. Protein Chem.* **5,** 255 (1986).
[4] T. Nakamura, T. Morita, and S. Iwanaga, *Eur. J. Biochem.* **154,** 511 (1986).

METHODS IN ENZYMOLOGY, VOL. 223

to estimate or quantify nanogram quantities of endotoxins in human body fluids and pharmaceutical products.[5] We have studied the *Limulus* clotting system to define the overall molecular events in the hemolymph clotting and to compare these reactions with those of mammalian blood coagulation. Studies[6-9] have shown that there are sequential activations of intracellular serine protease zymogens similar to those of the mammalian coagulation system. Among these protease zymogens, factor C is an endotoxin-sensitive, intracellular serine protease zymogen that initiates the coagulation cascade system in the *Limulus* hemolymph.[4,10,11]

The present article describes the purification and biochemical properties of *Limulus* clotting factor C. The cDNA and entire amino acid sequences of factor C have been published elsewhere.[12]

Assay Method

The amidase activity of factor C after activation in the presence of LPS is measured using Boc-Val-Pro-Arg-*p*-nitroanilide (pNA) as the substrate. This chromogenic substrate is commercially available from the Peptide Institute (Minoh, Osaka, Japan). A reaction mixture containing 10–20 μl of sample, 20 μl of 1 M Tris-HCl (pH 8.0) containing 50 mM MgCl$_2$, and 4 μl of LPS (10 μg/ml), in a total volume of 200 μl, is preincubated at 37° for 10 min. Then, 50 μl of 2 mM Boc-Val-Pro-Arg-pNA is added, and this mixture is incubated for another 7 min. Finally, 0.8 ml of 0.6 M acetic acid is added to terminate the reaction and the absorbance at 405 nm is read in a Hitachi 220A spectrophotometer. LPS prepared from *Escherichia coli* (*E. coli*) 0111 : B4 is a product of List Biological Laboratories, Inc.[5] One unit of the enzyme activity is defined as 1 μmol of *p*-nitroaniline liberated per minute. The amidase activity of factor $\overline{\text{C}}$ (active form of factor C) is measured as that described for factor C, with the exception that no LPS is added.

[5] S. Tanaka and S. Iwanaga, this volume [23].
[6] T. Nakamura, T. Horiuchi, T. Morita, and S. Iwanaga, *J. Biochem.* (*Tokyo*) **99**, 847 (1986).
[7] T. Nakamura, T. Morita, and S. Iwanaga, *J. Biochem.* (*Tokyo*) **97**, 1561 (1985).
[8] T. Miyata, M. Hiranaga, M. Umezu, and S. Iwanaga, *J. Biol. Chem.* **259**, 8924 (1984).
[9] T. Morita, S. Tanaka, T. Nakamura, and S. Iwanaga, *FEBS Lett.* **129**, 318 (1981).
[10] T. Nakamura, F. Tokunaga, T. Morita, S. Iwanaga, S. Kusumoto, T. Shiba, T. Kobayashi, and K. Inoue, *Eur. J. Biochem.* **176**, 89 (1988).
[11] T. Nakamura, F. Tokunaga, T. Morita, and S. Iwanaga, *J. Biochem.* (*Tokyo*) **103**, 370 (1988).
[12] T. Muta, T. Miyata, Y. Misumi, F. Tokunaga, T. Nakamura, Y. Toh, Y. Ikehara, and S. Iwanaga, *J. Biol. Chem.* **266**, 6554 (1991).

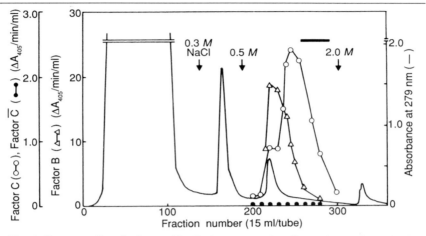

FIG. 1. Dextran sulfate-Sepharose CL-6B chromatography of horseshoe crab (*Tachypleus tridentatus*) hemocyte lysate.[12] The solid bar indicates fractions that were pooled and concentrated. Factor C (○), factor $\overline{\text{C}}$ (●), and factor B[6] (△) activities were assayed by using Boc-Leu-Gly-Arg-pNA as a substrate.

Purification Procedures

Materials

Horseshoe crabs (*Tachypleus tridentatus*), collected in Fukuoka prefecture, are bled by cardiac puncture in the presence of caffeine (10 mM).[4] The hemocytes are sedimented at 4000 rpm for 10 min at 4°, and the lysate is prepared according to the method previously described.[11] Sepharose CL-6B and CM-Sepharose CL-6B obtained are from Pharmacia Fine Chemicals (Uppsala, Sweden).

Isolation of Zymogen Factor C from Hemocyte Lysate

All the following procedures are performed under sterile conditions. The lysate (1150 ml) is applied to a dextran sulfate-Sepharose CL-6B column (5.0 × 23.5 cm), and the clotting factors are eluted as shown in Fig. 1. The detailed procedures concerning the *Limulus* proclotting enzyme have been described.[13] The factor C-containing fractions (solid bar, Fig. 1) are collected and concentrated by Diaflo ultrafiltration. The concentrated fraction is applied to a Sepharose CL-6B column (4.0 × 127 cm), equilibrated with 0.02 M sodium acetate (pH 5.0) containing 0.1 M NaCl. Elution is carried out at a flow rate of 60 ml/hr, and the fractions are

[13] T. Muta, T. Nakamura, R. Hashimoto, and S. Iwanaga, this volume [22].

TABLE I
PURIFICATION PROCEDURE FOR *Tachypleus* FACTOR C

Step	Volume (ml)	Total protein (mg)	Total activity (units)	Specific activity (units/mg)	Purification (−fold)	Yield (%)
Hemocyte lysate	1150	6095	—	—	—	—
Dextran sulfate-Sepharose CL-6B	525	98.4	767	7.8	1	100
Sepharose CL-6B	215	25.4	572	22.5	2.9	74.6
CM-Sepharose CL-6B	215	15.5	443	28.6	3.7	57.8

collected, 15.4 ml per tube, at 4°. Factor C activity appears in the first protein peak. Fractions 69–82 are collected for further purification. The pooled factor C fraction is applied to a CM-Sepharose CL-6B column (2.0 × 25.5 cm), equilibrated with 0.02 M sodium acetate–0.1 M NaCl (pH 5.0). Factor C is eluted with a linear NaCl gradient (0.1–0.35 M) in 0.02 M sodium acetate (pH 5.0) (500 ml/chamber) at a flow rate of 72 ml/hr. The fractions are collected, 15.4 ml per tube. One major protein peak appears and factor C activity coincides with this peak.

A summary of the procedures for purification of factor C is shown in Table I. The yield of purified factor C is 15.5 mg. The procedures for factor C are repeated three times using 1.2 liters of hemocyte lysate and reproducible results are obtained. The mean yield is 20.7 mg from 173 g of hemocytes (wet weight). Moreover, the same methods as described here are applicable to purification of the zymogen factor C from the hemocyte lysate of the American horseshoe crab, *Limulus polyphemus*.[14]

The purity of the final preparation is examined by PAGE in the presence and absence of SDS. On SDS–PAGE the purified zymogen gives a single band before reduction and two protein bands after reduction by 2-mercaptoethanol (Fig. 2A), suggesting that zymogen factor C consists of two polypeptide chains connected by a disulfide linkage(s). As shown in Fig. 2B, purified factor C gives a single protein band on disc PAGE at pH 4.3, and factor C activity is observed at the same position as the protein band stained with Coomassie brilliant blue. The zymogen factor C preparation does not exhibit any amidase activity toward either Boc-Leu-Gly-Arg-pNA or Boc-Val-Pro-Arg-pNA. Furthermore, no apparent molecular change in factor C is observed on SDS–PAGE following incubation at 37° for 3 hr under LPS-free conditions, indicating that the zymogen factor C

[14] F. Tokunaga, H. Nakajima, and S. Iwanaga, *J. Biochem.* (*Tokyo*) **109**, 150 (1991).

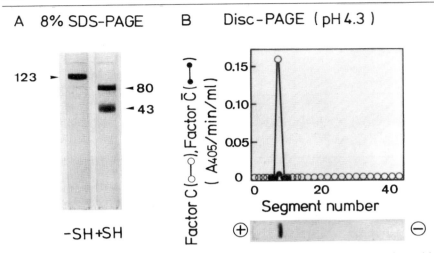

FIG. 2. PAGE of purified factor C.[12] (A) SDS–PAGE (13-μg sample) was performed in the presence (right lane) or absence (left lane) of 2-mercaptoethanol. (B) PAGE (7.5% gel) was performed at pH 4.3. One of the two gels was sliced into 1.5-mm segments and the segments were soaked in 0.4 ml of 0.5 M Tris-HCl buffer, pH 8.0, containing 0.05% bovine serum albumin, and kept at 4° for 24 hr. Aliquots of the extracts (50 μl) were assayed for factor C (○) and factor $\overline{C}$ (●) by using Boc-Val-Pro-Arg-pNA as a substrate. The other gel was stained with Coomassie brilliant blue R-250. Arrows show $M_r \times 10^{-3}$.

preparation does not contain factor $\overline{C}$ (the active form) and is not contaminated with LPS.

Preparation and Purification of Factor $\overline{C}$ from Hemocyte Lysate

The lysate (624 ml) prepared from 58 g (wet weight) of hemocytes is applied to a dextran sulfate-Sepharose CL-6B column (5.0 × 23.5 cm). The elution of factor C is performed by the procedure described in Fig. 1. The pooled and concentrated factor C sample is applied to a Sepharose CL-6B column (3.5 × 122 cm), equilibrated with 0.05 M Tris-HCl (pH 8.0) containing 0.2 M NaCl. Elution is carried out with the same buffer at a flow rate of 55 ml/hr. The fractions containing factor C activity are collected and NaCl is added to give a final concentration of 0.5 M. This sample is applied to a benzamidine-CH-Sepharose CL-6B column (1.5 × 14 cm), equilibrated with 0.02 M Tris-HCl–0.5 M NaCl (pH 8.0). The column is then washed with 100 ml of the equilibrating buffer. The zymogen factor C is eluted in the unadsorbed fraction of the column under these conditions. To the factor C pool, MgCl$_2$ and $E.$ $coli$ 0111 : B4 LPS are added to give a final concentration of 5 mM and 2 μg/ml, and the mixture is incubated at 37° for 4 hr to activate the zymogen factor C. This reaction mixture is applied to the same benzamidine-CH-Sepharose CL-6B column (1.5 × 14 cm), equilibrated with 0.02 M Tris-HCl–0.5 M NaCl (pH 8.0). The column is initially washed with the equilibrating buffer (100 ml) and subsequently with 0.02 M Tris-HCl–1.0 M NaCl (pH 8.0) (100 ml). The active factor $\overline{C}$ is then eluted from the column with 0.02 M Tris-HCl–1.0 M NaCl (pH 8.0) containing 1.0 M guanidine hydrochloride. Factor $\overline{C}$-containing fractions are collected and dialyzed against 0.02 M Tris-HCl–0.1 M NaCl (pH 8.0).

Properties

Physicochemical Properties

Limulus factor C (M_r of 123,000) glycoprotein consists of a heavy chain (M_r 80,000) and a light chain (M_r 43,000).[4] Factor C is converted to an activated form (M_r 123,000; designated factor $\overline{C}$) in the presence of LPS.[4] On activation, a single cleavage of the Phe-Ile bond in the light chain occurs, resulting in the accumulation of two new fragments of M_r 34,000 and 8500 on reduced SDS–PAGE.[15] A diisopropyl fluorophosphate (DFP)-sensitive active site is localized in the 34-kDa light chain of factor C.[4] We have determined the entire amino acid sequence of factor C using recombinant DNA technique.[16] The zymogen consists of 994 amino acid residues with a calculated molecular mass of 109,648 Da.[16] Most interestingly, factor C has five repeating units ("Sushi domain," or short consensus repeat) of about 60 amino acid residues each, which are found in many proteins participating in the mammalian complement system (Figs. 3 and 4). In addition to a typical serine protease domain in the carboxyl-terminal portion, characteristic segments with epidermal growth-factor-like, lectin-like, cysteine-rich, and proline-rich domains are also found, revealing a unique mosaic protein structure.[16] The serine protease domain is most analogous to human thrombin.[16] Furthermore, we identified a transcript, possibly derived by alternative splicing of factor C mRNA, which encodes a protein sharing the amino-terminal portion of factor C (Fig. 3).[16] The reconstitution experiments, using factor C, factor B,[6] proclotting enzyme,[7,16] and LPS, demonstrate that all of these proteins are essential for the endotoxin-mediated coagulation system.[4,6]

[15] F. Tokunaga, T. Miyata, T. Nakamura, T. Morita, K. Kuma, T. Miyata, and S. Iwanaga, *Eur. J. Biochem.* **167**, 405 (1987).
[16] T. Muta, R. Hashimoto, T. Miyata, H. Nishimura, Y. Toh, and S. Iwanaga, *J. Biol. Chem.* **265**, 22426 (1990).

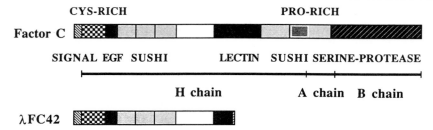

FIG. 3. The domain structure of factor C.[12] Factor C has a mosaic structure of several domains, including a signal peptide (-25 to -1), one Cys-rich region (1–76), one EGF-like domain (77–111), five Sushi domains (117–170, 174–229, 235–296, 551–609, and 615–723), one lectinlike domain (411–543), one Pro-rich region (618–665, located within the fifth Sushi domain), and one serine protease domain (738–994). The H, A, and B chains correspond to each domain as illustrated. A tentative alternative splicing product encoded by λFC42 contains the signal peptide, the Cys-rich region, the EGF-like domain, three Sushi domains, and the first half of the lectinlike domain.[12]

The optimum pH for activation of the zymogen factor C by LPS and lipid A[10] is 7.2 and the resulting active enzyme shows an optimum pH 8.5 when Boc-Val-Pro-Arg-pNA is used as the substrate.[4] The zymogen factor C and the active enzyme factor $\overline{C}$ are both stable at 0° for several weeks and may be stored at $-80°$ indefinitely.

Specific Enzymatic Properties

Zymogen factor C is rapidly activated not only by bacterial endotoxins but also by acylated ($\beta 1 \rightarrow 6$)-D-glucosamine disaccharide bisphosphate

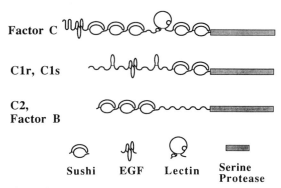

FIG. 4. Comparison of the structures of serine proteases containing Sushi domains. The domain structures of serine proteases with Sushi domains including *Limulus* factor C, C1r, C1s, C2, and complement factor B are illustrated. Reproduced from Muta *et al.*[12]

(synthetic *E. coli*-type lipid A) and the corresponding 4′-monophosphate analogs, as shown in Tables II and III.[10] However, the corresponding nonphosphorylated lipid A does not activate factor C, indicating that a phosphate ester group linked with the ($\beta 1 \rightarrow 6$)-D-glucosamine disaccharide backbone is required for the zymogen activation. During these studies we also found that the zymogen factor C is significantly activated by acidic phospholipids, such as phosphatidylinositol, phosphatidylglycerol, and cardiolipin, but not at all by neutral phospholipids (Table IV).[10] The rate of this activation, however, is affected markedly by the ionic strength of the reaction mixture, although such an effect is not observed in the lipid A-mediated activation of factor C. A variety of negatively charged surfaces, such as sulfatide, dextran sulfate, and ellagic acid, that are known

TABLE II
STRUCTURES OF CHEMICALLY SYNTHESIZED LIPID A ANALOGS

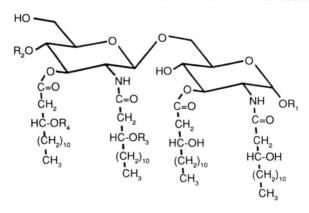

Compound number	Substituent			
	R_1	R_2	R_3	R_4
403	—H	—H		
404	—H	—PO(OH)$_2$		
405	—PO(OH)$_2$	—H	—H	—H
406	—PO(OH)$_2$	—PO(OH)$_2$		
503	—H	—H		
504	—H	—PO(OH)$_2$		
505	—PO(OH)$_2$	—H	—CO(CH$_2$)$_{10}$CH$_3$	—CO(CH$_2$)$_{12}$CH$_3$
506	—PO(OH)$_2$	—PO(OH)$_2$		

TABLE III
ACTIVATION OF FACTOR C BY LIPID A ANALOGS[a]

Compound number	Activation	Concentration of lipid A to activate 50% of factor C (ED_{50}) (nM)
403	−	
404	+	16
405	+	143
406	+	6
503	−	
504	+	66
505	+	220
506	+	55

[a] Synthetic lipid A analogs were dissolved in 0.025% triethyla-mine (1 mg/ml), sonicated at room temperature for 10 min, and diluted with 0.05 M Tris-HCl (pH 8.0) containing 0.5 mg/ml human serum albumin. Compounds 403 and 503 were not soluble in the above buffer, thus their sonicated emulsions were used alone for assay. The activation of factor C (7.7 nM) at various concentrations of lipid A analogs was examined, and the mini-mal concentration of each lipid A analog for 50% activation of factor C (ED_{50}) was estimated.

TABLE IV
ACTIVATION OF FACTOR C BY PHOSPHOLIPIDS[a]

Phospholipid	Activation	Concentration of phospholipid to activate 50% of factor C (ED_{50}) (nM)
Phosphatidic acid	+	>10,000
Phosphatidylglycerol	+	22
Phosphatidylserine	+	45
Phosphatidylinositol	+	8
Phosphatidylethanolamine	−	
Phosphatidylcholine	−	
Diphosphatidylglycerol (cardiolipin)	+	24

[a] Phospholipids were dissolved in water (1 mg/ml), sonicated at room temperature for 10 min, and diluted with 0.05 M Tris-HCl (pH 8.0) containing 0.5 mg/ml human serum albumin. Then the activation of factor C (7.7 nM) at various concentrations of phospholipids was examined and the minimal concentration of each phospholipid for 50% activation of factor C was estimated, as shown in Table III.

as typical initiators of activation of the mammalian intrinsic clotting system have no effect on the zymogen factor C activation. This suggests that lipid A is the most effective trigger to initiate the activation of the horseshoe crab hemolymph clotting system.[10]

In addition to the nonenzymatic activators described above, zymogen factor C is rapidly activated by α-chymotrypsin and rat mast cell chymase,[15] but not by trypsin, which activates many serine protease zymogens associated with mammalian coagulation, fibrinolysis, and complement systems. Thus, *Limulus* factor C seems to be a novel type of serine protease zymogen susceptible to α-chymotrypsin.[14]

The amidase activity of factor C is strongly inhibited by antithrombin III, but not by α_2-antiplasmin.[6] Soybean trypsin inhibitor and hirudin have no inhibitory effect on factor C activity.[4] DFP and D-Phe-Pro-Arg-chloromethyl ketone have a potent inhibitory effect.[4] Benzamidine and leupeptin also have inhibitory activity at high concentrations (5 and 0.23 mM, respectively), whereas p-chloromercuribenzoate has no apparent inhibitory effect.[4]

Subcellular Localization of Factor C

Using antifactor $\overline{C}$ polyclonal antibody, the subcellular localization of factor C has been analyzed by an immunohistochemical technique. The hemocytes are collected into a sterilized tube and immediately fixed with 4% paraformaldehyde (v/v) and 1.2% glutaraldehyde (v/v). The sections, mounted on a nickel grid, are incubated with factor $\overline{C}$ IgG after blocking with 1% bovine serum albumin. The sections are washed and then incubated with goat antirabbit IgG gold colloidal particles (20 nm) (E-Y Labs, Inc., San Mateo, CA). After extensive washing, the sections are doubly stained with 2% uranyl acetate and lead tartrate and examined under an electron microscope (Hitachi H-500). Many gold particle conjugates are observed, mainly in large granules in the cell, indicating that factor C antigen is localized in the large granules[12] and exocytosed by stimulation with LPS.[17]

---

[17] Y. Toh, A. Mizutani, F. Tokunaga, T. Muta, and S. Iwanaga, *Cell Tissue Res.* **266,** 137 (1991).

[21] *Limulus* Clotting Factor B

By Takanori Nakamura, Tatsushi Muta, Toshio Oda, Takashi Morita, and Sadaaki Iwanaga

Introduction

Horseshoe crab hemocytes are highly sensitive to gram-negative bacterial endotoxins, which are lipopolysaccharides (LPS); exposure to LPS results in activation of intracellular coagulation systems.[1-3] In the Japanese horseshoe crab, *Tachypleus tridentatus,* the coagulation system consists of at least three intracellular serine protease zymogens, named proclotting enzyme,[4] factor B,[5] and factor C,[6] in addition to the fibrinogen-like substance, coagulogen.[7-10] These zymogens and coagulogen constitute the cascade pathway responsible for the LPS-mediated activation,[11] as shown in Fig. 1. In the presence of LPS, a LPS-sensitive serine protease zymogen, factor C, is autocatalytically activated.[6] The active factor $\overline{C}$ then activates zymogen factor B to factor $\overline{B}$, which subsequently activates proclotting enzyme to clotting enzyme. The resulting clotting enzyme converts soluble coagulogen to an insoluble coagulin gel. Like the mammalian coagulation and complement systems, this cascade reaction is propagated by limited proteolysis.[4-6]

This article describes the purification and properties of *Limulus* clotting factor B. Methods for other clotting factors are described in [20] and [22] of this volume.

[1] J. Levin and F. B. Bang, *Bull. Johns Hopkins Hosp.* **115**, 265 (1964).

[2] E. H. Muller, J. Levin, and R. Holme, *J. Cell. Physiol.* **816**, 533 (1975).

[3] S. Iwanaga, T. Morita, T. Miyata, T. Nakamura, and J. Aketagawa, *J. Protein Chem.* **5**, 255 (1986).

[4] T. Nakamura, T. Morita, and S. Iwanaga, *J. Biochem.* (*Tokyo*) **97**, 1561 (1985).

[5] T. Nakamura, T. Horiuchi, T. Morita, and S. Iwanaga, *J. Biochem.* (*Tokyo*) **99**, 847 (1986).

[6] T. Nakamura, T. Morita, and S. Iwanaga, *Eur. J. Biochem.* **154**, 511 (1986).

[7] T. Miyata, M. Hiranaga, M. Umezu, and S. Iwanaga, *J. Biol. Chem.* **259**, 8924 (1984).

[8] T. Takagi, Y. Hokama, T. Miyata, and S. Iwanaga, *J. Biochem.* (*Tokyo*) **95**, 1445 (1984).

[9] T. Miyata, K. Usui, and S. Iwanaga, *J. Biochem.* (*Tokyo*) **95**, 1793 (1984).

[10] S. Srimal, T. Miyata, S. Kawabata, T. Miyata, and S. Iwanaga, *J. Biochem.* (*Tokyo*) **98**, 305 (1985).

[11] T. Morita, T. Nakamura, T. Miyata, and S. Iwanaga, *Prog. Clin. Biol. Res.* **189**, 53 (1985).

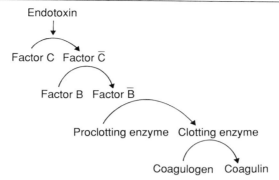

FIG. 1. A mechanism for the intracellular coagulation cascade found in the *Limulus* hemocytes.

Assay Methods

The methods for preparations of *Limulus* proclotting enzyme and factor $\overline{C}$ used in the assay of factor $\overline{B}$ activity have been described in [20] and [22] of this volume. The zymogen factor B activity is expressed as the amidase activity of the clotting enzyme generated from proclotting enzyme by activated factor $\overline{B}$. A 200-μ aliquot of reaction mixture containing sample, 0.1 M Tris-HCl (pH 8.0), human serum albumin (0.5 mg/ml), and factor $\overline{C}$ (1.9 μg/ml) is incubated at 37° for 10 min to activate the zymogen factor B. Then, 20 μl of 5 mM Boc-Leu-Gly-Arg-pNA and 30 μl of partially purified proclotting enzyme (960 μg/ml) are added and the mixture is incubated for an additional 4 min. Finally, 0.8 ml of 0.6 M acetic acid is added to terminate the reaction and the absorbance at 405 nm is read in a Hitachi 220A spectrophotometer. Factor $\overline{B}$ assay is the same as that of the zymogen factor B assay, except for no treatment with factor $\overline{C}$. One unit of the enzyme activity is defined as the amount of 1 μmol of *p*-nitroaniline liberated per min.

Purification Procedures

Materials

Horseshoe crabs (*T. tridentatus*) are collected in Imatsu, Fukuoka prefecture, and the hemocyte lysate is prepared by the method described previously.[6] Sepharose CL-6B, CH-Sepharose 4B, and Sephacryl S-200 are purchased from Pharmacia Fine Chemicals (Uppsala, Sweden). SP-5PW is obtained from TOSOH Manufacturing Co., Ltd. (1-11-39, Akasaka, Minato-ku, Tokyo-107, Japan). Fluorogenic and chromogenic peptide substrates are from Protein Research Foundation (Minoh, Osaka, Japan).

Methods

Purification of Zymogen Factor B from Hemocyte Lysate. All procedures for preparation of factor B except for the final two steps are performed under sterile conditions. The lysate (1000 ml), prepared from 135 g (wet weight) of *T. tridentatus* hemocytes, is first fractionated on a dextran sulfate-Sepharose CL-6B column (5 × 20 cm), according to the method described in [22] of this volume. Fraction B-containing factor B (Fig. 1 in Ref. 6) is collected and concentrated by Diaflo ultrafiltration. The concentrated fraction (43 ml) is applied to a Sepharose CL-6B column (4 × 127 cm) equilibrated with 0.02 M Tris-HCl–0.2 M NaCl (pH 8.0) and is eluted with the same buffer. Factor B-containing fractions are collected and diluted to give a concentration of 0.15 M NaCl. The diluted solution (330 ml) is then applied to a benzamidine-CH-Sepharose 4B column (2 × 14.5 cm) equilibrated with 0.02 M Tris-HCl–0.15 M NaCl (pH 8.0). Elution is performed in stepwise fashion with 0.02 M Tris-HCl–0.5 M NaCl (pH 8.0). The breakthrough fractions containing factor B activity are collected and concentrated by Diaflo ultrafiltration (Millipore, Ltd., MS.

The concentrated fraction (9 ml) is applied to a Sephacryl S-200 column (3.2 × 143 cm) equilibrated with 0.02 M Tris-HCl–1 M NaCl (pH 8.0) and is eluted with the same buffer. The factor B-containing fractions are collected and dialyzed overnight against 0.05 M sodium phosphate buffer (pH 5.5). The dialyzed solution (25 ml) is finally applied to an SP-5PW (0.6 × 5 cm) column (TOSOH Manufacturing Co.) equilibrated with 0.05 M sodium phosphate buffer (pH 5.5). After washing of the column with the same buffer, elution is performed with a linear gradient on NaCl (0–0.5 M) in the equilibration buffer, by using a Pharmacia fast protein liquid chromatography (FPLC) system. Factor B-containing fractions are collected and concentrated by Diaflo ultrafiltration. The yield of purified factor B is 1.1 mg from 135 g (wet weight) of the hemocytes. All the steps mentioned above were repeated three times using 500–1000 ml of the lysate and the results were confirmed to be reproducible. The final preparation does not show any amidase activity toward Boc-Met-Thr-Arg-MCA, which was found to be a good substrate for factor $\overline{\text{B}}$. This result indicates that no activated form of factor $\overline{\text{B}}$ coexists in the preparation. Moreover, the preparation was not contaminated by factor C,[6] factor G,[12] or proclotting enzyme.[4] On exposure of the purified preparation to LPS (*Escherichia*

---

[12] T. Morita, S. Tanaka, T. Nakamura, and S. Iwanaga, *FEBS Lett.* **129**, 318 (1981).

coli 0111-B4), it does not show any amidase activity due to the factor $\overline{B}$, indicating that the zymogen is essentially insensitive to LPS.[13]

Purification of Active Factor $\overline{B}$ from Hemocyte Lysate. The lysate (985 ml), prepared from 156 g (wet weight) of hemocytes, is applied to a dextran sulfate-Sepharose CL-6B column (5 × 23 cm). Elution of factor B is performed as described above. The pooled sample (150 ml, total A_{280} 459) obtained from the column is applied to a benzamidime-CH-Sepharose 4B column (2 × 9 cm) equilibrated with 0.02 M Tris-HCl–0.5 M NaCl (pH 8.0). The breakthrough fractions containing factor B activity are collected and concentrated by Diaflo ultrafiltration. The concentrated sample (21 ml, total A_{280} 57.8) is applied to a Sephadex G-100 column (3 × 126 cm) equilibrated with 0.02 M Tris-HCl–0.2 M NaCl (pH 8.0). Factor B-containing fractions are collected. The pooled fraction (93 ml, total A_{280} 13.5) is applied again to a dextran sulfate-Sepharose CL-6B column (1 × 16.5 cm) equilibrated with 0.02 M Tris-HCl–0.2 M NaCl (pH 8.0). Elution is performed with a linear gradient of NaCl (0.2–0.6 M) in the same buffer. Factor B-containing fractions are collected and concentrated by Diaflo ultrafiltration. The concentrated fraction (4.1 ml, total A_{280} 1.0) is incubated with 50 μl of factor $\overline{C}$ (224 μg/ml) at 37° for 60 min to activate fully the zymogen factor B. The reaction mixture after activation is applied to a benzamidine-CH-Sepharose CL-6B column (1.5 × 14 cm) equilibrated with 0.02 M Tris-HCl–0.2 M NaCl (pH 8.0). Elution is performed with a linear gradient of NaCl (0.2–0.6 M) in the same buffer. The active factor $\overline{B}$-containing fractions are collected and concentrated by Diaflo ultrafiltration.

Properties

Stability

The purified factor B is stable at pH 5.0–9.0 for 24 hr at 4° but less stable under more acidic conditions.[5] It may be stored at −80°.

Physicochemical Properties

Limulus clotting factor B is a single-chain glycoprotein with molecular weight of 64,000 (nonreduced SDS–PAGE). The purified factor B, however, gives three bands with apparent molecular weights of 64,000, 40,000 (designated heavy chain), and 25,000 (light chain) after reduction. There-

[13] T. Nakamura, F. Tokunaga, T. Morita, S. Iwanaga, S. Kusumoto, T. Shiba, T. Kobayashi, and K. Inoue, *Eur. J. Biochem.* **176,** 89 (1988).

fore, the purified preparation contains two molecular species: single-chain factor B and two-chain factor B bridged by a disulfide linkage. The two-chain species appears most likely to be a factor B zymogen cleaved in a disulfide loop that links the heavy and the light chains. On activation of the zymogen factor B by active factor $\overline{C}$, it is converted into factor $\overline{B}$ with the heavy (M_r 32,000) and light (M_r 25,000) chains, releasing an activation peptide.[5]

When the purified factor $\overline{B}$ is treated with [³H]DFP, one major peak with radioactivity before reduction is found at the position corresponding to the major protein band of factor $\overline{B}$, and two minor peaks with radioactivity are detected at positions corresponding to the M_r 120,000 and 32,000 chains, respectively. On reduction, the major radioactive peak migrates to the position corresponding to the mobility of the heavy chain (M_r 32,000) derived from factor $\overline{B}$ (Fig. 2), indicating that the DFP-reactive site is located in the heavy chain portion of factor $\overline{B}$. The minor peaks with radioactivity (less than 5% of the total radioactivity) seem to be an aggregate or a degraded form of factor $\overline{B}$.

We have succeeded in determining the cDNA sequence for the zymogen factor B.[14] It consists of 400 amino acid residues with 23 residues of signal sequence. The mature protein is composed of 377 amino acids, with the calculated molecular mass (40,570) of the protein. Three potential glycosylation sites for N-linked carbohydrate chain are found in the mature polypeptide sequence. The entire amino acid sequence has a similarity to that of the *Limulus* proclotting enzyme. Particularly, the sequence identity of the carboxyl-terminal serine protease domains between factor B and the proclotting enzyme[15] is 43.9%. Moreover, the amino-terminal regions up to 60th residue of both proteins share a similar structure with six cysteines, suggesting that the proteins have arisen from a gene duplication.

Specific Enzymatic Properties

Factor $\overline{B}$ efficiently hydrolyzes Boc-Met-Thr-Arg-4-methylcoumaryl-7-amide (MCA) and shows a relatively weak activity toward Boc-Leu-Gly-Arg-MCA, which is the best substrate for *Limulus* clotting enzyme.[4] A good substrate (Boc-Val-Pro-Arg-MCA) for factor $\overline{C}$ is also weakly hydrolyzed by factor $\overline{B}$. Factor $\overline{B}$ shows an apparent affinity for hydroxylamino acids (Thr or Ser) at the P2 site (nomenclature of Schechter and Berger[16]). As regards natural substrates, it does not activate mammalian

[14] T. Oda, T. Muta, and S. Iwanaga, *Seikagaku* **63,** 969 (abstr.) (1991).
[15] T. Muta, R. Hashimoto, T. Miyata, H. Nishimura, Y. Toh, and S. Iwanaga, *J. Biol. Chem.* **265,** 22426 (1990).
[16] T. Schechter and A. Berger, *Biochem. Biophys. Res. Commun.* **27,** 157 (1967).

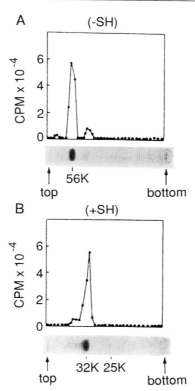

FIG. 2. Distribution of radioactivity of [³H]DFP-labeled factor $\overline{B}$ on SDS–PAGE. A 180-μl aliquot of factor $\overline{B}$ (15 μg) was incubated with 44 μM [³H]DFP at 37° for 30 min. After incubation, [³H]DFP-treated factor $\overline{B}$ (7.1 μg) was subjected to SDS–PAGE (8% gel) before reduction (A) or after reduction (B). Then, the gels were stained and sliced, and the radioactivity of the gel slices was counted.

plasma coagulation factors such as factor IX, factor X, prothrombin, plasminogen, protein C, and prekallikrein. It also does not convert fibrinogen to fibrin gel. Moreover, factor $\overline{B}$ does not catalyze the activation of the zymogen factor C or the conversion of coagulogen to coagulin.[5]

The amidase activity of factor $\overline{B}$ is strongly inhibited by a_2-plasmin inhibitor, but is not inhibited by antithrombin III, which is a potent inhibitor of factor $\overline{C}$ and *Limulus* clotting enzyme. Among four synthetic inhibitors, DFP has the strongest inhibitory effect. Benzamidine, leupeptin, and PCMB partially inhibit the factor $\overline{B}$ activity.[5]

[22] *Limulus* Proclotting Enzyme

By Tatsushi Muta, Takanori Nakamura, Ryuji Hashimoto, Takashi Morita, and Sadaaki Iwanaga

Introduction

Exposure of the *Limulus* hemocytes to bacterial endotoxins, which are lipopolysaccharides (LPS), results in the activation of the intracellular clotting system leading to gel formation.[1-4] This phenomenon is known to be one of the self-defense systems of the horseshoe crab against invading microorganisms. To elucidate the overall molecular events of the LPS-induced activation of the *Limulus* clotting system, we have been isolating clotting factors from the hemocyte lysate.[5-10] These studies have indicated that there are at least three intracellular protease zymogens, designated factor C, factor B, and proclotting enzyme, and one clottable protein, coagulogen, all of which are closely associated with the LPS-mediated clotting system, and that factor C is an LPS-sensitive trigger for this system.[11-13] This chapter concentrates on the purification and properties of *Limulus* proclotting enzyme. Other *Limulus* clotting factors, including factor C and factor B, have been described in other issues of this series.

Materials and Reagents

Horseshoe crabs (*Tachypleus tridentatus*), collected in Imatsu and Fukuyoshi, Fukuoka prefecture, are bled by cardiac puncture in the pres-

[1] J. Levin and F. B. Bang, *Bull. Johns Hopkins Hosp.* **115,** 265 (1964).

[2] E. H. Muller, J. Levin, and R. Holme, *J. Cell. Physiol.* **816,** 533 (1975).

[3] J. Y. Tai and T.-Y. Liu, *J. Biol. Chem.* **252,** 2178 (1977).

[4] J. Y. Tai, R. C. Seid, R. D. Hunh, and T.-Y. Liu, *J. Biol. Chem.* **252,** 4773 (1977).

[5] S. Iwanaga, T. Morita, T. Miyata, T. Nakamura, and J. Aketagawa, *J. Protein Chem.* **5,** 255 (1986).

[6] T. Nakamura, T. Morita, and S. Iwanaga, *Eur. J. Biochem.* **154,** 511 (1986).

[7] T. Nakamura, T. Horiuchi, T. Morita, and S. Iwanaga, *J. Biochem.* (*Tokyo*) **99,** 847 (1986).

[8] T. Nakamura, T. Morita, and S. Iwanaga, *J. Biochem.* (*Tokyo*) **97,** 1561 (1985).

[9] T. Miyata, M. Hiranaga, M. Umezu, and S. Iwanaga, *J. Biol. Chem.* **259,** 8924 (1984).

[10] T. Morita, S. Tanaka, T. Nakamura, and S. Iwanaga, *FEBS Lett.* **129,** 318 (1981).

[11] T. Nakamura, F. Tokunaga, T. Morita, S. Iwanaga, S. Kusumoto, T. Shiba, T. Kobayashi, and K. Inoue, *Eur. J. Biochem.* **176,** 89 (1988).

[12] T. Nakamura, F. Tokunaga, T. Morita, and S. Iwanaga, *J. Biochem.* (*Tokyo*) **103,** 370 (1988).

[13] T. Muta, T. Miyata, Y. Misumi, F. Tokunaga, T. Nakamura, Y. Toh, Y. Ikehara, and S. Iwanaga, *J. Biol. Chem.* **266,** 6554 (1991).

ence of caffeine (10 mM), and the hemolymph is centrifuged at 1000 rpm for 10 min at 4°. The hemocytes are washed with sterilized 0.02 M Tris-HCl buffer, pH 8.0, containing 0.154 M NaCl. A chromogenic substrate, Boc-Leu-Gly-Arg-p-nitroanilide (pNA), is obtained from the Peptide Research Foundation (Minoh, Osaka, Japan).

Preparation of Hemocyte Lysate

All glassware, metalware, and buffer solutions used for the preparation are sterilized by heating at 200° or autoclaving. The blade of a Physcotron (Nihon-Seimitsu Kogyo, Ltd., Tokyo) is sterilized overnight by soaking it in 95% ethanol containing 0.2 M NaOH. About 53 g (wet weight) of hemocytes collected from 1.2 liters of hemolymph is suspended in 250 ml of 0.02 M Tris-HCl buffer, pH 8.0, containing 50 mM NaCl, 1 mM benzamidine hydrochloride, and 1 mM EDTA in a metal tube, homogenized in the Physcotron, and centrifuged at 8000 rpm for 30 min in a Hitachi 20 PR-52 centrifuge. The pellet is reextracted twice with 200 ml of the same buffer and the supernatant from the first centrifugation is pooled with the two 200-ml extractions to give 640 ml of lysate.

Preparation of Dextran Sulfate-Sepharose CL-6B

Dextran sulfate-Sepharose CL-6B is prepared by a modification of the procedure of Andersson *et al.*[14] and Pepper and Prowse.[15] Dextran sulfate sodium salt (12.5 g) dissolved in 500 ml of cold water is mixed with 250 ml of Sepharose CL-6B and then 62.5 g of cyanogen bromide in 90 ml of acetonitrile is added to the mixture. Throughout the coupling stage, the pH is adjusted to 10.5 with 10 N NaOH, and the temperature is held between 4 and 10°. After coupling, the dextran sulfate-Sepharose CL-6B is filtered off and suspended in 500 ml of 0.5 M Tris-HCl buffer, pH 8.5, and stirred at 4° for 15 hr. The incubation allows for hydrolyzing the remaining CNBr groups. The final product is thoroughly washed with pyrogen-free water, resuspended in 1 liter of 0.02 M Tris-HCl buffer, pH 8.0, autoclaved at 120° for 30 min, and stored at 4°.

Preparation of Activated Factor $\bar{B}$ Used for Assay
of Proclotting Enzyme

A factor B-containing fraction (0.1 mg/ml) obtained from the lysate by using a heparin-Sepharose column[16] is mixed with 250 μl of 1 M Tris-

[14] L-O. Andersson, H. Borg, and M. Miller-Andersson, *Thromb. Res.* **7**, 451 (1975).
[15] D. S. Pepper and C. Prowse, *Thromb. Res.* **11**, 689 (1977).
[16] M. Ohki, T. Nakamura, T. Morita, and S. Iwanaga, *FEBS Lett.* **120**, 217 (1980).

HCl buffer, pH 8.0, containing 50 mM MgCl$_2$ and 50 μl of *E. coli* 0111-B4 LPS (10 μg/ml), and the mixture is incubated at 37° for 10 min. An aliquot of this mixture is used for the activation of the proclotting enzyme.

Assay Method

The amidase activity of the proclotting enzyme after activation with activated factor $\overline{\text{B}}$ is measured by using Boc-Leu-Gly-Arg-pNA as a substrate. A reaction mixture containing 5 μl of sample, 100 μl of activated factor $\overline{\text{B}}$, 20 μl of 1 M Tris-HCl buffer, pH 8.0, 10 μl of 1% BSA, and 65 μl of distilled water, in a total volume of 200 μl, is preincubated at 37° for 30 min. Then, 50 μl of 2 mM Boc-Leu-Gly-Arg-pNA is added, and the whole mixture is incubated for an additional 1–3 min. Then, 0.8 ml of 0.6 M acetic acid is added to terminate the reaction and the absorbance at 405 nm is read in a Hitachi 220A spectrophotometer. One unit of the enzyme activity is defined as 1 μmol of *p*-nitroaniline liberated per min.

Purification Procedure

Step 1: Chromatography on Dextran Sulfate-Sepharose CL-6B. All procedures for step 1 are performed under sterile conditions. All the buffers used for the following procedures are 0.02 M Tris-HCl buffer, pH 8.0, containing 1 mM benzamidine hydrochloride and 1 mM EDTA (Tris–benzamidine–EDTA) unless otherwise stated. The lysate (630 ml) is centrifuged at 8000 rpm for 1 hr before use and the supernatant is applied to a dextran sulfate-Sepharose CL-6B column (5.0 × 23.5 cm) equilibrated with 3.5 liters of Tris–benzamidine–EDTA containing 50 mM NaCl. The column is then washed with 750 ml of the equilibration buffer and 750 ml of the same buffer without 1 mM benzamidine-HCl and 1 mM EDTA. As shown in Fig. 1, the proclotting enzyme passes through the column under these conditions. Other components adsorbed on the column are eluted in a stepwise fashion with 1.5 liters of Tris–benzamidine–EDTA containing 0.3 M NaCl for coagulogen and factor G,[17] with 1.3 liters of Tris–benzamidine–EDTA containing 0.5 M NaCl for factor B,[7] and finally with 1.5 liters of Tris–benzamidine–EDTA containing 2.0 M NaCl for anti-LPS factor.[18] Elution is performed at a flow rate of 105 ml/hr at 4°, and the transmittance of the eluate at 279 nm is monitored with an LKB 2138 Uvicord S (LKB-Produkter AB). The breakthrough fractions containing

[17] T. Morita, S. Tanaka, T. Nakamura, and S. Iwanaga, *FEBS Lett.* **129,** 318 (1981).
[18] J. Aketagawa, T. Miyata, R. Ohtsubo, T. Nakamura, T. Morita, H. Hayashida, T. Miyata, S. Iwanaga, T. Takao, and Y. Shimonishi, *J. Biol. Chem.* **261,** 7357 (1988).

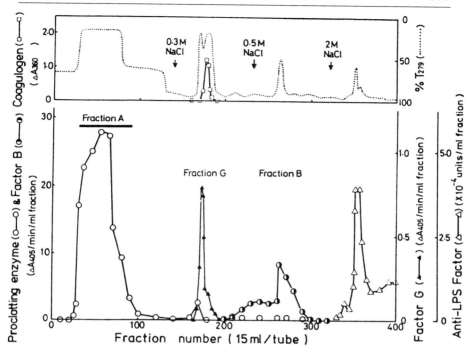

FIG. 1. Dextran sulfate-Sepharose CL-6B chromatography of horseshoe crab (*Tachypleus tridentatus*) hemocyte lysate.[8] The activities of proclotting enzyme (○), factor B (◑), and coagulogen (□) were assayed by methods described in Refs. 6–8; factor G[16] (▲) and anti-LPS factor[18] (△) activities were measured by methods described in Ref. 5.

proclotting enzyme (fraction A) are collected and diluted twofold with 935 ml of cold Tris–benzamidine–EDTA. The following procedures are carried out under nonsterile conditions.

Step 2: Chromatography on DEAE-Sepharose CL-6B. The breakthrough fraction from step 1 is applied to a DEAE-Sepharose CL-6B column (2.2 × 21.5 cm) equilibrated with 3.0 liters of Tris–benzamidine–EDTA containing 25 mM NaCl. The column is then washed with 1.4 liters of the same buffer. Proteins are eluted by a linear salt gradient formed from 500 ml each of Tris–benzamidine–EDTA containing 0.025 M NaCl and the same buffer, containing 0.30 M NaCl, at a flow rate of 100 ml/hr. The proclotting enzyme activity appears in the early fractions (15 ml/tube) with the salt gradient. The fractions are collected and concentrated by Diaflo ultrafiltration.

Step 3: Gel Filtration on Sephadex G-150. The concentrated fraction from step 2 is applied to a Sephadex G-150 column (3.4 × 127 cm) equilibrated with tris–benzamidine–EDTA containing 0.2 M NaCl. Elution is carried out with the equilibration buffer at a flow rate of 50 ml/hr. The major proclotting enzyme activity is detected in tubes 50 to 65 (9.4 ml/tube) and these fractions are collected and dialyzed overnight against 2 liters of Tris–benzamidine–EDTA.

Step 4: Chromatography on Benzamidine-CH-Sepharose CL-6B. the dialyzed solution from step 3 is applied to a benzamidine-CH-Sepharose column (1.5 × 13 cm) equilibrated with the buffer used for dialysis. Elution is performed with a linear concentration gradient of NaCl from the equilibration buffer to 0.25 M at a flow rate of 50 ml/hr. The proclotting enzyme fractions (10 ml/tube) are combined and concentrated to about 6 ml with the Diaflo ultrafiltration apparatus.

Step 5: Gel Filtration on Sephacryl S-300. The final purification is performed by gel filtration on a Sephacryl S-300 column (2.6 × 140 cm) equilibrated with 0.05 M Tris-HCl buffer, pH 8.0, containing 0.2 M NaCl. The concentrated solution from step 4 is applied to the column and eluted with the equilibration buffer at a flow rate of 45 ml/hr. Two protein peaks appear and the first major peak contains the proclotting enzyme activity. These fractions (5.0 ml/tube) are collected.

A summary of the procedures for purification of the proclotting enzyme is shown in Table I. The yield of purified protein is about 20% and approximately 300-fold purification over the proclotting enzyme activity of the lysate is achieved. All the steps mentioned above are repeated several times using 300–700 ml of the lysate and the results have been found to be reproducible.

The purified zymogen gives a single protein band on disc PAGE (8% gel, w/v) at pH 8.3. Moreover, a single peak with the proclotting enzyme activity is observed at the same position as the protein band stained with Coomassie brilliant blue. On SDS–PAGE (8% gel, w/v) with and without 2-mercaptoethanol, the preparation gives a single molecular species, sug-

TABLE I
PURIFICATION PROCEDURE FOR *Tachypleus* PROCLOTTING ENZYME

Step	Volume (ml)	Total protein (mg)	Total activity (units)	Specific activity (units/mg)	Purification (fold)	Yield (%)
Hemocyte lysate	630	2026	3119	1.54	1.0	100
Dextran sulfate-Sepharose CL-6B	945	318	2552	8.03	5.2	81.8
DEAE-Sepharose CL-6B	355	65.0	2414	37.1	24.1	77.4
Sephadex G-150	150	26.9	1830	68.0	44.2	58.7
Benzamidine-CH-Sepharose CL-6B	90	8.91	1242	139	90.3	39.8
Sephacryl S-300	29	1.36	609	448	291	19.5

gesting that the protein consists of a single polypeptide chain. Also, the preparation is not contaminated by factor B,[7] factor C,[6,19] factor G,[17] or anti-LPS factor,[18] all of which are associated with the hemolymph coagulation system. On exposure of the preparation to bacterial endotoxin (*Escherichia coli* 0111-B4), it does not show any amidase activity due to the clotting enzyme, indicating that the zymogen is essentially insensitive to endotoxin.

Properties

Stability

The purified proclotting enzyme is stable at pH 5.0–9.0 for 24 hr at room temperature but is less stable under more acidic conditions. The zymogen is stable at 0° for a week and may be stored at −80° indefinitely.

Physicochemical Properties

The proclotting enzyme is a glycoprotein with an apparent molecular weight (M_r) of 54,000[8] (SDS–PAGE), and no γ-carboxyglutamic acid is detected.[8] It is converted to the active form by purified factor $\overline{\text{B}}$ or by trypsin. The resulting clotting enzyme has a molecular weight of 54,000, consisting of a heavy chain of M_r 31,000 and a light chain of M_r 25,000, the former of which contains a serine active site triad.[8] The isolated cDNA for proclotting enzyme consists of 1501 base pairs.[20] The open reading frame of 1125 base pairs encodes a sequence comprising 29 amino acid residues of prepro sequence and 346 residues of the mature protein with a molecular mass of 38,194 Da.[20] Three potential glycosylation sites for N-linked carbohydrate chains are confirmed to be glycosylated. Moreover the zymogen contains six O-linked carbohydrate chains in the amino-terminal light chain generated after activation.[20] The cleavage site that accompanies activation catalyzed by trypsinlike active factor $\overline{\text{B}}$ is an Arg-Ile bond.[20] The resulting carboxyl-terminal heavy chain is composed of a typical serine protease domain, with a sequence similar to that of human coagulation factor XIa (34.5%) or factor Xa (34.1%).[20] The light chain has a unique disulfide-knotted domain that shows a significant sequence similarity to that of the *Drosophila* serine protease precursor, easter.[21]

[19] F. Tokunaga, T. Miyata, T. Nakamura, T. Morita, K. Kuma, T. Miyata, and S. Iwanaga, *Eur. J. Biochem.* **167**, 405 (1987).
[20] T. Muta, R. Hashimoto, T. Miyata, H. Nishimura, Y. Toh, and S. Iwanaga, *J. Biol. Chem.* **265**, 22426 (1990).
[21] R. Chasan and K. V. Anderson, *Cell (Cambridge, Mass.)* **56**, 391 (1989).

Substrate Specificities and Inhibitors

The active clotting enzyme specifically hydrolyzes Bz-Ile-Glu-Gly-Arg-pNA, which is a good chromogenic substrate for the mammalian coagulation factor, factor Xa. Boc-Leu-Gly-Arg-pNA and Boc-Leu-Gly-Arg-4-methylcoumaryl-7-amide (MCA) are the most effective substrates for *Limulus* clotting enzyme, the former of which is employed for the chromogenic substrate assay of bacterial endotoxins, using the hemocyte lysate. Thus, the clotting enzyme has a very similar specificity to factor Xa, but not to α-thrombin, although it catalyzes the transformation of coagulogen to coagulin gel, as α-thrombin does. This enzyme is maximally active at pH 8.5–8.8

The amidase activity of the clotting enzyme is strongly inhibited by antithrombin III in the presence of heparin and α_2-plasmin inhibitor, but is not inhibited by hirudin, aprotinin, and soybean trypsin inhibitor.[7,22] Diisopropyl fluorophosphate (DFP) and benzamidine have the strongest inhibitory effect.[22]

Subcellular Localization

The *Limulus* hemocytes contain two types of granules, large but less dense granules and smaller but denser ones.[23] Immunohistochemical studies show that the proclotting enzyme is localized exclusively in the large granules and is apparently released through exocytosis on LPS stimulation of the hemocytes.[23]

[22] S. Nakamura, S. Iwanaga, T. Harada, and M. Niwa, *J. Biochem. (Tokyo)* **80,** 1011 (1976).
[23] Y. Toh, A. Mizutani, F. Tokunaga, T. Muta, and S. Iwanaga, *Cell Tissue Res.* **266,** 137 (1991).

[23] *Limulus* Test for Detecting Bacterial Endotoxins

By SHIGENORI TANAKA and SADAAKI IWANAGA

Introduction

Limulus test, a test for detecting bacterial endotoxins, was invented by Levin and Bang based on their finding that a trace amount of endotoxin coagulates hemocyte (amebocyte) lysate of the horseshoe crab, *Limulus*

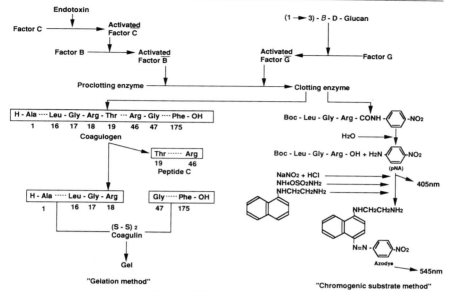

FIG. 1. Principle of *Limulus* test.

polyphemus.[1-3] This gelation reaction has been widely employed as a simple and very sensitive assay method for endotoxins.[4,5] The original method is qualitative or semiquantitative; the presence of endotoxin is determined by reading the formation of gel clot after incubation of a sample with the lysate at 37° for 1 hr (*Limulus* gelation test).

During the past decade we studied the molecular mechanism of hemolymph coagulation in *Limulus* and established a protease cascade. This cascade, as shown in Fig. 1, is based on three kinds of serine protease zymogens, factor C,[6] factor B,[7] and proclotting enzyme,[8] and one clottable

[1] J. Levin and F. B. Bang, *Thromb. Diath. Haemorrh.* **19,** 186 (1968).

[2] N. S. Young, J. Levin, and R. A. Prendergast, *J. Clin. Invest.* **51,** 1790 (1972).

[3] E. T. Yin, C. Galanos, S. Kinsky, R. A. Bradshaw, S. Wessler, O. Luderitz, and M. E. Sarmiento, *Biochim. Biophys. Acta* **261,** 284 (1972).

[4] H. D. Hochstein, R. J. Ein, J. F. Cooper, E. B. Seligmam, and S. M. Wolff, *Bull. Parenter. Drug. Assoc.* **27,** 139 (1973).

[5] J. F. Cooper, J. D. Hochstein, and R. B. Seligman, *J. Lab. Clin. Med.* **78,** 138 (1971).

[6] T. Muta, T. Miyata, Y. Misumi, F. Tokunaga, T. Nakamura, Y. Toh, Y. Ikehara, and S. Iwanaga, *J. Biol. Chem.* **266,** 6554 (1991).

[7] T. Nakamura, T. Horiuchi, T. Morita, and S. Iwanaga, *J. Biochem.* (*Tokyo*) **99,** 847 (1986).

[8] T. Muta, R. Hashimoto, T. Miyata, H. Nishimura, Y. Toh, and S. Iwanaga, *J. Biol. Chem.* **265,** 22426 (1990).

protein, coagulogen.[9] Endotoxin (bacterial lipopolysaccharide, LPS) activates the zymogen factor C to the active form, factor C̄. Factor C̄ then activates factor B to factor B̄, which in turn converts the proclotting enzyme to the active clotting enzyme. Each activation proceeds by limited proteolysis. The resulting clotting enzyme cleaves two bonds in coagulogen, which is a fibrinogen-like molecule in arthropods, to yield an insoluble coagulin gel. The coagulation cascade is also activated by $(1 \rightarrow 3)$-β-D-glucan. In the presence of $(1 \rightarrow 3)$-β-D-glucan, a protease that is tentatively named factor G is activated, leading to activation of the proclotting enzyme.[10] Because the *Limulus* lysate contains all the enzymes described above, the *Limulus* test reacts with $(1 \rightarrow 3)$-β-D-glucan as well as endotoxin. The latter activates factor C, whereas the former activates factor G; both pathways converge on proclotting enzyme, ensuing its activation and release of pNA (Fig. 1). In addition to linear $(1 \rightarrow 3)$-β-D-glucan, fungal polysaccharides with (1,6) branches, such as lentinan and schizophyllan, and the rinse through the cellulosic dialyzer (used for renal disease) also activate factor G.[11]

Our purpose in this chapter is to describe the specific assay for bacterial endotoxins by using a chromogenic substrate of Boc-Leu-Gly-Arg-*p*-nitro-anilide (pNA). The sequence of this substrate originates from the sequences located close to the site cleaved during the gelation of coagulogen by *Limulus* clotting enzyme (Fig. 1).[12] The chromogenic substrate is hydrolyzed by clotting enzyme to release pNA. By measuring the absorbance of released pNA at 405 nm, endotoxin concentration in the samples can be determined. Endotoxin concentration can also be determined by measuring the absorbance at 545 nm after the diazo coupling of pNA, when a yellowish color in samples interferes with the measurement at 405 nm.

Reagents: Endotoxin-free

Endospecy: Endospecy is a mixture of a lyophilized powder of *Limulus* coagulation factors and a substrate, and Boc-Leu-Gly-Arg-(pNA). This reagent is now commercially available from Seikagaku Corp., Ltd. (2-1-5, Nihonbashi-Honcho, Chuo-ku, Tokyo-103, Japan)
Assay buffer: 0.2 M Tris-HCl buffer, pH 8.0

[9] T. Miyata, H. Matsumoto, M. Hattori, Y. Sakaki, and S. Iwanaga, *J. Biochem. (Tokyo)* **100**, 213 (1986).
[10] T. Morita, S. Tanaka, T. Nakamura, and S. Iwanaga, *FEBS Lett.* **129**, 318 (1981).
[11] T. Obayashi, H. Tamura, S. Tanaka, M. Ohki, S. Takahashi, M. Arai, M. Matsuda, and T. Kawai, *Prog. Clin. Biol. Res.* **231**, 357 (1987).
[12] S. Iwanaga, T. Morita, T. Harada, S. Nakamura, M. Niwa, K. Takada, T. Kimura, and S. Sakakibara, *Haemostasis* **7**, 183 (1978).

Other Reagents

Acetic acid (AcOH): 0.6 M and 0.8 M

Sodium nitrite: 0.04% (w/v) in 0.48 M HCl and 1.0 M HCl

Ammonium sulfamate: 0.3% (w/v)

N-1-Naphthylethylenediamine dihydrochloride: 0.07% (w/v) in 14% (v/v) N-methyl-2-pyrrolidone

Standard endotoxin: *Escherichia coli* 0111: B4 endotoxin (Westphal type); 1 ng of this endotoxin equals 2.9 USP (definition by U.S. Pharmacopeial Forum) reference standard endotoxin unit (EU). EU is a unit that represents the approximate threshold pyrogen dose for humans and rabbits.[12]

Glassware and Plasticware

It is essential that the assay procedure be performed completely in an endotoxin-free system until completion of the main reaction. All glassware should be heated in an oven at 250° for 2 hr. Commercial endotoxin-free glassware may be used. Endotoxin-free plasticware (Seikagaku Corp.) is also available commercially and should be used. The plasticware is preferred over glassware.

Endotoxin-Specific Chromogenic Limulus Test and Its Assay Procedures

As described above, because a conventional *Limulus* test using whole hemocyte lysate is not specific to endotoxin, we tried to develop an endotoxin-specific *Limulus* reagent by removing factor G from the lysate. This is accomplished by fractionating the lysate with a dextran sulfate-Sepharose CL-6B column and by reconstituting those clotting factors that make up the endotoxin-sensitive pathway.[13,14] The new reagent, called Endospecy, consisting of the reconstituted lysate, Mg^{2+}, and Boc-Leu-Gly-Arg-pNA, reacts only with endotoxins, including lipid A and its analogs.[15]

Test Tube Method 1: A_{405} nm. A 0.1-ml aliquot of samples, a blank (distilled water), and serial twofold dilutions of the standard endotoxin are dispensed in a test tube placed in a thawing ice bath. A 0.1-ml aliquot of Endospecy dissolved in the assay buffer is added to each tube. After being shaken well, the mixture is incubated at 37° for 30 min. The reaction

[13] T. Nakamura, T. Morita, and S. Iwanaga, *J. Biochem. (Tokyo)* **97**, 1561 (1985).

[14] T. Obayashi, H. Tamura, S. Tanaka, M. Ohki, S. Takahashi, M. Arai, M. Matsuda, and T. Kawai, *Clin. Chim. Acta* **149**, 55 (1985).

[15] T. Obayashi, *Adv. Exp. Med. Biol.* **256**, 215 (1990).

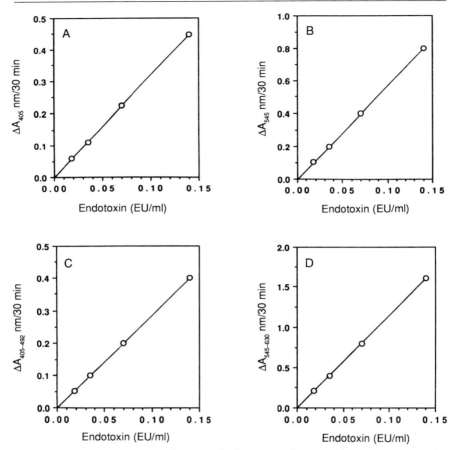

FIG. 2. Calibration curves of reagent Endospecy against *E. coli* 0111 : B4 endotoxin. (A) Test tube assay at 405 nm. (B) Test tube assay at 545 nm with diazo coupling. (C) Microplate assay at 405 and 492 nm by dual wavelength. (D) Microplate assay at 545 and 630 nm with diazo coupling. Each of the plots are quite linear to 0.02 EU/ml. This is the limit of sensitivity under the conditions used.

is stopped by adding a 0.4-ml portion of 0.8 M acetic acid and the absorbance is measured at 405 nm. The endotoxin concentration is calculated from the standard curve plotted against the serial dilutions of the standard endotoxin (Fig. 2A).

Test Tube Method 2: A_{545} nm with Diazocoupling. The procedure is similar to that above except a 0.5-ml portion of 0.04% sodium nitrate (0.48 M HCl solution) is used to stop the reaction instead of acetic acid. Then, a 0.5-ml aliquot of 0.3% ammonium sulfamate and 0.07% *N*-1-naphthylethylenediamine dihydrochloride is added successively to diazocouple the released pNA. Absorbance is measured at 545 nm. The endotoxin concentration is calculated from the standard curve (Fig. 2B). This method has an advantage over the preceding one for samples containing yellow-brownish pigments.

Microplate Method by Dual-Wavelength Spectrophotometry: A$_{405}$ and A$_{492}$ nm. A 0.05-ml aliquot of samples, a blank (distilled water), and serial twofold dilutions of the standard endotoxin are dispensed in a well of a 96-well microplate (Toxipet plate 96 F, Seikagaku Corp.). A 0.05-ml aliquot of Endospecy dissolved in the assay buffer is added to each well. With an accompanying plastic lid placed on top, the microplate is incubated at 37° for 30 min on a dry microplate incubator (Hotplate CT-961, Seikagaku Corp.). The reaction is stopped with 0.6 *M* acetic acid (0.2 ml) and the absorbance is measured with a microplate reader at 405 and 492 nm simultaneously, the latter being used as a reference. The endotoxin concentration is calculated from the standard curve (Fig. 2C). The dual-wavelength spectrophotometry corrects nonspecific absorption derived from the bend of the microplate and from the turbidity of the samples.

Microplate Method by Dual-Wavelength Spectrophotometry: A$_{545}$ and A$_{630}$ nm with Diazocoupling. The procedure is similar to that above except a 0.05-ml portion of 0.04% sodium nitrate (1.0 *M* HCl solution) is used to stop the reaction instead of acetic acid. Released pNA is then diazocoupled with 0.05 ml each of 0.3% ammonium sulfamate and 0.07% N-1-naphthylethylenediamine dihydrochloride in 14% N-methyl-2-pyrrolidone solution. Absorbance is measured with a microplate reader at 545 and 630 nm simultaneously, the latter being used as a reference. The endotoxin concentration is calculated from the standard curve (Fig. 2D). If a 545-nm filter is not available, a 540- or 550-nm filter will suffice.

Kinetic Assay Procedure

A 0.05-ml aliquot of samples, a blank (distilled water), and serial two-fold dilutions of the standard endotoxin are dispensed in a well (Toxipet plate 96F). A 0.05-ml aliquot of Endospecy dissolved in the assay buffer is added to each well. After being covered with a plastic lid, the microplate is placed on a microplate incubator equipped with a reader (Wellreader SK601, Seikagaku Corp.) and heated at 37°. The absorbance is measured every 15 sec automatically at 405 and 492 nm, with the latter as a reference. The maximum rate of increase in absorbance is read and the endotoxin concentration is calculated from the standard curve (logarithmic equation) obtained simultaneously (Fig. 3). This method allows accurate measurement, without diazocoupling, of samples containing yellow-brownish pigments.

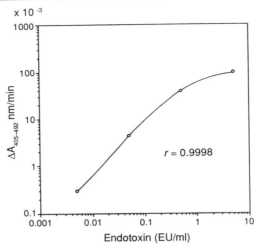

FIG. 3. Calibration curve of *E. coli* 0111 : B4 endotoxin measured by kinetic assay procedure.

General Comments

The methods described above are a 100 times more sensitive than the *Limulus* gelation test and are very reproducible. If this technique is to be applied to blood samples, however, the activities of *Limulus* test-interfering factors in the samples, such as thrombin, factor Xa, and α_1-antitrypsin, need to be abolished. To remove such interferences, various methods have been studied and applied to blood samples, such as pretreatment with chloroform, ether, acid, or alkali and heating.[15–17] Unfortunately, these techniques still present some unsolved problems.

[16] T. Ikeda, K. Hirata, K. Tabuchi, H. Tamura, and S. Tanaka, *Circ. Shock* **23**, 263 (1987).
[17] H. Tamura, S. Tanaka, T. Obayashi, M. Yoshida, and T. Kawai, *Clin. Chim. Acta* **200**, 35 (1991).

[24] Snake Venom Hemorrhagic and Nonhemorrhagic Metalloendopeptidases

By Hiroyuki Takeya, Toshiyuki Miyata, Norikazu Nishino, Tamotsu Omori-Satoh, and Sadaaki Iwanaga

Introduction

Hemorrhage is a common manifestation in the victim following the bite of crotalid and viperid snakes.[1] Various components, such as hemorrhagic factors and metalloendopeptidases,[2] of these venoms cause localized hemorrhage by direct actions on the blood vessel walls. Electron microscopic studies indicate that erythrocytes are leaked in a one-by-one fashion through widened interendothelial gaps when capillaries are exposed to these hemorrhagic proteins.[1] The enzymes may disrupt the pericellular basement membrane through proteolytic activity, with subsequent damage to the integrity of the vessel wall, after which hemorrhage occurs.[3] On the other hand, crotalid and viperid venoms contain many metalloendopeptidases, which are completely free from hemorrhagic activity.[4,5] Some of them were originally reported as anticoagulant proteases with fibrinogenolytic or fibrinolytic activity.[6–8] Additional groups of venom metalloendopeptidases are procoagulants with very strict substrate specificities. Two well-known examples of this type are Russell's viper venom factor X activator (RVV-X) and *Echis carinatus* venom prothrombin activator (ecarin), which have been detailed in an earlier volume.[9,10]

The proteolytic specificity for various venom hemorrhagic and nonhemorrhagic metalloendopeptidases has been studied, commonly using oxidized insulin B chain as substrate.[2] Most of metalloendopeptidases,

---

[1] A. Ohsaka, *in* "Handbook of Experimental Pharmacology" (C.-Y. Lee, ed.), Vol. 52, p. 480. Springer-Verlag, Berlin and New York, 1979.

[2] A. T. Tu, *in* "Handbook of Natural Toxins" (A. T. Tu, ed.), Vol. 5, p. 297. Marcel Dekker, New York, 1991.

[3] A. Ohsaka, M. Just, and E. Habermann, *Biochim. Biophys. Acta* **323,** 415 (1973).

[4] T. Takahashi and A. Ohsaka, *Biochim. Biophys. Acta* **198,** 293 (1970).

[5] S. Iwanaga, G. Oshima, and T. Suzuki, this series, Vol. 45, p. 459.

[6] B. V. Pandya and A. Z. Budzynski, *Biochemistry* **23,** 460 (1984).

[7] T. W. Willis and A. T. Tu, *Biochemistry* **27,** 4769 (1988).

[8] F. S. Markland, Jr., *Thromb. Haemostasis* **65,** 438 (1991).

[9] B. C. Furie and B. Furie, this series, Vol. 45, p. 191.

[10] T. Morita and S. Iwanaga, this series, Vol. 80, p. 303.

including those from *Trimeresurus flavoviridis* and *Crotalus atrox*,[4,7,11-13] hydrolyze mainly X-Leu peptide bonds of the insulin B chain. This observation strengthens the concept that venom proteinase hydrolyzes various substrates with specificity toward hydrophobic P_1' residue.[5] No significant difference of substrate specificity has been evaluated between hemorrhagic and nonhemorrhagic metalloendopeptidases using oxidized insulin B chain as substrate.

The hemorrhagic and nonhemorrhagic metalloendopeptidases isolated from *T. flavoviridis* and *C. atrox* have been well characterized and their amino acid sequences have been established recently.[14-18] This chapter describes these well-studied venom metalloendopeptidases.

Assay Methods

Principle

Hemorrhagic activity is defined as the ability to cause hemorrhagic necrosis after injection of venom fraction into the depilated skin of rabbits.[19] The proteolytic activity is generally determined using casein as substrate as described in this series.[5] Several hemorrhagic factors have been shown to be nonproteolytic using this method. However, when dimethylcasein instead of casein is used as substrate,[20] the hemorrhagic factors hydrolyze this substrate extensively, and after proteolysis newly evident terminal amino groups are detected with 2,4,6-trinitrobenzenesulfonic acid (TNBS).[21] Another assay method for proteolytic activity is a more convenient and highly sensitive system employing new synthetic

[11] J. B. Bjarnason and J. W. Fox, *Biochemistry* **22**, 3770 (1983).

[12] T. Nikai, M. Niikawa, Y. Komori, S. Sekoguchi, and H. Sugihara, *Int. J. Biochem.* **19**, 221 (1987).

[13] J. B. Bjarnason and J. W. Fox, *in* "Hemostasis and Animal Venoms" (H. Pirkle and F. S. Markland, Jr., eds.), p. 457. Marcel Dekker, New York, 1988.

[14] T. Miyata, H. Takeya, Y. Ozeki, M. Arakawa, F. Tokunaga, S. Iwanaga, and T. Omori-Satoh, *J. Biochem.* (*Tokyo*) **105**, 847 (1989).

[15] H. Takeya, M. Arakawa, T. Miyata, S. Iwanaga, and T. Omori-Satoh, *J. Biochem.* (*Tokyo*) **106**, 151 (1989).

[16] J. D. Shannon, E. N. Baramova, J. B. Bjarnason, and J. W. Fox, *J. Biol. Chem.* **264**, 11575 (1989).

[17] K. Yonaha, personal communication (1991).

[18] H. Takeya, K. Oda, T. Miyata, T. Omori-Satoh, and S. Iwanaga, *J. Biol. Chem.* **265**, 16068 (1990).

[19] H. Kondo, S. Kondo, H. Ikezawa, R. Murata, and A. Ohsaka, *Jpn. J. Med. Sci. Biol.* **13**, 43 (1960).

[20] Y. Lin, G. E. Means, and R. E. Feeney, *J. Biol. Chem.* **244**, 789 (1969).

[21] R. Fields, this series, Vol. 25, p. 464.

peptide substrates.[22] Because the venom metalloendopeptidases exhibit primary specificity toward the amino group of the scissile peptide bond (P_1' residue), commonly used 4-nitroanilide or 4-methylcoumaryl-7-amide substrates are useless.[23] The intermolecularly quenched fluorogenic peptide substrates have recently been developed.[22] The substrates contain a 2-aminobenzoyl group (Abz) as fluorophore and 2,4-dinitroanilinoethylamide (Dna) as a quenching group. These aromatic groups, linked via amide bonds to the N- and C-terminal ends of the synthetic peptides, are separated on enzymatic hydrolysis with a resultant increase in fluorescence that is utilized to measure hydrolysis rates.

Hemorrhagic Activity Assay[19]

Procedure

Skin on the back of an albino rabbit weighing 2–2.5 kg is depilated after treatment with a paste of 20% barium sulfide; the depilated area is thoroughly washed with warm water. One day after depilation, 0.1 ml of a sample, diluted serially threefold with physiological saline, is injected intracutaneously into the depilated skin. One day later, the rabbits are killed by ether inhalation and the skin is removed immediately and fixed on a glass plate. The cross diameter of each hemorrhagic spot is measured and hemorrhagic activity (minimum hemorrhagic dose, MHD) is expressed as micrograms of protein required to produce a hemorrhagic spot 10 mm in diameter.

Proteolytic Activity

Dimethylcasein Method[20,21]

Reagents

Dimethylcasein (0.2%): Dimethylcasein is obtained from Sigma Chemical Co. (St. Louis, MO) or is prepared as follows: casein (1.5 g) in 150 ml of 0.1 M borate buffer, pH 9.0, is rapidly stirred, and 300 mg of sodium borohydride is added. A few drops of 2-octanol are also added to prevent any subsequent tendency to foam. Formaldehyde

[22] N. Nishino, Y. Makinose, and T. Fujimoto, *Chem. Lett.* p. 77 (1992).
[23] S. Kawabata, T. Miura, T. Morita, H. Kato, K. Fujikawa, S. Iwanaga, K. Takada, T. Kimura, and S. Sakakibara, *Eur. J. Biochem.* **172,** 17 (1988).

(3 ml) is then added in 100-μl increments over a period of 30 min. The solution is acidified to pH 6 by the addition of 50% acetic acid and is dialyzed against water. The lyophilized dimethylcasein is dissolved in 10 mM borate buffer, pH 8.5, to give a final concentration of 0.2%.

Procedure. Dimethylcasein solution, 50 μl, is digested for 30 min at 37° with 50 μl of a suitably diluted enzyme solution. The reaction is terminated by immersing the sample briefly in a boiling water bath. Then, 200 μl of 0.2 M borate buffer, pH 10.5, and 20 μl of 1.1 M TNBS are added and the solution is rapidly mixed. After 10 min, the reaction is stopped by adding 1 ml of 0.2 M NaH$_2$PO$_4$ containing 1.5 mM Na$_2$SO$_3$ (made fresh daily), and the absorbance at 420 nm is determined relative to a blank incubated with all of the components present in the sample except active enzyme.

Synthetic Substrate Method[22]

Reagents

Peptide substrates: Abz-Gly-Phe-Arg-Leu-Dna, Abz-Gly-Phe-Arg-Leu-Leu-Dna, Abz-Ala-Gly-Leu-Ala-Dna, Abz-Ala-Gly-Pro-Ala-Dna, Abz-Ala-Pro-Glu(OBzl)-Ala-Dna, Abz-Gly-Pro-Leu-Gly-Pro-Dna, Abz-Ser-Pro-Met-Leu-Dna, Abz-Phe-Ser-Pro-Met-Leu-Dna, Abz-Thr-Glu-Lys-Leu-Val-Dna, Abz-Ala-Thr-Asp-Ile-Val-Dna, Abz-Asn-Ala-Pro-Leu-Ala-Dna, and Abz-Asn-Ala-Pro-Ser-Ala-Dna; these are synthesized by conventional methods according to the procedure reported earlier,[24] and by Kaiser's oxime resin solid-phase methods.[25] For instance, Abz-Gly-Phe-Arg-Leu-Dna and Abz-Gly-Phe-Arg-Leu-Leu-Dna are synthesized as follows: *tert*-butyloxycarbonyl (Boc)-Leu (4 mmol) is reacted with the oxime resin (4.0 g) using dicyclohexylcarbodiimide (DCC). The incorporated amount is determined by picrate assay to be 0.48 mol/g resin. The Boc-Leu-resin (4.36 g) is treated with 25% (w/v) trifluoroacetic acid (TFA) in CH$_2$Cl$_2$ (60 ml). Boc-Arg(tosyl) (3 equivalents) is coupled with DCC/1-hydroxybenzotriazole (HOBt). The third amino acid, Boc-Phe, is coupled by the symmetrical anhydride method to avoid spontaneous cleavage of the dipeptide, and the further elongation of peptide with Boc-Gly and Boc-Abz is performed using DCC/HOBt. Finally, the efficient amidation cleavage of the peptide assembled on the resin is performed with Dna and Leu-Dna (4 equivalents) in the presence of acetic acid (4

---

[24] N. Nishino and J. C. Powers, *J. Biol. Chem.* **255**, 3482 (1980).

[25] S. H. Nakagawa and E. T. Kaiser, *J. Org. Chem.* **48**, 678 (1983).

equivalents) in 30 ml of dimethylformamide (DMF). Finally, they are treated with anhydrous HF and purified by reversed-phase HPLC.

Procedure. Abz-Ser-Pro-Met-Leu-Dna, the best substrate for the venom metalloendopeptidases so far tested,[26] is dissolved in 0.95 ml of 50 mM Tris-HCl buffer containing 2.5% DMF. To a cuvette containing this solution, 50 μl of enzyme solution is added at 25°, and the fluorescence change is monitored continuously in a fluorometer connected to a chart recorder. The excitation and emission wavelengths are 320 and 425 nm, respectively. The increase in fluorescence on complete hydrolysis is affected by the concentration of substrate and is determined at each substrate concentration utilized. For the routine assays of proteases during purification, a substrate concentration of 20 μM is recommended.

Purification

Isolation of High-Molecular-Mass Hemorrhagic Metalloendopeptidases from Venom of Trimeresurus flavoviridis[27]

The venom used is taken from the specimens of Habu (*T. flavoviridis*) collected in the Amami Oshima Islands, or is purchased from Sigma Chemical Co. or Latoxan (Rosans, France). The venom solution (10%) in 5 mM Tris-HCl buffer, pH 8.5, containing 0.15 M NaCl is applied to a column (5 × 95 cm) of Sephadex G-100. Elution is carried out with the same buffer at a flow rate of 20 ml/hr, and 15-ml fractions are collected. The high-molecular-mass fraction (tubes 42–52), containing 63% of the hemorrhagic activity, is separated from the low-molecular-mass fraction (tubes 65–85). The high-molecular-mass fraction, after concentration by lyophilization and dialysis overnight against 6 mM borax–HCl buffer (pH 9.0), is applied to a column (2.5 × 25 cm) of DEAE-Sephadex A-50. The column is eluted with linear gradient of 0.05–0.25 M NaCl in 2 liters of buffer. A fraction of hemorrhagic activity (HR1 fraction), emerging at a NaCl concentration of 0.2 M, is applied to a column (5 × 95 cm) of Sephadex G-200 superfine. Protein is eluted with 0.05 M Tris-HCl buffer (pH 8.5) containing 0.1 M NaCl at a flow rate of 20 ml/hr, and 3 ml of fractions are collected. The hemorrhagic activity is resolved into two peaks, designated as HR1A and HR1B. Each fraction is concentrated and dialyzed overnight against 0.01 M sodium acetate, and then applied to a column (2 × 30 cm) of TEAE-cellulose. The column is eluted with a linear gradient of 0.01–0.2 M sodium acetate in 500 ml. Each active fraction, emerging at a sodium acetate

---

[26] H. Takeya, S. Nishida, N. Nishino, Y. Makinose, T. Omori-Satoh, T. Nikai, H. Sugihara, and S. Iwanaga, *J. Biochem. (Tokyo)* **113**, 473 (1993).

[27] T. Omori-Satoh and S. Sadahiro, *Biochim. Biophys. Acta* **580**, 392 (1979).

concentration of 0.13 M (HR1A) and 0.07 M (HR1B), is rechromato-graphed under the conditions described above, except that 0.075 M sodium acetate is used for HR1A as the starting buffer. The purified HR1A and HR1B yield a single band on sodium dodecyl sulfate polyacrylamide gel electrophoresis (SDS–PAGE) and their apparent molecular weights are both estimated to be 60,000. The MHDs of HR1A and HR1B are 0.016 and 0.010 μg, respectively, whereas that of crude venom is 0.29 μg. From 10 g of crude venom as starting material, 42 mg of HR1A and 38 mg of HR1B are obtained.

Isolation of Low-Molecular-Mass Hemorrhagic and Nonhemorrhagic Metalloendopeptidases[4,28]

The low-molecular-mass fraction obtained by gel filtration of the venom described above is dialyzed against 5 mM borax–NaOH buffer (pH 9.3) containing 2 mM Ca^{2+} and is then concentrated. The solution is applied to a column (1.6 × 42 cm) of Bio-Rex 70 (Bio-Rad, Richmond, CA) and eluted with a linear gradient of 0–0.25 M NaCl in 2.2 liters of buffer. The major proteinase, H$_2$-proteinase, which is completely free of hemorrhagic activity, emerges at a NaCl concentration of 0.13 M. The main part of the hemorrhagic activity is eluted at 0.17–0.20 M NaCl in two peaks, designated as HR2a and HR2b. The fractions of HR2a and HR2b are concentrated by lyophilization and dialyzed overnight against 5 mM bo-rax–NaOH buffer (pH 9.3) containing 2 mM Ca^{2+}. The HR2a fraction is rechromatographed under the conditions described above. The purified HR2a yields a single band on SDS–PAGE, with an apparent molecular weight of 25,000. HR2b contaminated with HR2a at this stage in the procedure is further purified as described elsewhere.[29] From 1 g of crude venom, 15.1 mg of HR2a is obtained and the MHD is 0.066 μg.

Isolation of Hemorrhagic Toxins c and d from Venom of Crotalus atrox[30]

Lyophilized crude venom, purchased from Miami Serpentarium Labo-ratories, is dissolved in 50 ml of distilled water and dialyzed overnight against 10 mM borate, pH 9.0, containing 0.1 M NaCl and 2 mM CaCl$_2$. The venom is applied to a column (2.5 × 35 cm) of DEAE-cellulose. Four protein fractions are eluted continuously with the same buffer used for

[28] T. Takahashi and A. Ohsaka, *Biochim. Biophys. Acta* **207**, 65 (1970).
[29] K. Yonaha, M. Iha, Y. Tomihara, M. Nozaki, M. Yamakawa, T. Kamura, and S. Toyama, *Toxicon* **26**, 1205 (1988).
[30] J. B. Bjarnason and A. T. Tu, *Biochemistry* **17**, 3395 (1978).

dialysis, and three fractions (1, 2, and 4) have hemorrhagic activity. Bound proteins are finally eluted with 0.4 M NaCl. Fraction 1 yields hemorrhagic toxin a (Ht-a) and Ht-b, and nonhemorrhagic fibrinolytic metalloproteinase, atroxase.[7,30] Fraction 2 yields Ht-c and Ht-d, and fraction 4 yields Ht-e. The following purification procedure is described for Ht-c and Ht-d, because details on their properties, including the amino acid sequence of Ht-d, have been established. Further purification methods of Ht-a, Ht-b, and Ht-e and atroxase are described in the literature.[7,30]

Fraction 2 is lyophilized and dissolved in distilled water and applied to a column (5 × 90 cm) of Sephadex G-25 superfine equilibrated with 5 mM Tris-HCl (pH 8.5) containing 0.1 M NaCl and 2 mM CaCl$_2$. The active fraction is lyophilized and equilibrated against 10 mM Tris-HCl (pH 8.5) containing 0.04 M NaCl and 2 mM CaCl$_2$ and is applied to a column (1.5 × 40 cm) of DEAE-cellulose. The column is eluted with linear gradient of 0.04–0.1 M NaCl. Two hemorrhagic fractions, Ht-c and Ht-d, are rechromatographed separately under the conditions described above. The yields for Ht-c and Ht-d are 34 and 62 mg, respectively, from 20 g of venom.

Properties

Various properties and amino acid compositions of hemorrhagic and nonhemorrhagic metalloendopeptidases isolated from the venoms of *T. flavoviridis* and *C. atrox* are summarized in Tables I and II. Their hemorrhagic and proteolytic activities are inhibited by EDTA or 1,10-phenanthroline. The detailed properties of these enzymes isolated from the venom of *Agkistrodon halys blomhoffii* have also been described in an earlier volume.[5]

Venom metalloendopeptidases are divided into two classes: high-molecular-mass (60,000–68,000) hemorrhagic factors such as HR1A, HR1B, and Ht-a (Table I), and low-molecular-mass (23,000–26,000) hemorrhagic factors. The most characteristic aspect of the amino acid compositions of HR1A, HR1B, and Ht-a is the very high levels of cysteine residues (Table II). High-molecular-mass hemorrhagic factors isolated from other snake species, all of which express 10–25 times higher hemorrhagic activities than low-molecular-mass ones, have also been shown to be cysteine rich.[31–33] It seems very likely that all of these factors contain a cysteine-

[31] D. J. Civello, H. L. Duong, and C. R. Geren, *Biochemistry* **22**, 749 (1983).

[32] G. Ohshima, T. Omori-Satoh, S. Iwanaga, and T. Suzuki, *J. Biochem. (Tokyo)* **72**, 1483 (1972).

[33] G. Oshima, S. Iwanaga, and T. Suzuki, *Biochim. Biophys. Acta* **250**, 416 (1971).

TABLE I

BIOCHEMICAL PROPERTIES OF METALLOENDOPEPTIDASES ISOLATED FROM VENOMS OF *Trimeresurus flavoviridis* AND *Crotalus atrox*[a]

Property	Trimeresurus flavoviridis					Crotalus atrox					
	HR1A	HR1B	HR2a	HR2b	H₂	Ht-a	Ht-b	Ht-c	Ht-d	Ht-e	Atroxase
Molecular weight	60,000[b]	60,000[b]	23,015[c]	23,318[c]	22,991[c]	68,000[b]	24,000[b]	24,000[b]	23,234[c]	25,700[b]	23,500[b]
Carbohydrate content (%)	17–18	17–18	–	–	–	N.D.	N.D.	N.D.	–	–	–
Moles of zinc per molecule	N.D.	N.D.	0.78	N.D.	0.78	0.99	0.82	0.86	0.86	1.03	1
Hemorrhagic activity	+	+	+	+	–	+	+	+	+	+	–
Moles of free sulfhydryl per molecule	N.D.	+	–	N.D.	+	N.D.	N.D.	–	–	–	–
Isoelectric point	4.4	4.4	8.6	9.2	Basic	Acidic	Basic	6.1	6.2	5.6	9.6
References	d	d, e	f–h	g–i	j, k	l	l	m	m, n	l	o

[a] H₂, H₂-proteinase; N.D., not determined; +, positive; –, negative.
[b] Estimated from SDS–PAGE.
[c] Calculated from amino acid sequence.
[d] Omori-Satoh and Sadahiro.[29]
[e] Takeya et al.[18]
[f] Miyata et al.[14]
[g] Takahashi and Ohsaka.[28]
[h] Nikai et al.[12]
[i] Yonaha.[17]
[j] Takeya et al.[15]
[k] Takahashi and Ohsaka.[4]
[l] Bjarnason and Tu.[30]
[m] Bjarnason and Fox.[13]
[n] Shannon et al.[16]
[o] Willis and Tu.[7]

TABLE II

AMINO ACID COMPOSITIONS OF METALLOENDOPEPTIDASES ISOLATED FROM VENOMS OF *Trimeresurus flavoviridis* AND *Crotalus atrox*

Amino acid	Trimeresurus flavoviridis					Crotalus atrox					
	HR1A[a]	HR1B[b,c]	HR2a[b,d]	HR2b[b,e]	H₂[b,f,g]	Ht-a[h]	Ht-b[h]	Ht-c[h]	Ht-d[h,i]	Ht-e[h]	Atroxase[j]
Asp	54	54	27	27	32	85	26	33	29	30	27
Thr	22	19	13	13	10	36	7	11	10	10	9
Ser	25	28	13	13	13	35	14	16	14	14	23
Glu	39	39	16	16	17	55	16	24	22	26	23
Pro	16	20	6	6	7	27	6	6	7	8	6
Gly	32	26	9	9	7	43	13	9	11	14	9
Ala	24	21	12	12	9	36	8	10	9	7	7
Cys	38	35	6	6	7	66	4	2	4	8	12
Val	22	23	13	13	14	35	11	11	9	12	12
Met	5	7	10	10	8	11	6	5	7	8	4
Ile	14	23	15	15	12	36	12	17	14	21	11
Leu	28	25	14	14	19	49	19	25	25	14	17
Tyr	19	21	7	7	8	21	7	7	7	11	6
Phe	13	15	7	7	5	17	7	7	6	11	8
Lys	18	27	16	16	18	27	10	7	8	9	8
His	11	13	8	8	7	18	8	8	9	9	8
Trp	4	4	3	3	2	6	4	5	2	4	3
Arg	23	16	7	7	6	33	12	10	11	8	11
Total	407	416	202	204	201	635	200	213	203	209	204

[a] H. Takeya, J. Kubo, and S. Iwanaga, unpublished results (1991).
[b] Calculated from the amino acid sequence.
[c] Takeya *et al.*[18]
[d] Miyata *et al.*[14]
[e] Yonaha.[17]

[f] Takeya *et al.*[15]
[g] H₂, H₂-proteinase.
[h] Bjarnason and Tu.[30]
[i] Shannon *et al.*[16]
[j] Willis and Tu.[7]

rich domain in their structures similar to that found in HR1B (described later).

Proteolytic Specificity

Human fibrinogen and fibrin have often been used as substrates for various venom proteinases.[8] The fibrinogen cleavage site by HR2a is the Pro-Met bond of the Ser-Pro-Met-Leu sequence (residues 515–518) of the α chain.[26] The new synthetic peptide substrate, Abz-Ser-Pro-Met-Leu-Dna, which refers to the cleavage site of fibrinogen, is cleaved by HR1A, HR1B, and H_2-proteinase as well as HR2a at the Pro-Met bond having the highest k_{cat}/K_m value, whereas bacterial metalloproteinase thermolysin is inert toward this substrate.[26]

Hemorrhagic metalloendopeptidases degrade basement membrane preparation,[3] as well as the isolated components, including type IV collagen, laminin, nidogen, and fibronectin.[34] The initial cleavage site of type IV collagen by Ht-e has been identified to be the Ala^{258}-Gln^{259} bond located in a pepsin-susceptible triplet interruption region of the $\alpha1(IV)$ chain and the Gly^{191}-Leu^{192} bond located in a triple-helical region of the $\alpha2(IV)$ chain.[35]

Structure and Function

The amino acid sequences and disulfide bridge locations have been determined for low-molecular-mass hemorrhagic and nonhemorrhagic

---

FIG. 1. Sequence comparison of snake venom metalloendopeptidases. Low-molecular-mass metalloendopeptidases and the amino-terminal half corresponding to the proteinase domain of high-molecular-mass HR1B are shown. These proteins have been isolated, respectively, from the following snake venoms; HR1B, HR2a, HR2b, and H_2-proteinase from *T. flavoviridis,* HT-2 from *C. ruber ruber,* Ht-d from *C. atrox,* J protease from *Bothrops jararaca,* and AC1 proteinase from *Agkistrodon acutus.* Identical residues shared by all proteins are shaded. The residues common to only hemorrhagic metalloendopeptidases, such as HR1B, HR2a, HR2b, HT-2, and Ht-d, are boxed. The putative zinc ligands (▲) and active site (△) are indicated, as are sugar chains (●) linked to Asn residues of HR1B. Three disulfide bridges link Cys-117 to Cys-197, Cys-157 to Cys-181, and Cys-159 to Cys-164 in HR2a and H2-proteinase, whereas two disulfide bridges link Cys-117 to Cys-197 and Cys-157 to Cys-164 in HT-2 and Ht-d. The residue numbering is that of HR1B. <E denotes 5-pyrrolidone-2-carboxylic acid. The data are taken from the following references: HR1B from Takeya *et al.*[18]; HR2a from Miyata *et al.*[14]; HR2b from Yonaha[17]; HT-2 from Takeya *et al.*[36]; Ht-d from Shannon *et al.*[16]; H_2-proteinase from Takeya *et al.*[15]; J proteinase from Tanizaki *et al.*[37]; AC1 proteinase from Nakagawa *et al.*[38]

[34] E. N. Baramova, J. D. Shannon, J. B. Bjarnason, and J. W. Fox, *Arch. Biochem. Biophys.* **275,** 63 (1989).

[35] E. N. Baramova, J. D. Shannon, J. B. Bjarnason, and J. W. Fox, *Matrix* **10,** 91 (1990).

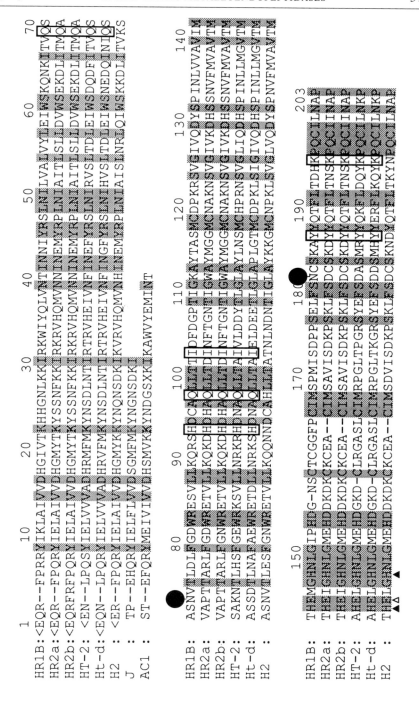

metalloendopeptidases, HR2a, HR2b, and H_2-proteinase from *T. flavoviridis*, HT-2 from *Crotalus ruber ruber,* and Ht-d from *C. atrox.*[14-17,36] As shown in Fig. 1, they have a very similar structure consisting of 201–204 amino acid residues.[37,38] Moreover, the sequence of the amino-terminal half (203 residues) of HR1B[18] represents a metalloproteinase structure similar to those of the low-molecular-mass metalloendopeptidases, HR2a, HR2b, H_2-proteinase, HT-2, and Ht-d (Fig. 1). They have peptide segments His-Glu-X-X-His (residues 143–147) consisting of the putative zinc-chelating (His-143, His-147) and catalytic (Glu-144) residues, identified by homology to the zinc-chelating sequence of thermolysin. However, there is no significant sequence similarity with thermolysin or any other known metalloproteinase, except for this His-Glu-X-X-His sequence.

HR1B expresses LD_{50} values of 4.9 μg/mouse and shows 10 times higher hemorrhagic activities than does HR2a or HR2b, thereby indicating the major lethal factor in *T. flavoviridis* venom. In addition to these pathological functions, HR1 (mixture of HR1A and HR1B) as well as the crude venom inhibit ADP-stimulated platelet aggregation.[39] The most interesting structural feature of HR1B is that the cysteine-rich middle region (residues 204–300) shows a striking similarity to disintegrins, Arg-Gly-Asp-containing platelet aggregation inhibitors found in several viper venoms (Fig. 2).[40-47] Interestingly, however, this region of HR1B does not contain

[36] H. Takeya, A. Onikura, T. Nikai, H. Sugihara, and S. Iwanaga, *J. Biochem.* (*Tokyo*) **108**, 711 (1990).

[37] M. M. Tanizaki, R. B. Zingall, H. Kawazaki, S. Imajoh, S. Yamazaki, and K. Suzuki, *Toxicon* **27**, 747 (1989).

[38] H. Nakagawa, R. A. Miller, A. T. Tu, T. Nikai, and H. Sugihara, *Biochem. Arch.* **5**, 243 (1989).

[39] M. Yamanaka, M. Matsuda, J. Isobe, A. Ohsaka, T. Takahashi, and T. Omori-Satoh, *in* "Platelets, Thrombosis, and Inhibitors" (P. Didisheim, ed.), p. 335. Schattauer Verlag, Stuttgart, and New York, 1974.

[40] T.-F. Huang, J. C. Holt, E. P. Kirby, and S. Niewiarowski, *Biochemistry* **28**, 661 (1989).

[41] Z.-R. Gan, R. J. Gould, J. W. Jacobs, P. A. Friedman, and M. A. Polokoff, *J. Biol. Chem.* **263**, 19827 (1988).

[42] B. H. Chao, J. A. Jakubowski, B. Savage, E. P. Chow, U. M. Marzec, L. A. Harker, and J. M. Maraganove, *Proc. Natl. Acad. Sci. U.S.A.* **86**, 8050 (1989).

[43] R. J. Shebuski, D. R. Ramjit, G. H. Bencen, and M. A. Polokoff, *J. Biol. Chem.* **264**, 21550 (1989).

[44] M. S. Dennis, W. J. Henzel, R. M. Pitti, M. T. Lipari, M. A. Napier, T. A. Deisher, S. Bunting, and R. A. Lazarus, *Proc. Natl. Acad. Sci. U.S.A.* **87**, 2471 (1989).

[45] R. J. Gould, M. A. Polokoff, P. A. Friedman, T. F. Huang, J. C. Holt, J. J. Cook, and S. Niewiarowski, *Proc. Soc. Exp. Biol. Med.* **195**, 168 (1990).

[46] J. J. Calvete, W. Schafer, T. Soszka, W. Lu, J. J. Cook, B. A. Jameson, and S. Niewiarowski, *Biochemistry* **30**, 5225 (1991).

[47] V. Saudek, R. A. Atkinson, and J. T. Pelton, *Biochemistry* **30**, 7369 (1991).

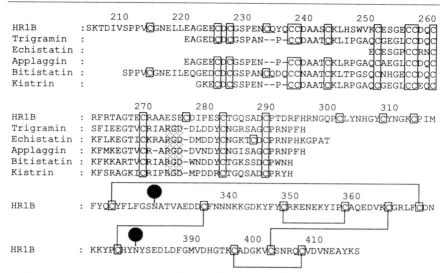

FIG. 2. Sequence comparison of the carboxyl-terminal portion of HR1B with several disintegrins. Trigramin, echistatin, applaggin, bitistatin, and kistrin are all viper venom platelet aggregation inhibitors, called disintegrins. Cysteine residues are boxed and five disulfide bridges assigned in the C-terminal region of HR1B (residues 317–416) are also shown. The Arg-Gly-Asp sequences common to disintegrins are shaded; sugar chains (●) are linked to Asn residues. The data are taken from following references: HR1B from Takeya et al.[18]; trigramin from Huang et al.[40]; echistatin from Gan et al.[41]; applaggin from Chao et al.[42]; bitistatin from Shebuski et al.[43]; kistrin from Dennis et al.[44]; the disulfide bridge locations of HR1B are from H. Takeya and S. Iwanaga (unpublished results). For more information on disintegrins the reader is referred to Ref. 45. Recently, the disulfide bridge locations of one of the disintegrins, albolabrin, and the three-dimensional structure of echistatin have been reported.[46,47]

the Arg-Gly-Asp sequence, which is known to be a putative disintegrin binding site for the platelet fibrinogen receptor, the glycoprotein IIb/IIIa complex. The remaining region at the carboxyl-terminal end is unique in possessing a cysteine-rich sequence. These results suggest that the middle portion of HR1B, which shows structural similarities to that of disintegrin, may be important in synergistically stimulating the hemorrhagic activity with the amino-terminal metalloproteinase domain.

The gross structures of venom metalloendopeptidases are shown in Fig. 3. Because these features indicate the structural and evolutional relationship among these proteins, the venom metalloendopeptidases maybe belong to a new gene family. It is also expected that many kinds of venom metalloendopeptidases, originally isolated with different biological

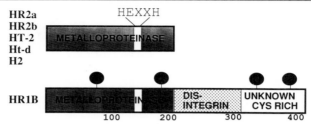

FIG. 3. Gross structure of snake venom metalloendopeptidases. HR2a, HR2b, HT-2, Ht-d and H_2-proteinase are the venom low-molecular-mass hemorrhagic and nonhemorrhagic metalloendopeptidases. HR1B is composed of three domains: the metalloproteinase domain, the disintegrin-like domain, and an unknown cysteine-rich region. The putative zinc ligands and active site are shown (HEXXH). Sugar chains (●) linked to Asn-residues are indicated.

activities, such as fibrinogenase,[8] serpin inactivator,[48,49] and several procoagulants,[9,10] are members of this newly identified venom metalloendopeptidase family.

[48] T. Kurecki, M. Laskowski, Sr., and L. F. Kress, *J. Biol. Chem.* **253**, 8340 (1978).
[49] L. F. Kress, T. Kurecki, S. K. Chan, and M. Laskowski, *J. Biol. Chem.* **254**, 5317 (1979).

[25] Horseshoe Crab Transglutaminase

By FUMINORI TOKUNAGA and SADAAKI IWANAGA

Introduction

The hemocytes circulating in horseshoe crab (*Limulus*) hemolymph contain a coagulation system that participates both in hemostasis and in defense against invading microorganisms.[1] This coagulation system is highly sensitive to gram-negative bacterial endotoxins, which are lipopolysaccharides (LPS).[2-4] In 1973, Lorand and Campbell-Wilkes discovered

[1] J. Levin and F. B. Bang, *Bull. Johns Hopkins Hosp.* **115**, 265 (1964).
[2] S. Iwanaga, T. Morita, T. Miyata, T. Nakamura, and J. Aketagawa, *J. Protein Chem.* **5**, 255 (1986).
[3] T. Muta, R. Hashimoto, T. Miyata, H. Nishimura, Y. Toh, and S. Iwanaga, *J. Biol. Chem.* **265**, 22426 (1990).
[4] T. Muta, T. Miyata, Y. Misumi, F. Tokunaga, T. Nakamura, Y. Toh, Y. Ikehara, and S. Iwanaga, *J. Biol. Chem.* **266**, 6554 (1991).

the existence of Ca^{2+}-dependent transglutaminase (TGase) in *Limulus polyphemus* amebocytes and showed the TGase-catalyzed incorporation of [^{14}C]putrescine into β-lactoglobulin and α-casein.[5] Later, Chung *et al.*[6] confirmed this fact and suggested that TGase participates in the cross-linking of coagulin gel. However, no further studies have been made on the purification and characterization of *Limulus* TGase, because of an extremely unstable enzyme in the lysate. In 1989, Roth *et al.*[7] reported that TGase (factor XIII-like) activity in either native or gelled *Limulus* lysate was negative, and that stable gels formed after coagulation of *limulus* lysate by bacterial endotoxin were not covalently cross-linked.

TGase [EC 2.3.2.13] constitutes a family of Ca^{2+}-dependent enzymes that catalyze an acyl transfer reaction in which the γ-carboxamide group of the peptide-bound Gln residue acts as acyl donor and the primary amino group of either the peptide-bound Lys residue or the free amine can serve as acyl acceptor.[8] In mammals, TGase is widely distributed in blood plasma, platelets, placenta, keratinocyte, and tissues such as liver, prostate gland, and epidermis.[8] TGase is classified into three groups—plasma TGase (factor XIII),[9] type I membrane-bound TGase (keratinocyte type),[10] and type II cytosolic TGase (liver type).[11] Moreover, one of the major erythrocyte membrane proteins, band 4.2, is related to the TGase family, although it does not show any TGase activity.[12] Recently, we have succeeded in isolating TGase from the hemocyte lysate of the Japanese horseshoe crab (*Tachypleus tridentatus*), providing the first demonstration of invertebrate TGase.[13]

This chapter describes the purification, properties, and cDNA sequence of *Limulus* TGase.

[5] Campbell-Wilkes, Ph.D. Dissertation (Examiner, L. Lorand), Northwestern University, Univ. Microfilms, 73-30, 763, Ann Arbor, MI (1973).

[6] S. I. Chung, R. C. Seid, and T.-Y. Liu, *Thromb. Haemostasis* **38**, 182 (1977).

[7] R. I. Roth, J. C.-R. Chen, and J. Levin, *Thromb. Res.* **55**, 25 (1989).

[8] J. E. Folk and J. S. Finlayson, *Adv. Protein Chem.* **31**, 1 (1977).

[9] A. Ichinose, B. A. McMullen, K. Fujikawa, and E. W. Davie, *Biochemistry* **25**, 4633 (1986).

[10] M. A. Phillips, B. E. Stewart, Q. Qin, R. Chakravarty, E. E. Floyd, A. M. Jetten, and R. H. Rice, *Proc. Natl. Acad. Sci. U.S.A.* **87**, 9333 (1990).

[11] K. Ikura, T. Nasu, H. Yokota, Y. Tsuchiya, R. Sasaki, and H. Chiba, *Biochemistry* **27**, 2898 (1988).

[12] L. A. Sung, S. Chien, L.-S. Chang, K. Lambert, S. A. Bliss, E. E. Bouhassira, R. L. Nagel, R. S. Schwartz, and A. Rybicki, *Proc. Natl. Acad. Sci. U.S.A.* **87**, 955 (1990).

[13] F. Tokunaga, M. Yamada, T. Muta, M. Hiranaga-Kawabata, S. Iwanaga, A. Ichinose, and E. W. Davie, *Thromb. Haemostasis* **65**, 936 (1991).

Assay Method

Principle

TGase activity was determined by the fluorescence of monodansylca-daverine (DCA) incorporated into N,N-dimethylcasein as described in the method of Lorand and Gotoh.[14]

Reagents and Procedure

The reaction mixture, containing 20 μl of 50 mM Tris-HCl buffer, pH 7.5, 20 μl of 100 mM CaCl$_2$, 20 μl of 100 mM dithiothreitol (DTT), 100 μl of 2 mM DCA (Sigma Chemical Co., St. Louis, MO), 200 μl of 0.4% (w/v) dimethylcasein (Sigma), and the sample, in a total of 400 μl, is incubated at 37° for 30 min. After incubation, 400 μl of 10% trichloroacetic acid is added to terminate the reaction and the mixture is centrifuged at 15,000 rpm for 15 min. The precipitate is washed three times with 1 ml of ethanol/ethyl ether (1 : 1, v/v) and dried by Speed-Vac concentrator (Savant Instruments, NY). The dried precipitate is dissolved in 1 ml of 50 mM Tris-HCl, pH 8.0, containing 8 M urea and 0.5% SDS. Fluorescent intensity is then measured with excitation at 355 nm and emission at 525 nm. DCA dissolved in the same solution to give a concentration of 10 μmol/liter is used as a standard. One unit of the enzyme activity, defined as amine incorporation unit per minute (AIU), is calculated as described.[14]

Purification Procedure

Japanese horseshoe crab (*T. tridentatus*) hemocytes are collected as described.[15] Briefly, the fresh hemocyte pellet is homogenized three times with 300 ml of 50 mM Tris–acetate buffer, pH 7.5, containing 1 mM EDTA (TAE) using a Physcotron homogenizer (Nihon-Seimitsu Kogyo, Ltd., Tokyo) in a sterilized centrifugation tube; it is then centrifuged at 8000 rpm for 30 min. The lysate (890 ml) thus obtained from 32.4 g of hemocytes (wet weight) is dialyzed at 4° for 12 hr against 20 liters of 50 mM TAE buffer. This procedure is very important to stabilize the *Limulus* TGase. The dialyzed lysate is centrifuged at 8000 rpm for 1 hr and the supernatant is then applied to a CM-Sepharose CL-6B column (3.5 × 22 cm) equili-brated with 50 mM TAE buffer. The column is washed with 700 ml of the equilibrated buffer and proteins are eluted with TAE buffer containing 2

[14] L. Lorand and T. Gotoh, this series, Vol. 19, p. 770.
[15] T. Nakamura, T. Morita, and S. Iwanaga, *J. Biochem.* (*Tokyo*) **97**, 1561 (1985).

M NaCl. The breakthrough fractions containing TGase activity are pooled and dialyzed overnight against 50 m*M* TAE buffer. The dialyzate is brought to 50% saturation of ammonium sulfate and the resulting precipitate is collected by centrifugation at 8000 rpm for 30 min. The pellet thus obtained is dissolved in 460 ml of 50 m*M* TAE buffer and dialyzed overnight against same buffer. The dialyzed solution is applied to a DEAE-cellulose (DE-52) column (2.0 × 14.5 cm) equilibrated with 50 m*M* TAE buffer. The column is washed with 300 ml of the equilibration buffer and the elution is carried out with a linear salt gradient with 500 ml each of 50 m*M* TAE buffer and the same buffer containing 0.3 *M* NaCl. As shown in Fig. 1, TGase is eluted in the fractions with a conductivity of 8 mmho. These fractions (indicated by a bar) are collected, dialyzed overnight against 50 m*M* TAE buffer, and concentrated by Diaflo ultrafiltration (Amicon, Danvers, MA). The concentrated sample is applied to a Sephacryl S-300 column (3.0 × 94 cm) equilibrated with 50 m*M* TAE buffer and eluted

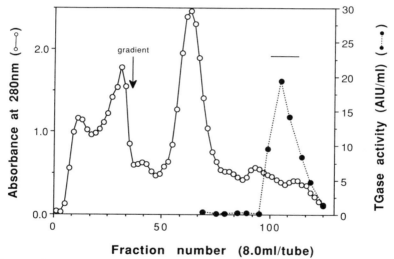

Fraction number (8.0ml/tube)

Fig. 1. Chromatography of *Limulus* TGase fraction on a DEAE-cellulose column (2.0 × 14.5 cm). The breakthrough fractions obtained from a CM-Sepharose CL-6B column were pooled and fractionated by 50% saturated $(NH_4)_2SO_4$, and the resulting precipitate was dissolved in 50 mM TAE buffer and dialyzed overnight against the same buffer. The dialyzed solution was then applied to a DEAE-cellulose column equilibrated with 50 mM TAE buffer. The elution was performed with a linear salt gradient from 0 to 0.3 *M* NaCl in the equilibration buffer. Fractions of 8 ml were collected at 4°, and the fractions indicated by a bar were pooled. The absorbance of the eluate at 280 nm (○) and the TGase activity (●) are shown.

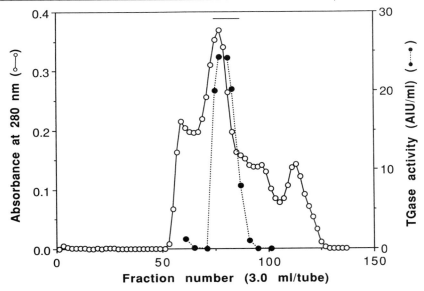

FIG. 2. Gel filtration of TGase fraction on a Sephacryl S-300 column (3.0 × 94 cm). The pooled fraction obtained from Fig. 1 was concentrated and applied to a Sephacryl S-300 column. The elution was carried out with 50 mM TAE buffer and fractions of 3.4 ml were collected at 4°, and the fractions indicated by a bar were pooled. The absorbance of eluate at 280 nm (○) and the TGase activity (●) are shown.

with the same buffer (Fig. 2). The pooled TGase-containing fraction (indicated by a bar) is applied to a DEAE-Cosmogel (7.5 × 50 mm) column (Nakalai Tesque Co.) linked to high-performance liquid chromatography. The column is equilibrated with 50 mM Tris–acetate, pH 7.5, and the elution is performed with a linear gradient from 0 to 0.4 M NaCl in the same buffers for 120 min at a flow rate of 1.0 ml/min, as shown in Fig. 3. The pooled TGase-containing fraction is dialyzed against 50 mM Tris–acetate, pH 7.5, and the dialyzed fraction is finally applied to a zinc-chelating Sepharose 6B (1.6 × 25 cm) column equilibrated with 50 mM Tris–acetate, pH 7.5. The proteins are eluted with a linear gradient of histidine (0–30 mM) in 50 mM Tris–acetate, pH 7.5. In each of the procedures, the protein concentrations are determined by a dye-binding method using bovine serum albumin as a standard (Bio-Rad, Richmond, CA).

Table I shows the summary of the purification of TGase from the hemocyte lysate. The yield of purified TGase is about 1.6 mg from 32.4 g of hemocytes (wet weight, collected from 10 horseshoe crabs) and 156-fold purification is achieved. The purified TGase shows a single band

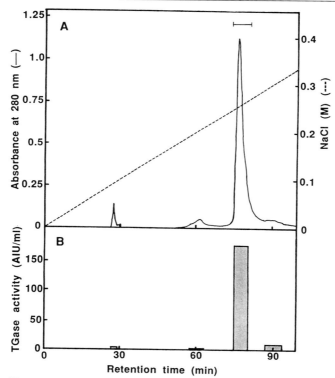

FIG. 3. Chromatography of TGase fraction on a DEAE-Cosmogel column (7.5 × 50 mm). The pooled fraction obtained from Fig. 2 was dialyzed against 50 mM TAE buffer and the dialyzed solution was applied to a DEAE-Cosmogel column connected to high-performance liquid chromatography. The elution was performed with a linear salt gradient to 0.4 M NaCl in 50 mM Tris–acetate, pH 7.5, at a flow rate of 1.0 ml/min. The fractions indicated by a bar were collected manually. (A) Absorbance of eluate at 280 nm; (B) shaded bars indicate the TGase activity.

of 86 kDa under nonreduced SDS–PAGE (Fig. 4), and it gave higher molecular mass in the presence of 2-mercaptoethanol.

Properties

The extinction coefficient of *Limulus* TGase (1% solution in 50 mM TAE buffer) is estimated to be 21.7 at 280 nm. The sample should be stored at 4°, not in a frozen state, because freeze-thawing procedures unstabilize this enzyme. *Limulus* TGase shows Ca^{2+}-dependent activity like mammalian tissue TGase and 10 mM Ca^{2+} is required for maximum

TABLE I
PURIFICATION OF *Limulus* TGASE

Step	Volume (ml)	Total protein (mg)	Total activity (AIU)[a]	Specific activity (AIU/mg)	Purification (fold)	Yield (%)
Hemocyte lysate	890	1335	8055	6.0	1.0	100
CM-Sepharose CL-6B	1000	462	8425	18.3	3.0	105
Ammonium sulfate precipitation	225	191	2690	14.0	2.3	33.3
DEAE-cellulose	175	17.1	4541	266	44.3	56.4
Sephacryl S-300	27.0	9.2	3109	340	56.6	38.6
DEAE-Cosmogel	12.0	2.7	2305	851	142	28.6
Zinc-chelating Sepharose 6B	13.0	1.6	1528	938	156	19.0

[a] Amine incorporation unit.[8]

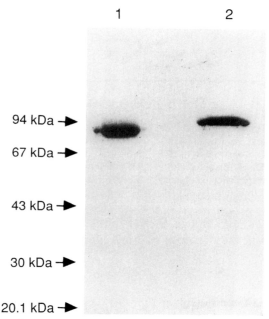

FIG. 4. SDS–PAGE of purified *Limulus* TGase. The purified *Limulus* TGase was electrophoresed in the absence (lane 1) and presence (lane 2) of 2-mercaptoethanol. SDS–PAGE was carried out using a 6–15% gradient gel. After electrophoresis, the gel was stained with Coomassie brilliant blue.

TABLE II
EFFECT OF VARIOUS INHIBITORS ON ACTIVITY
OF *Limulus* TGASE

Inhibitor	Remaining activity (%)
None	100
100 mM hydroxylamine	0
1 mM ICH$_2$COOH	6
1 mM N-ethylmaleimide	2
1 mM CuCl$_2$	0
5 mM EDTA	0
5 mM diisopropyl fluorophosphate	100

activity.[5] Interestingly, however, dithiothreitol, which enhances tissue TGase activity, shows little effect on *Limulus* TGase activity. Table II shows the inhibitor spectrum of *Limulus* TGase. At concentrations of 1 to 5 mM, CuCl$_2$, EDTA, and SH-blocking reagents such as iodoacetic acid and N-ethylmaleimide strongly inhibit the *Limulus* TGase activity. Furthermore, hydroxylamine shows inhibitory effect, but diisopropyl fluorophosphate (DFP) has no effect. The *Limulus* TGase activity is also inhibited by high concentrations ($>0.5\ M$) of NaCl. A micromolar amount of GTP is known to inhibit guinea pig liver TGase,[16] but nucleoside 5'-triphosphates such as ATP, CTP, GTP, and UTP have no effect on the *Limulus* TGase activity up to final concentrations of 1 mM in the presence of 5 mM CaCl$_2$.

cDNA and Amino Acid Sequence

We have succeeded in cloning the cDNA encoding *Limulus* TGase and have determined the entire amino acid sequence using recombinant DNA technique. As shown in Fig. 5, *Limulus* TGase consists of 764 amino acid residues with a calculated molecular mass of 87,110 Da, which is in good agreement with the value estimated on SDS–PAGE. Based on a sequence similarity with other mammalian tissue TGases, Cys-343 is deduced to be a catalytic site. Although there are five potential carbohydrate attachment sites (Asn-X-Ser/Thr) at positions of 12, 196, 206, 578, and 611, no hexosamines are detected by amino acid analysis, indicating that *Limulus* TGase is a simple protein.

[16] K. E. Achyuthan and C. S. Greenberg, *J. Biol. Chem.* **262,** 1901 (1987).

```
                                                    -115  CTCACCGAAGACGAGTATACCAGA                         -91
                                                                                                           -1
CGAGCAACTTAGTTAGGGTTATTAGACCCTTTGTCTTCTGAACTTGTGAGGACAGTCAACGTTAAACTGAAGACAGTATC

ATGTATGGTTTTGGAAGAGGTAACATGTTCCGAAACAGTAACAGAGGAGTAACAGAGAAGAGGCCCAGGTACGAGGCTGAGAATTATCATTCC    90
  M  Y  G  F  G  R  G  N  M  F  R  N  R  S  T  R  Y  R  R  P  R  Y  R  A  E  N  Y  H  S
  1

TACATGTTAGATTTGTTGGAGAATATGAATGAAGAGTTCGGAAGAAATTGGTGGGGGACCCCGAATCACCAGCCTGCAGTGGCCCA    180
  Y  M  L  D  L  L  E  N  M  N  E  E  F  G  R  N  W  G  T  P  E  S  H  Q  P  P  D  S  G  P
 31

TCATCACTACAAGTGGAAAGTGTGGAACTGTACACAAGAGATAACGCTCGCGAACACAACACTTTCATGTATGATCTTGTGTGATGGTACC    270
  S  S  L  Q  V  E  S  V  E  L  Y  T  R  D  N  A  R  E  H  N  T  F  M  Y  D  L  V  D  G  T
 61

AAGCCTGTCCTGATTCTTCGTCGTGGTCAGCCCTTCAGCATAGCAATTCGATTTAAAAGAAACTACAATCCACAGCAAGACCGCCTGAAG    360
  K  P  V  L  I  L  R  R  G  Q  P  F  S  I  A  I  R  F  K  R  N  Y  N  P  Q  Q  D  R  L  K
 91

CTGGAAATTGGTTTTGGGCAACAACCACTAATTACTAAAGGCACTTTGATAATGTTGCCCGTCTCTGGCAGCGATACGTTTACAAAGGAT    450
  L  E  I  G  F  G  Q  Q  P  L  I  T  K  G  T  L  I  M  L  P  V  S  G  S  D  T  F  T  K  D
121

AAAACCCAATGGGACGTTCGATTACGTCAACATGATGCGCCTGTCGATCACTCTTGAAATACAGATTCCTGCTGCTGTTGCAGTAGGAGTT    540
  K  T  Q  W  D  V  R  L  R  Q  H  D  G  A  V  I  T  L  E  I  Q  I  P  A  A  V  A  V  G  V
151

TGGAAAATGAAAATTGTATCCCAGCTGACTTCAGAAGAACAACCAATGTCTGCACCCATGAGTGTAAAATAAGACATATATA    630
  W  K  M  K  I  V  S  Q  L  T  S  E  E  Q  P  N  V  S  A  V  T  H  E  C  K  N  K  T  Y  I
181

CTGTTTAATCCATGGTGTAAACAGGATTCAGTGTATATGGAGGATGAACAATGGAGAAAGGAATATGTTCTGAGTGATGTAGGAAAGATA    720
  L  F  N  P  W  C  K  Q  D  S  V  Y  M  E  D  E  Q  W  R  K  E  Y  V  L  S  D  V  G  K  I
211

TTCACTGGTTCCTTCAAACAACCGGTTGGACGAAGATGGATTTTTGGACAGTTTACAGATTCCAGCCTGTATGTTGATCTTG    810
  F  T  G  S  F  K  Q  P  V  G  R  R  W  I  F  G  Q  F  T  D  S  V  L  P  A  C  M  L  I  L
241

GAACGTTCTGGACTTGACTACACTGCTAGGTCCAACCCGATAAAAGTAGTCCGAGCTATATCAGCGATGGTAAACAACATAGACGATGAA    900
  E  R  S  G  L  D  Y  T  A  R  S  N  P  I  K  V  V  R  A  I  S  A  M  V  N  N  I  D  D  E
271

GGGGTTCTAGAAGGAAGATGGCAAGAACCTTATGATGGTGTTGCACCTTGGATGTGGACCGGAAGTTCTGCCATCTTGGAGAAATAC    990
  G  V  L  E  G  R  W  Q  E  P  Y  D  D  G  V  A  P  W  M  T  G  S  S  A  I  L  E  K  Y
301

CTGAAAACTCGAGGAGTTCCAGTGAAATACGGACGTGCTGGCGTTGTTCGCTGGCAATACTGTGTCTCGAGCCTGGGCATCCCC    1080
  L  K  T  R  G  V  P  V  K  Y  G  Q (C) W  V  F  A  G  V  A  N  T  V  S  R  A  L  G  I  P...
331

AGCAGGACCGTGACCAATTATGATTCAGCCCATGACACGGATGACACCTTGACCATTGACAAATGGTTCGACAAAATGGAGATAAAATT    1170
  S  R  T  V  T  N  Y  D  S  A  H  D  T  D  D  T  L  T  I  D  K  W  F  D  K  N  G  D  K  I
361

GAGGATGCTACAAGTGATTCAATTTGGAATTTCCATGTGTGGAATGACTGTTGGATGGCAGACCTGATTTACCTACAGGTTACGGAGGA    1260
  E  D  A  T  S  D  S  I  W  N  F  H  V  W  N  D  C  W  M  A  R  P  D  L  P  T  G  Y  G  G
391
```

```
421  TGGCAAGCATATGATTCTACACCACAGGAGACCAGTGAAGGCCGTGTACCAAACAGGACCTGCTTCAGTACTTGCTGTCAAAGAGGAGAA  1350
     W ... Q  A  Y  D  S  T  P  Q  E  T  S  E  G  V  Y  Q  T  G  P  A  S  V  L  A  V  Q  R  G  E

451  ATCGGTTACATGTTTGACTCTCCTTTGTTTCTCCGAAGTGAAGTCAACGCAGACGTTGTACATTGGCAAGAGATGATAGTGAGACAGGC  1440
     I  G  Y  M  F  D  S  P  F  V  F  S  E  V  N  A  D  V  V  H  W  Q  E  D  D  S  S  E  T  G
                    CGT                                                                      TGT
                    Arg                                                                      Cys

481  TACAAAAAGCTGAAAATCGACAGCTATCGTGTGGGTCGGCTTCTTCTCACCAAGAAAATTGGAGTAGATGACTTTGGAGATGCTGAC  1530
     Y  K  K  L  K  I  D  S  Y  R  V  G  R  L  L  L  T  K  K  I  G  V  D  D  D  F  G  D  A  D
                          AGC
                          Ser

511  GCAGAAGATATCACAGACCAGTACAAAAACAAAGAGGGAACAGATGAAGAGAATGTCTGTACTAAATGCAGCTAGAAGCAGTGTTTTC  1620
     A  E  D  I  T  D  Q  Y  K  N  K  E  G  T  D  E  E  R  M  S  V  L  N  A  A  R  S  S  G  F

541  AATTACGCATTCAATCTGCCCTCCCAGAGAAAGAGAGATGTTTATTTTAATTTGTTGACATCGAGAAAATAAAGATTGGTCAACCCTTC  1710
     N  Y  A  F  N  L  P  S  P  E  K  E  D  V  Y  F  N  L  L  D  I  E  K  I  K  I  G  Q  P  F

571  CACGTGACACTGAACATGAAAAGCAAAGTAGCGACAGAAGAGTCAGCTGTCCTCTCAGCCAGTAGTATTTACTACACAGGTATA  1800
     H  V  T  V  N  I  E  N  Q  S  S  E  T  R  R  V  S  A  V  L  S  A  S  S ... I  Y  Y  T  G  I

601  ACTGGAAGCAAAATAAAACGGGAAAATGGAAACTTTCATTACAACCACATCAAAAGGAAGTATTATCTCATCGAGGTGACCCCAGATGAA  1890
     T  G  R  K  I  K  R  E  N  G  N  F  S  L  Q  P  H  Q  K  E  V  L  S  I  E  V  T  P  D  E

631  TATCTGGAGAAACTGGTTGATTACGCAATGATCAAGCTTTATGCTATAGCCAACGACGTGGTCAGAAGAAGAC  1980
     Y  L  E  K  L  V  D  Y  A  M  I  K  L  Y  A  I  A  T  V  K  E  T  Q  Q  T  W  S  E  E  D

661  GATTTCATGGTTGAGAAACCGAATCTAGAACTTGAGATCCGTGGAAATTTGCAGGTGGGAACACATTGTACTTGCAATCAGTTTGACC  2070
     D  F  M  V  E  K  P  N  L  E  L  E  I  R  G  N  L  Q  V  G  T  P  P  F  V  L  A  I  S  L  T

691  AACCCCACTGAAGCCGTGTGTCTGGACAACTGTTTCTTCACGATCGAAGCTCCAGGGGTGACTGGAGCGTTCCGGGTTACCAATAGGGATATT  2160
     N  P  L  K  R  V  L  D  N  C  F  F  T  I  E  A  P  G  V  T  G  A  F  R  V  T  N  R  D  I

721  CAACCTGAAGAAGTGGCTGTGCCACACCGTGTCGGCTCATCCCTCAAAAACCAGGTCCAAGAGAAAGATCGTGTGCTACCTTCAGTTCCGACAA  2250
     Q  P  E  E  V  A  V  H  T  V  R  L  I  P  Q  K  P  G  P  R  K  I  V  A  T ... F  S  S ... R  Q ...

751  TTGATACAGGTAGTTGGATCTAAGCAAGTCGAAGTGCTGGACTAGGTTGACTTCTAGACAAAAGGAACATTGGAATGTTTTATAGAGTTG  2340
     L  I  Q ... V  V  G  S  K  Q  V  E  V  L  D

     TCACCTTCTTTCACGTTAAATATACATTTATTTGTGTCAGTTATGTCTCAAAGAACACACATTTAGTGTTTCTAAGTATATCCAAGTAGT  2430
     TTGCTTTTCTTTATAAAATATGCGACGTTTATTAAATGTACTATTTTATTTAGTATTGCTACTAAACCACCTGATTCACTTTAATTACTTGT  2520
     TAGGTTTAGTAAAGTTTGTATTTTCCTCATAATTAACAGAGAATTATCACGTCACGTGTTTTATTTAATTATTTCATCATAAAAGG  2610
     TTTGTATTCGTTCTCTCATTCATAATAATGCTTTAGTAGTTTCTAATGTGTTGTTTTATTTTAATTATTTCGTTATTTTTGAAGTATTAT  2700
     TTCAATTTTACCAGGTTCCTTTCCAGTAATTGTTGTCTTGAAGTGGTCAACCAATAAAAATATAGCAC  2769
```

The entire sequence of *Limulus* TGase shows a significant similarity to mammalian TGase family sequences as follows: guinea pig liver TGase (32.7%),[11] human factor XIIIa subunit (36.9%),[9] human keratinocyte TGase (39.9%),[10] and human erythrocyte band 4.2 (23.7%).[12] However, *Limulus* TGase has a unique NH_2-terminal cationic 60-residue extension with no homology to the mammalian TGases. The function of this extension is not known.

FIG. 5. The composite nucleotide sequence (upper) and the deduced amino acid sequence (lower) of *Limulus* TGase. A polyadenylation signal (AATAAA) is double underlined. The amino acid residues that have been confirmed by sequencing of purified peptides are single underlined. The amino acid residue suspected to be an active Cys residue is circled.

Author Index

Numbers in parentheses are footnote reference numbers and indicate that an author's work is referred to although the name is not cited in the text.

W

Subject Index

A

A549 cells, *see* Tumor cells, lung carcinoma

N-Acetyl-L-arginine methyl ester, 79

N-Acetyl-β-D-glucosaminidase, in oligosaccharide analysis of recombinant plasminogen, 177, 182–184

N-Acetylglycine-L-lysine methyl ester, 79

Acetyltyrosine ethyl ester, 102–103

Affinity chromatography
  complement C1 inhibitor, 110–111
  complement C3, 58–60
  complement C4, C4A, and C4B, 55–58
  α₂-plasmin inhibitor, 191–192
  urokinase plasminogen activator receptor, 215–216

Agkistrodon acutus, AC1 protease, 374

Agkistrodon halys blomhoffi, metalloendopeptidase, 371

Alteplase, 8

Amblyomma americanum, factor Xa inhibitor, 309

Anion-exchange chromatography, high-pH, *see* High-pH anion-exchange chromatography

Anisoylated plasminogen streptokinase activator complex, clinical applications, 7–8

α₂-Antichymotrypsin, structure–function relationships, 117

α₂-Antiplasmin, 6, 273

Antistatin
  domain structure, 303
  inhibitory characteristics, 304–305
  isoforms
    amino acid sequence comparisons, 297–299
    HPLC comparison with recombinant antistatin, 297, 299
  properties, 292

purification from leech salivary glands, 295–300
reactive site, 303
recombinant, 297–300
  amino acid sequence, 302–304
  HPLC comparison with isoforms, 297, 299
  physical properties, 302–304
  purification, 300–305
  purified
    analytical analysis of, 300–302
    properties, 302–305
  purity, 302
  structure, 302–304
  in vitro plasma-based clotting assay, 305–306

Antithrombin III
  hinge region, 115
  inhibition of *Limulus* proclotting enzyme, 358
  reactive center, 115
    mutations, 116

α₂-Antitrypsin
  hinge region, 115–116
  reactive center, 115–116
  structure–function relationships, 117

Apolipoprotein receptors, low-density, module in factor I, 19–21

APSAC, *see* Anisoylated plasminogen streptokinase activator complex

Armyworm, *see Spodoptera frugiperda* cells

Arthrobacter ureafaciens, neuraminidase, in oligosaccharide analysis of recombinant plasminogen, 177, 182–184

Arthropod, α-macroglobulin, 125

Aspergillus phoenicis, α-mannosidase, in oligosaccharide analysis of recombinant plasminogen, 177–178

ATEE, *see* Acetyltyrosine ethyl ester

Atherosclerosis, lipoprotein (a) in, 272

T

ISBN 0-12-182124-2

90038